2D Metals

2D Metals: Fundamentals, Emerging Applications, and Challenges delves into the state-of-the-art advancements in utilizing 2D metals for emerging applications, encompassing a comprehensive overview of synthetic methodologies and characterization techniques provided by leading experts in the field.

2D nanomaterials have emerged as highly promising candidates for a diverse array of cutting-edge applications, spanning energy and biomedicine, owing to their adjustable electrochemical properties, versatility, and exceptional mechanical resilience. Notably, carbon-based 2D materials have already demonstrated extensive utility across various domains. Meanwhile, 2D metals, often referred to as Metallenes, represent a burgeoning class of materials with broad-reaching potential. In contrast to alternative 2D materials like graphene and transition metal chalcogenides, as well as bulk metals, 2D metals exhibit remarkable conductivity, expansive surface area, and customizable electronic and optoelectronic characteristics.

This book explores the influence of structural modifications on the properties of 2D metals and addresses the myriad challenges associated with their burgeoning applications. Each chapter, authored by esteemed specialists from across the globe, offers invaluable insights, rendering this book an indispensable resource for students while furnishing researchers and industry professionals with novel guidance and perspectives.

Ram K. Gupta is Professor of Chemistry at Pittsburg State University (PSU). He also serves as Associate Vice-President for Research and Support at PSU and Director of Research at the National Institute for Materials Advancement (NIMA). Dr. Gupta has been recently named by Stanford University as being among the top 2% of research scientists worldwide. Dr. Gupta's research spans a range of subjects critical to current and future societal needs including, semiconducting materials and devices, biopolymers, flame-retardant polymers, green energy production and storage using nanostructured materials and conducting polymers, electrocatalysts, optoelectronics and photovoltaics devices, organic–inorganic heterojunctions for sensors, nanomagnetism, biocompatible nanofibers for tissue regeneration, scaffold and antibacterial applications, and biodegradable metallic implants.

Dr. Gupta has published over 350 peer-reviewed journal articles, made over 500 national/international/regional presentations, chaired and organized multiple sessions at national/international meetings, written several book chapters (120+), edited over 70 books and received several million dollars for research and educational activities from external agencies. He also serves as editor, associate editor, guest editor, and editorial board member for various journals.

2D Metals

Fundamentals, Emerging Applications, and Challenges

Edited by
Ram K. Gupta

CRC Press
Taylor & Francis Group
Boca Raton London New York

CRC Press is an imprint of the
Taylor & Francis Group, an **informa** business

Designed cover image: Shutterstock_2174519929

First edition published 2025
by CRC Press
2385 NW Executive Center Drive, Suite 320, Boca Raton FL 33431

and by CRC Press
4 Park Square, Milton Park, Abingdon, Oxon, OX14 4RN

CRC Press is an imprint of Taylor & Francis Group, LLC

© 2025 Ram K. Gupta

Reasonable efforts have been made to publish reliable data and information, but the author and publisher cannot assume responsibility for the validity of all materials or the consequences of their use. The authors and publishers have attempted to trace the copyright holders of all material reproduced in this publication and apologize to copyright holders if permission to publish in this form has not been obtained. If any copyright material has not been acknowledged please write and let us know so we may rectify in any future reprint.

Except as permitted under U.S. Copyright Law, no part of this book may be reprinted, reproduced, transmitted, or utilized in any form by any electronic, mechanical, or other means, now known or hereafter invented, including photocopying, microfilming, and recording, or in any information storage or retrieval system, without written permission from the publishers.

For permission to photocopy or use material electronically from this work, access www.copyright.com or contact the Copyright Clearance Center, Inc. (CCC), 222 Rosewood Drive, Danvers, MA 01923, 978-750-8400. For works that are not available on CCC please contact mpkbookspermissions@tandf.co.uk

Trademark notice: Product or corporate names may be trademarks or registered trademarks and are used only for identification and explanation without intent to infringe.

ISBN: 9781032638553 (hbk)
ISBN: 9781032644981 (pbk)
ISBN: 9781032645001 (ebk)

DOI: 10.1201/9781032645001

Typeset in Minion
by Deanta Global Publishing Services, Chennai, India

Contents

Contributors, vii

CHAPTER 1 ■ Introduction to 2D Metals (Metallenes)   1

SUNIL KUMAR BABURAO MANE AND NAGHMA SHAISHTA

CHAPTER 2 ■ Fundamentals of 2D Metals   16

PINTU BARMAN, APURBA DAS, AND ANINDITA DEKA

CHAPTER 3 ■ Synthesis and Characterization of 2D Metals   32

HSU-SHENG TSAI, JING LI, AND ZHENGGUANG SHI

CHAPTER 4 ■ Structure, Stability, and Properties of 2D Metallenes   47

AKANKSHA A. SANGOLKAR, RAMAKRISHNA KADIYAM, AND RAVINDER PAWAR

CHAPTER 5 ■ Properties and Functionalization of 2D Metals   63

HARINI G. SAMPATKUMAR, G. BANUPRAKASH, AND SIDDAPPA A. PATIL

CHAPTER 6 ■ 2D Metals in Electronics   76

REZA RAHIGHI, SEYED MORTEZA HOSSEINI-HOSSEINABAD, AMIRMAHMOUD BAKHSHAYESH, AND SOMAYEH GHOLIPOUR

CHAPTER 7 ■ 2D Metals (Metallenes) in Spintronics   91

PUJA KUMARI AND SOUMYA J. RAY

CHAPTER 8 ■ 2D Metals in Optoelectronics   107

SWIKRUTI SUPRIYA AND RAMAKANTA NAIK

CHAPTER 9 ■ 2D Metals as Photocatalysts   122

SUBHADEEP BISWAS AND ANJALI PAL

vi ■ Contents

CHAPTER 10 ■ 2D Metal-based Electrocatalysts: Properties and Applications   138

PANDIAN LAKSHMANAN, KRISHNA KUMAR M, RADHA D PYARASANI, AND JOHN AMALRAJ

CHAPTER 11 ■ 2D Metals as Photocatalysts   156

ASMA RAFIQ, MISBAH NAZ, SHEHNILA ALTAF, AND SAIRA RIAZ

CHAPTER 12 ■ Advancement in 2D Metals as Photocatalysts   173

AZAM ASLANI, HADISEH MASOUMI, AHAD GHAEMI, AND RAM K. GUPTA

CHAPTER 13 ■ Two-Dimensional Metallenes for Photocatalysis Applications   189

SOUMITA SAMAJDAR AND SRABANTI GHOSH

CHAPTER 14 ■ 2D Metals for Fuel Cells   203

ASHWANI KUMA, JYOTI BALA, AND MOHD. SHKIR

CHAPTER 15 ■ 2D Metals for Methanol/Ethanol Oxidation   218

YULI MA AND JUNYU LANG

CHAPTER 16 ■ 2D Metals for Energy Storage Applications   234

ANIT JOSEPH, SUDHA PRIYANGA, AND TIJU THOMAS

CHAPTER 17 ■ 2D Metals for Energy Conversion Applications   253

FATEMEH BAHMANZADGAN, FERESHTEH POURESMAEIL, SHANLI NEZAMI, AND AHAD GHAEMI

CHAPTER 18 ■ 2D Metals for Sensors and Actuators   267

HADISEH MASOUMI, AZAM ASLANI, AHAD GHAEMI, AND RAM K. GUPTA

CHAPTER 19 ■ 2D Metals for Biomedical Applications   283

ELNAZ FEKRI AND MIR SAEED SEYED DORRAJI

INDEX, 299

Contributors

Shehnila Altaf
University of Engineering and Technology
Lahore, Pakistan

John Amalraj
Universidad de Talca
Talca, Chile

Azam Aslani
University of Guilan
Rasht, Iran

Fatemeh Bahmanzadgan
Iran University of Science and Technology
(IUST)
Tehran, Iran

Jyoti Bala
Goswami Ganesh Dutta Sanatan Dharma
College
Chandigarh, India

G. Banuprakash
SJB Institute of Technology
Bangalore, India

Amirmahmoud Bakhshayesh
Iran University of Science and Technology
(IUST)
Tehran, Iran

Pintu Barman
Kamrup College
Assam, India

Subhadeep Biswas
National Institute of Technology, Silchar
Assam, India

Apurba Das
Handique Girls College
Assam, India

Anindita Deka
Indian Institute of Technology, Guwahati
Assam, India

Mir Saeed Seyed Dorraji
University of Zanjan
Zanjan, Iran

Elnaz Fekri
University of Zanjan
Zanjan, Iran

Ahad Ghaemi
Iran University of Science and Technology
(IUST)
Tehran, Iran

Somayeh Gholipour
Adolphe Merkle Institute
Fribourg, Switzerland

Srabanti Ghosh
Academy of Scientific & Innovative Research
(AcSIR)
Ghaziabad, India

Ram K. Gupta
Pittsburg State University
Pittsburg, Kansas

Seyed Morteza Hosseini-Hosseinabad
Agency for Science, Technology, and Research (A*STAR), Innovis
Fusionopolis, Singapore

Anit Joseph
Indian Institute of Technology Madras (IITM)
Chennai, India

Ramakrishna Kadiyam
National Institute of Technology, Warangal
Telangana, India

Ashwani Kuma
Goswami Ganesh Dutta Sanatan Dharma College
Chandigarh, India

Puja Kumari
Indian Institute of Technology, Patna
Bihta, India

Krishna Kumar M
CHRIST (Deemed to be University)
Karnataka, India

Pandian Lakshmanan
Inha University
Incheon, South Korea

Junyu Lang
ShanghaiTech University
Shanghai, China

Jing Li
Harbin Institute of Technology
Heilongjiang Province, China

Sunil Kumar Baburao Mane
Khaja Bandanawaz University
Karnataka, India

Yuli Ma
Shanghai Jiao Tong University
Shanghai, China

Hadiseh Masoumi
Iran University of Science and Technology (IUST)
Tehran, Iran

Ramakanta Naik
Institute of Chemical Technology – Indian Oil Odisha Campus
Bhubaneswar, India

Misbah Naz
University of Education
Lahore, Pakistan

Shanli Nezami
Iran University of Science and Technology (IUST)
Tehran, Iran

Anjali Pal
Indian Institute of Technology, Kharagpur
West Bengal, India

Siddappa A. Patil
Jain (Deemed-to-be University)
Bangalore, India

Ravinder Pawar
National Institute of Technology, Warangal
Telangana, India

Fereshteh Pouresmaeil
Iran University of Science and Technology (IUST)
Tehran, Iran

Sudha Priyanga
Indian Institute of Technology, Madras (IITM)
Chennai, India

Radha D. Pyarasani
Universidad católica de Maule
Talca, Chile

Asma Rafiq
University of Naples
Naples, Italy

Reza Rahighi
Sungkyunkwan University
Suwon, Republic of Korea

Soumya J. Ray
Indian Institute of Technology, Patna
Bihta, India

Saira Riaz
University of the Punjab
Lahore, Pakistan

Soumita Samajdar
CSIR – Central Glass and Ceramic Research
  Institute
Jadavpur, India

Harini G. Sampatkumar
Jain (Deemed-to-be University)
Bangalore, India

Akanksha A. Sangolkar
National Institute of Technology, Warangal
Telangana, India

Naghma Shaishta
Khaja Bandanawaz University
Karnataka, India

Zhengguang Shi
Harbin Institute of Technology
Heilongjiang Province, China

Mohd. Shkir
King Khalid University
Abha, Saudi Arabia

Swikruti Supriya
Institute of Chemical Technology – Indian
  Oil Odisha Campus
Bhubaneswar, India

Tiju Thomas
Indian Institute of Technology, Madras
  (IITM)
Chennai, India

Hsu-Sheng Tsai
Harbin Institute of Technology
Heilongjiang Province, China

CHAPTER 1

Introduction to 2D Metals (Metallenes)

Sunil Kumar Baburao Mane and Naghma Shaishta

1.1 INTRODUCTION

Metal-based catalysts have always been essential for heterogeneous catalysis. The peculiar electron confinement in two-dimensional (2D) materials, a feature not available in other classes of nanomaterials (NMs) or their bulk counterparts, causes materials with ultra-thin 2D framework to demonstrate extraordinary and unusual chemical, electronic, and catalytic characteristics. It is widely recognized that the topologies and contents of these materials have a significant impact on their physicochemical characteristics and catalytic activity [1]. The majority of accessible active locations and greater responsiveness may be provided by metallic entities with larger surface-to-volume ratios and surface energies, which in turn leads to specific catalytic features that can be distinguished from that of their bulk counterparts. Thus, to achieve outstanding activity, specificity, and durability regarding different heterogeneous catalytic processes, substantial and in-depth studies are conducted in regulated fabrication of improved catalysts containing metallic nanoparticles (NPs), nanowires (NWs), and nanosheets (NSs) [2–4].

The study of two-dimensional materials (2DMs) has expanded significantly since the initial finding of graphene in 2004. These materials include elemental 2D materials such as graphene, hexagonal boron nitride (h-BN), transition metal sulfides (TMDs), layered double hydroxides (LDHs), 2D metal-organic frameworks (MOFs), and their derivatives [5, 6]. These 2D materials are incredibly useful in the disciplines of biomedical science, detectors, optics and electronics, heterogeneous catalysis, energy conversion and preservation, and plentiful defect sites and exterior active sites. They offer large specific surface areas, outstanding conductivity, and even surface varieties. Though elemental 2DMs above graphene remain uncommon, despite their unique electronic and structural characteristics and wide

DOI: 10.1201/9781032645001-1

2 ■ 2D Metals

range of potential applications, the introduction of 2DMs such as graphene, TMDs, LDHs, and 2D MOFs has been provided in some inspiring reviews in detail thus far [7, 8].

Metallenes (METs) are recent2D metallic material with a thickness of only a few atomic layers. Compared to nonmetallic elemental 2DMs such as graphene, phosphorene, silicone, and borophene, METs can provide high-proportioned surface unsaturated metal locations, which typically serve as the active sites. This makes them especially well suited in the area of heterogeneous catalysis [9]. The atoms in METs exhibit greater connection as compared to traditional fine metal NPs, which may result in increased electron transfer throughout catalysis, particularly electrocatalysis. Furthermore, METs have more versatile uses because they can stand alone without the need for support. Nonetheless, the production of atomically thin METs continues to be extremely difficult since their surface energy sharply rises with decreasing thickness.

In this chapter, we highlight the latest developments in METs for heterogeneous catalysis. We begin by providing a quick overview of METs, including their definition, categorization, and some unique characteristics. The essential procedures for MET synthesis are then outlined. We then go into great detail about the uses of METs in a range of heterogeneous catalytic processes, particularly in electrocatalysis. In conclusion, we outline the outstanding issues and suggest related directions for the continued advancement of METs in heterogeneous catalysis. Lastly, we outline the remaining difficulties and future directions for MET growth in the direction of heterogeneous catalysis. It is our goal that this study will inspire additional research on the production and uses of METs.

1.2 DETAILED DESCRIPTION OF METS

1.2.1 Definition

A single atomic layer made up of metal atoms is, in theory, referred to as a well-defined metallene. The term MET has an analogous genesis to graphene with monolayer-thick metal sheet and has a great degree of flexibility. Nevertheless, obtaining independent monolayer metal sheets is highly challenging due to inadequate production options and thermodynamic unpredictability [9]. Because of its high surface free energy, it can exhibit surface curvature while unsupported or atomical surface uniformity when placed on a substrate. When metal decreases to a monolayer thickness, MET and graphene have similar electrical characteristics, which is another reason why "ene" is used for metal [10].

1.2.2 Features of METs

METs have various unique characteristics that make them particularly suitable for use in heterogeneous catalysis concerning other non-2DMs and their bulk equivalents. METs' ultrahigh surface-to-volume percentages are mainly due to their atomically thin frameworks, which also allow for the exposure of ultrahigh dimensions of coordinatively unsaturated metal atoms at their edges and surface which are typically thought of as active centers possibly enhancing their catalytic performance [11]. The perimeter atoms have more defects, dislocations, and other coordination unsaturation than the surface atoms, which may indicate enhanced catalytic activity. Perhaps more significantly, METs' fully exposed active sites enable easy functionalization like ligand alterations, alloying, doping, and imperfections which is highly advantageous for adjusting the materials' security, electronic frameworks, physicochemical characteristics, and ultimately catalytic success [4].

In the meantime, the adsorption of reactants onto METs can be significantly enhanced by their large specific surface areas and enhanced surface energy of the metal atoms. Furthermore, METs' ultrathin properties can greatly shorten the distances over which reactants and products diffuse as well as the distances over which charges transfer, speeding up the transfer of mass and charges throughout the catalytic procedure [12]. In addition, several METs exhibit excellent application in photothermal hyperthermia, photocatalysis, photothermal catalysis, and photodynamic treatment due to their wide light absorption spectrum from the ultraviolet to the near-infrared area [13]. However, because of their zero bandgap characteristics, most of them are unable to serve effectively as photocatalysts in terms of photocatalysis. But while they are semiconductors with thickness-dependent bandgaps, bismuth and antimony-based METs are the anomalies [14].

1.2.3 Classification

The unique characteristics of METs are mostly dictated by the inherent properties of metals themselves, albeit they can be adjusted by adjusting compositions, diameters, thicknesses, and surface chemistry. As a result, it's essential to classify METs appropriately. The METs that have been discovered thus far are mostly divided into three categories, such as (1) transition metal-based, which includes noble (such as Au, Pd, Pt, Ru, Ir, Rh, Ag) and non-noble (such as Fe, Co, Ni, Cu, Zn) METs; (2) main group METs, such as gallenene (Ga-ene), germanene (Ge-ene), stanene (Sn-ene), plumbene (Pb-ene), antimonene (Sb-ene), and bismuthene (Bi-ene); and (3) alloy-based metallenes [15, 16].

1.2.4 Synthesis of METs

To investigate the physicochemical characteristics and catalytic efficiency of METs, it is necessary to first prepare them in a controlled manner with the appropriate thickness, content, and crystal phases. Numerous methods have been established for the production of 2D metal NMs with regulated thickness, layout, and crystal arrangement, owing to the collaborative efforts of multiple organizations. The techniques for MET synthesis that are commonly used are briefly introduced and summarized in this section. These techniques can be broadly classified into three categories: (1) bottom-up approaches, such as the different wet chemical methods, (2) top-down strategies, such as liquid-phase and mechanical exfoliation, and (3) topotactic metallization methods.

1.2.4.1 Bottom-up Approaches

Anisotropic metal development at the nanoscale is the main cause for worry when using bottom-up approaches to generate MET. The traditional LaMer model states that to cause the ensuing creation of an ultrathin metal nanostructure, asymmetrical nucleation of metal atoms for the birth of 2D nuclei is essential. To make this ultrathin nanostructure thermodynamically advantageous, a well-selected ligand selectively linked to the bottom planes may accomplish passive expansion perpendicular to the nanostructure's bottom planes, thereby confining the thickness of the 2D nanostructure. The investigators showed that to create ultrathin Pd METs, the ligands required for anisotropic confinement can be either organic surfactants or inorganic compounds like CO. Anisotropy can also be produced via laminar frameworks, where metal growth is tightly controlled within the interlayer space of the layered structures, much like in ligand-confined development. A space-confined development of PdCo single layers in the angstrom-sized interlamination of layered crystalline clay material montmorillonite can yield PdCo single atomic layers [17, 18, 19].

Apart from the restricted development methods, template-mediated anisotropy can also thermodynamically favor 2D nanostructure. To create ultrathin metallic layers, metal atoms can be selectively formed on the surface of a template that has an atomically flat 2D surface. Graphene has been utilized as an instinctively chosen template to prepare METs. According to Zhang et al., Au atoms can be deposited on graphene to produce Au MET with an unusual hexagonal close-packed (hcp) crystal arrangement [20]. It is noteworthy that the free-standing Fe monolayer floating within graphene holes was obtained by stabilizing the thermodynamically unfavorable MET in the graphene holes [21]. In addition to graphene, metallic templates are frequently employed in the atomic layer-by-layer deposition method for the epitaxial development of metal atoms. This technique allows for the properly managed deposition of a range of metal atoms on templates' surfaces, opening the door to the production of MET with more complicated compositions. Lu et al., for instance, described the synthesis of MET, which is made up of high-entropy alloys, by use of a specific etching process that involves the galvanic substitution process and co-reduction of different metal ions on the surface of an Ag pattern [22].

In addition to the previously described techniques, various novel bottom-up processes such as pulsed laser deposition, pyrolysis, electrodeposition, and photoreduction can also be utilized to produce MET [23]. It is commonly recognized that laminar substances with weak interlayer van der Waals interactions and strong in-plane covalent bonds, including graphite, transition metal dichalcogenides, and layered double hydroxides, can be exfoliated into one or more monolayers [24, 25]. Similarly, metallenes made of layered elements (such as Ga, Sb, and Bi) can be easily extracted from their bulk equivalents using a variety of exfoliation methods [23]. Of these, liquid-phase exfoliation is the most popular because of it is easy to apply and also because of its efficiency.

1.2.4.2 Top-down Techniques

It has been demonstrated that layered materials with weak interlayer van der Waals interactions and strong in-plane covalent bonds, such as graphene, TMDs, and LDHs, can be readily exfoliated to form mono- or few-layer structures. Corresponding to this, METs with layered structures, such as Sb-ene and Bi-ene, can be easily extracted from bulk forms using a variety of exfoliation methods and it has also been reported that ion intercalation, accompanied by exfoliation powered by ultrasonication, can provide free-standing layers of Bi [26]. Apart from liquid-phase exfoliation, mechanical exfoliation from their bulk equivalents can also be used to generate 2D materials like graphene; this method has been applied in the production of Sb METs. By implementing a mechanical exfoliation method that involves pregrinding-induced shear force and ultrasonic exfoliation, it is possible to manufacture Sb plates and Sb METs with thicknesses of single and multiple layers [27]. Moreover, the production of METs also makes use of a few other top-down techniques, such as dealloying, solid-melt exfoliation, hot-pressing exfoliation, and electrochemical exfoliation. For example, solid-melt exfoliation was used by Vidya and colleagues to effectively prepare Ga metallenes from bulk Ga. Because bulk Ga has a low melting point (about 302.85 K), it is possible to remove a few layers of Ga from liquid Ga on various substrates [28]. A further popular instance involves the in situ dealloying of a bulk Au-Ag alloy using an electron beam in a transmission electron microscope (TEM) to produce a free-standing Au monolayer framed in bulk crystals [29].

1.2.4.3 Techniques for Topotactic Metallization

In addition to the several techniques previously discussed, topotactic metallization of the layered compounds comprising metals is an additional simple and efficient technique that helps to achieve controlled production of METs due to the unique architecture and content. In particular, under reductive circumstances, the nonmetallic groups (OH^-, O^{2-}) present in the interlayers and/or in-planes of the layered metal compounds are removed whereas their layered assemblies are preserved, resulting in the creation of METs. According to Kuang et al., for example, $Ni(OH)_2$ nanosheets (NSs) can be reduced in situ under solvothermal parameters to create ultrathin Ni NS arrays with a thickness of 2.2 nm. In this process, the Ni NSs displayed a single-crystalline lamellar morphology with an uncovered (111) facet [30].

Using the same procedure, the precursor $NiMoO_4$ NS arrays were converted into ultrathin Ni-Mo alloy NSs with a thickness of 2.0–2.1 nm [31]. Furthermore, it is important to remember that the topotactic metallization of layered ruthenate in an H_2 atmosphere may be used to create mono-layered Ru NSs [32]. Similarly, RuO_2 NSs were reduced with H_2 to create the two-layer Ru NSs. Additionally, under a reductive environment, Mo-incorporated Ni NSs and Co NSs have been effectively synthesized from their hydroxide counterparts [33–35]. Additionally, the in-plane/interlayered nonmetallic moieties from the laminar metal mixtures can be extracted under reductive circumstances to yield ultrathin layers consisting solely of metals without affecting the initial layered assemblies. Katsutoshi et al. stated that a Ru monolayer can be created, for example, by topotactic metallization of layered RuO_2 in an H_2 environment [36]. Furthermore, reductive synthesis of non-noble MET consisting of Co or Ni has been accomplished from their hydroxide counterparts [15].

1.3 CONSTRUCTING MET FOR ELECTROCHEMICAL REACTIONS

Modifying the physicochemical characteristics of METs for electrocatalytic applications is possible thanks to the variety of synthetic methodologies. As per the Sabatier principle, the relationship among the surface atoms of a heterogeneous catalyst and the intermediates participating in the catalytic reaction's rate-determining phase greatly influences the catalyst's effectiveness [37]. The development and occupation of bonding and antibonding orbitals among the intermediate and the catalytic sites are the basic root causes of the aforementioned interaction. This concept states that for a catalyst with an ideal activity, the

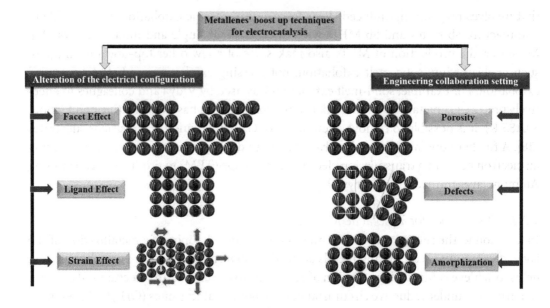

FIGURE 1.1 Diagrammatic representation of the design idea for enhancing metallenes' electrocatalytic qualities. The two main categories of optimization methodologies are as follows: (1) electronic framework manipulation through various pathways such as facet, ligand, and strain effects; (2) creating flaws, amorphization, or surface porosity to engineer the coordination environment.

capacity for adsorption of intermediate species ought to be at a medium value, meaning it must be neither too strong nor too weak. According to the d-band hypothesis, a shift in the electronic (d-band) configuration of transition metals can fundamentally modify the coupling strength among the adsorbate and the d-band, which will vary the metals' electrocatalytic activity [38].

Lattice stress, elemental composition, and surface atomic organization are the main factors that affect a metal's electrical behavior. Furthermore, the adsorption and dissociation energies of reactant entities on a metal surface can be intrinsically changed by the coordination number (CN) of active sites, which can change the catalytic activity of metal electrocatalysts [39]. Taking into consideration the fundamental principles of catalyst technology, two main approaches are chosen for the controlled production of METs. The first is electronic structure modification, which involves adjusting the surface strain, surface content, and surface facet. The alternative approach involves manipulating the surface atoms' coordination context (Figure 1.1).

1.4 APPLICATIONS OF METS

1.4.1 As Heterogeneous Catalysis

METs are thought to be the best heterogeneous catalysts due to their exceptional physicochemical characteristics, which include a large specific surface area, ultrahigh accessibility of unsaturated metal sites, and faulty sites. In order to assist readers comprehend the advantages of METs over traditional metal-based catalysts, we will briefly discuss their uses in a variety of heterogeneous catalysis processes here, such as heterogeneous organic catalysis, photocatalysis, and electrocatalysis. Among these, the uses of METs in electro-catalysis will be emphasizeddue to its promising future for energy conversion and storage.

1.4.2 Heterogeneous Organic Catalyst

The synthesis of fine compounds depends critically on conventional noble metal-based catalysts engaged in heterogeneous organic catalysis. Yet they frequently have poor cata-lytic properties in terms of operation, selectivity, or stability, due to the few active sites, intrinsic activity, unpredictable kinetic procedure, and poor structural stability. On the other hand, METs are anticipated to excel in these reactions due to their distinct physico-chemical characteristics, which set them apart from their bulk counterparts. For instance, the ultrathin Au NSs demonstrated much enhanced catalytic activity and stability relative to the Au NPs for the particular and solvent-free oxidation of C–H bonds due to the good exposure of the low-coordinated edge sites [40, 41].

1.4.3 Photo Catalysis

Because the redox processes sparked by photogenerated electron-hole pairs are the funda-mental component of photocatalysis, most metals cannot be employed effectively as pho-tocatalysts due to their intrinsic zero-gap nature. As a substitute, one of the best ways to enhance semiconductor photocatalysts' photocatalytic efficiency is to use METs with par-ticular geometries and contents as cocatalysts or substrates. Xu et al. presented a $CsPbBr_3$ nanocrystal/Pd NS composite photocatalyst as a proof of concept, in which the Pd NSs both enhances and speeds up the reaction kinetics and separate photo-excited electron-hole pairs more quickly [42]. Consequently, the addition of Pd NSs increases the rate of electron consumption in photocatalytic CO_2RR by a factor of 2.43. In a different investi-gation, Zhu et al. found that the density of edge sites on the Pd NS cocatalyst affects the photocatalytic activity of TiO2-Pd hybrid frameworks. Reduced Pd NS size will result in a rise in edge-site density, which will boost the relevant hybrid architectures' activity as well as selectivity in photocatalytic CO_2RR [43].

Additionally, He et al. created a vertically aligned $NiCo_2O_4$.Au–$NiCo_2O_4$ sandwich-type heterojunction photocatalyst using the Au NSs as both a cocatalyst and a substrate [14]. The efficiency of OER is significantly enhanced because it facilitates the separation and transmission of photo-excited carriers. Another crucial objective is to investigate effective non-noble MET-based cocatalysts for photocatalysis in addition to noble MET.

1.4.4 Electrocatalysis

The overuse of nonrenewable fossil fuels exacerbates energy and environmental issues, which puts society's ability to develop sustainably in jeopardy. Consequently, there is a strong need to create renewable and fossil-free energy conversion methods. Surprisingly, electrocatalysis coupled with renewable energy is thought to be a potential method for con-verting common feedstocks like water, CO_2, and atmospheric nitrogen into useful com-pounds and fuels.

8 ■ 2D Metals

1.4.4.1 Electrocatalytic Water Splitting

One of the greatest viable replacements for fossil fuels is hydrogen, which has an elevated density of energy and zero carbon emissions. The most effective technique to make real "green" hydrogen appears to be electrocatalytic water splitting in an atmospheric surrounding, as opposed to the "gray" hydrogen generated by conventional thermocatalysis techniques. It is widely recognized that cathodic hydrogen evolution process (HER) and anodic OER are involved in the electrocatalytic water splitting mechanism. Yet, the widespread use of water splitting methods is now hindered by the unavailability of highly efficient but reasonably priced electrocatalysts.

According to both theory and experimentation, noble metal-based materials – that is, Pt/Ru-based materials for HER and Ir/Ru-based materials for OER, are the most efficient electrocatalysts for water splitting. Unfortunately, their limited availability and high cost make widespread deployment very difficult. Thankfully, 2D engineering offers a solid platform to improve those noble metal-based electrocatalysts' intrinsic activity and atomic consumption, thereby lowering their usage [44].

1.4.4.2 Hydrogen Evaluation Reaction

Although it's still quite difficult, photocatalytic generation of hydrogen is a green, pollution-free technology that helps ease the fossil fuel issue. The HER, a crucial side process for the creation of hydrogen by water splitting, is often aided by noble metals including Pt, Pd, Ir, and Rh. A range of noble metallenes have been used in the HER with exceptional results, taking advantage of the 2D morphology. Here, 2D ultrathin palladium MET (Pd-ene) is firstly produced as cocatalysts for hydrogen evolution of g-C_3N_4 nanosheets by Qian et al [45]. Using acetylacetonate palladium as a Pd source, hydrothermal and deposition methods were used to create the Pd-ene and Pd-ene/g-C_3N_4 photocatalysts (Figure 1.2). The Pd-ene, with a size of 100 nm, was evenly distributed on the surface of g-C_3N_4 nanosheets, establishing a 2D/2D structure. With a photocatalytic hydrogen production rate of 31,292 μmol/h/g, the Pd-ene/g-C_3N_4 loading with 2 wt% Pd-ene demonstrated the highest rate, surpassing the PdNC/g-C_3N_4 loading with Pd nanocluster by almost 98 times.

1.4.4.3 Oxygen Reduction Reaction

Formate is the fuel used in direct formate fuel cells (DFFCs), which are energy conversion technologies. However, COads produced by formate dehydration build-up on the Pd surface and result in catalyst poisoning and deactivation. In order to improve the anti-CO poisoning capability, Hongjie et al. created a hydrogen intercalated palladium–copper MET (PdCuH MET) [46]. The mass and specific activity of PdCuH MET/C for formate oxidation were 3.4 and 3.2 times that of Pd/C, respectively (Figure 1.3). PdCuH MET's mass and specific activity in ORR were 12.4 and 13.9 times greater than Pt/C. The peak power and limiting current density of PdCuH MET/C || PdCuH metallene/C were 79.5% and 159.7% greater than those of Pt/C || Pd/C, respectively, when they built PdCuH metallene/C as an electrode material for DFFC.

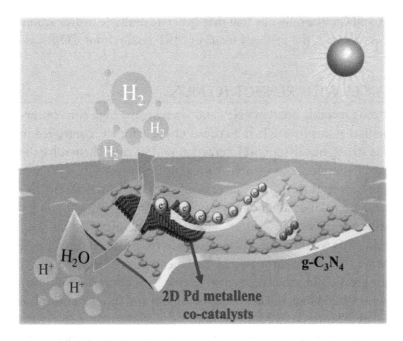

FIGURE 1.2 Graphical abstract Ultrathin Pd metallenes as novel cocatalysts for efficient photocatalytic hydrogen production. Adapted with permission [45]. Copyright © 2023, Elsevier Ltd.

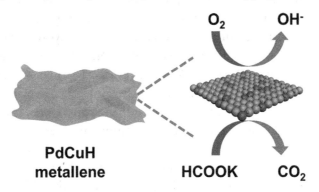

FIGURE 1.3 Graphical abstract showing the Palladium-copper metallene's resistance to CO poisoning was strengthened by hydrogen intercalation for direct formate fuel cells. Adapted with permission [46]. Copyright © 2023, Elsevier Ltd.

1.4.4.4 Carbon Dioxide Reduction Reaction (CO₂RR)

One of the most exciting ways to address the increasing CO_2 levels in the environment is through the electrochemical reduction of CO_2 into useful fuels (e.g., CO, formate, CH_4, C_2H_4, alcohols). This process can accomplish CO_2 fixing and energy storage at the same time. The enormous promise that MET holds for enhancing CO_2RR activity, product selectivity, and stability has led to a recent surge in interest in these areas. On the basis of the key CO_2RR products, earlier studies have determined that monometallic catalysts can be empirically classified into four main subgroups: CO-selective metals (such as Au, Ag, and Zn), formate-selective metals (such as Sn, Bi, Sb, and In), H_2-selective metals (such as Fe,

Co, Ni, and Pt), and Cu (the only one that may generate multiple hydrocarbon and carboxylic products). As a result, the primary focus of MET research for CO_2RR is on products [47].

1.5 CHALLENGES WITH RESPECT TO METs

Breaking the strong metallic bond in nNPs and lowering the high surface energy of unsaturated coordination atoms in METs is a crucial obstacle for the controlled production of METs. Using the strong metal-support interaction (SMSI) could help solve these problems and is anticipated to develop into a viable method of forming METs. For starters, the high surface energy of the metal NPs can be successfully reduced by the SMSI. The compromise between the catalytic balance and activity of conventional SMSI was going to be broken once the chosen encourages had greater contact with the metal atoms than the metal–metal attachment. This would cause the metal atoms to spread on the support to form 2D metal METs rather than three-dimensional (3D) metal NPs. More crucially, the use of 2D metal MET can greatly aid structure-sensitive processes because of its distinctive atom organization, which facilitated quicker mass movement and simpler desorption of particular intermediate in the interfacial configuration.

The present investigation uses the semi-hydrogenation of alkynes to alkenes as an example reaction since it is an important structurally sensitive reaction that is used in the commercial production of vitamins, polymers, and medicines. It is known that the over-hydrogenation of alkene, which results from the competing adsorption of alkyne and alkene, mostly happens at the edge sites, whereas the semi-hydrogenation of alkyne to alkene occurs predominantly at the plane sites of Pd NPs. When alkyne is semi-hydrogenated to alkene, ultrathin 2D MET NMs that reveal particular aspects like (111) and (100) plane sites ought to be preferred. Consequently, it is a crucial yet difficult topic to use SMSI to induce the in situ growth of Pd METs on the support in order to accomplish the semi-hydrogenation of alkynes in an excellent operation, high selectivity, and high stability method. The primary difficulty in production is the thermodynamically undesirable creation of atomically thin 2DMs because atoms favor bulk close-packed configurations.

Likewise, it has been documented that the edge defects on the MET layers may exhibit greater intrinsic activity in comparison to the facet atoms. Consequently, it can be said that decreasing the lateral size of 2D electrocatalytic materials may be just as crucial for enhancing efficiency. However, research on MET quantum dots (QDs) for electrocatalysis has not yet been published, possibly because of the significant difficulty in producing controlled production [4].

1.6 CONCLUSION

METs, the latest generation of 2DMs with atomically thin layers have stimulated a lot of interest in energy and catalytic uses because of their unique physicochemical and electrical characteristics. Specifically, the extremely high accessibility of the poorly coupled metal atoms in METs provides them with great catalytic potential, enhancing atom consumption and intrinsic activity. The production of METs has advanced significantly, leveraging the swift growth of synthetic methods at the nanoscale. This chapter has highlighted the

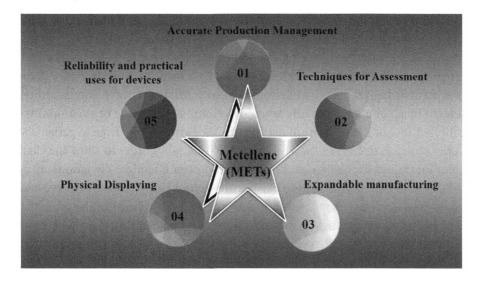

FIGURE 1.4 Prospects and possible avenues for future MET-related material studies.

synthesis, characterization, and design concepts of METs with regard to electrocatalysis for energy conversion, providing an overview of current developments in this field. In this section we address a few obstacles and potential avenues for further study to spur additional important contributions to this area (Figure 1.4).

1.6.1 Accurate Production Management

Even though METs with varying thicknesses, compositions, lattice constants, and porosities have been successfully synthesized, accurate oversight of the production of them at the atomic scale remains tough because of the complexity of high-level tampering when taking into account anisotropic development of metals in order to retain their 2D shape with atomic thickness. Synthetic regulate accuracy encompasses various aspects, such as scaling down thickness to an unfamiliar level with a stable and independent framework, creating anisotropic strain across the surface, accurately distributing elements in multielement METs and creating new crystal periods like body-centered cubic (bcc) or orthorhombic buildings with atomic thickness. To satisfy the demands of electrocatalysis for different energy conversions, all the MET-related factors must generally be carefully regulated and adjusted.

1.6.2 Techniques for Assessment

Despite significant progress in regulating the production of many METs through bottom-up growth, a thorough understanding of the development strategy of these special 2D metals still requires the application of both theoretical models and laboratory techniques. Because of this, detailed analysis of the development procedure for METs from nucleation to anisotropic growth needs sophisticated assessment approaches. The development of in situ characterization instruments, like as in situ transmission electron microscope (TEM), x-ray diffraction (XRD), Raman, and X-ray absorption spectroscopy (XAS), will facilitate the revelation of various production techniques, which will aid in the direction and forecasting of synthesis. Furthermore, characterizing atomic arrangements with standard methods is often challenging due to the sensitive nature of METs. For instance, METs are prone to disintegrating when viewed under a transmission electron microscope (TEM) due to their high sensitivity to electron beams. MET electron beam volatility must be considered in account, particularly when describing METs made of solution mixtures. Specifically, because of their ultrathin nature, the METs need more time to gather a strong enough energy-dispersive X-ray spectroscopy (EDS) signal in order to show the elemental dispersion. This could exacerbate architectural and even chemical deviations caused by electron beams. It may be possible to resolve this problem by employing cryo-electron microscopy. Furthermore, when used in practical and electrochemical uses, the thermodynamically adverse arrangement is more likely to undergo conformational alterations. As a result, creating new in situ characterization instruments is crucial to monitoring these potential changes in various MET uses.

1.6.3 Expandable Manufacturing

The majority of energy-associated uses encounter the difficulty of producing METs in huge amounts without sacrificing control over parameters including permeability, surface aspects, thickness, and content. Growing the volume of the reaction solution is a common tactic used in wet chemical bottom-up process synthesis to boost productivity. For instance, with strict control over multiple variables, the yield of PdMo METs can be raised by 40 times by raising the volume of reaction from 5 mL to over 200 mL. Yet, as the reaction volume is progressively ramped up, it becomes extremely difficult to get METs of excellent grade due to the inherent thermal and compositional heterogeneity over a wide volume. Thus, in order to guarantee the asymmetrical development of METs, different strategies must be established. By separating the product of the reaction into droplets, connectors with a millimeter scale diameter to preserve the uniformity of the reaction fluid, continuous-flow reactors, for instance, were utilized to tackle the heterogeneity problem related to the reaction volume. With this innovative method, scalable NM production can be achieved with well-controlled reaction conditions without sacrificing product quality.

1.6.4 Physical Displaying

To demonstrate the value of innovative architectures in a range of uses, physical modeling is essential. The appropriate study on materials connected to MET remains in its early stages, nevertheless. As previously mentioned, there is still a restricted amount of elements in the periodic table that can be obtained by experimentation or theoretical calculations related to METs. Thus far, increasing the theoretical modeling of 2D metal materials with atomic thickness has been crucial to increasing the appeal of this kind of materials to various groups for possible uses in medicine, electronics, the field of optics and catalysis.

1.6.5 Reliability and Practical Uses for Devices

Despite recent advancements in the production and usage of METs as electrocatalysts, several obstacles must continue to be overcome before this kind of NMs may be put to practical use. In fact, a large number of the electrocatalytic uses were conducted in laboratories as proof of concept. For usage in practical gadgets, METs' expenses, processing, and viability of scaling up must all be considered. In these everyday uses, the majority of METs may also lack durability and pollution tolerance. These issues could be resolved by directed construction of this material group, which would render the METs strong and useful in practical devices. Thus, much work needs to be done before this particular group of 2D metals may be applied to actual energy conversion, particularly in situations where extreme operating conditions are involved.

REFERENCES

1. Chen Y, Fan Z, Zhang Z, Niu W, Li C, Yang N, Chen B, Zhang H. Two-dimensional metal nanomaterials: synthesis, properties, and applications. *Chem. Rev.* **2018**, 118, 6409–6455.
2. Zhu QL, Xu Q. Immobilization of ultrafine metal nanoparticles to high-surface-area materials and their catalytic applications. *Chemistry.* **2016**, 1, 220–245.
3. Jiqing J, Rui L, Shoujie L, Weng-Chon C, Chao Z, Zheng C, Yuan P, Jianguo T, Konglin W, Sung FH, Hao MC, Lirong Z, Qi L, Xuan Y, Bingjun X, Hai X, Jun L, Dingsheng W, Qing P, Chen C, Yadong L. Copper atom-pair catalyst anchored on alloy nanowires for selective and efficient electrochemical reduction of CO_2. *Nat. Chem.* **2019**, 11, 222–228.
4. Prabhu P, Lee JM. Metallenes as functional materials in electrocatalysis. *Chem. Soc. Rev.* **2021**, 50, 6700–6719.
5. Jin H, Song T, Paik U, Qiao SZ. Metastable two-dimensional materials for electrocatalytic energy conversions. *Acc. Mater. Res.* **2021**, 2, 559–573.
6. Chakraborty G, Park IH, Medishetty R, Vittal JJ. Two-dimensional metal-organic framework materials: synthesis, structures, properties and applications. *Chem. Rev.* **2021**, 121 3751–3891.
7. Georgakilas V, Tiwari JN, Kemp KC, Perman JA, Bourlinos AB, Kim KS, Zboril R. Noncovalent functionalization of graphene and graphene oxide for energy materials, biosensing, catalytic, and biomedical applications. *Chem. Rev.* **2016**, 116 5464–5519.
8. Wei T, Na K, Xiaoyuan J, Yupeng Z, Amit S, Jiang O, Baowen Q, Junqing W, Ni X, Chulhun K, Han Z, Omid CF, Jong SK. Emerging two-dimensional monoelemental materials (Xenes) for biomedical applications. *Chem. Soc. Rev.* **2019**, 48, 2891–2912.
9. Liu Y, Dinh KN, Dai Z, Yan Q. Metallenes: recent advances and opportunities in energy storage and conversion applications. *ACS Mater. Lett.* **2020**, 2, 1148–1172.
10. Tang C, Qiao SZ. 2D atomically thin electrocatalysts: from graphene to metallene. *Matter.* **2019**, 1, 1454–1455.
11. Cao C, Ma D-D, Gu J-F, Xie X, Zeng G, Li X, Han S-G, Zhu Q-L, Wu X-T, Xu Q. Metal–organic layers leading to atomically thin bismuthene for efficient carbon dioxide electroreduction to liquid fuel. *Angew. Chem. Int. Ed.* **2020**, 59, 15014–15020.
12. Gao S, Lin Y, Jiao X, Sun Y, Luo Q, Zhang W, Li D, Yang J, Xie Y. Partially oxidized atomic cobalt layers for carbon dioxide electroreduction to liquid fuel. *Nature.* **2016,** 529, 68–71.
13. Kong X, Liu Q, Zhang C, Peng Z, Chen Q. Elemental two-dimensional nanosheets beyond graphene. *Chem. Soc. Rev.* **2017**, 46, 2127–2157.
14. He S, ChaiJ, Lu S, Mu X, Liu R, Wang Q, Chen F, Li Y, Wang J, Wang B. Solution-phase vertical growth of aligned $NiCo_2O_4$ nanosheet arrays on Au nanosheets with weakened oxygen-hydrogen bonds for photocatalytic oxygen evolution. *Nanoscale.* **2020**, 12, 6195–6203.
15. Changsheng C, Qiang X, Qi-Long Z. Ultrathin two-dimensional metallenes for heterogeneous catalysis. *Chem. Catal.* **2022,** 2, 693–723.
16. Teng Z, Li M, Li Z, Liu Z, Fu G, Tang Y. Facile synthesis of channel-rich ultrathin palladium-silver nanosheets for highly efficient formic acid electrooxidation. *Mater. Today Energy.* **2021**, 19, 100596.

14 ■ 2D Metals

17. Wenjin Z, Lei Z, Piaoping Y, Congling H, Zhibin L, Xiaoxia C, Zhi-Jian Z, Jinlong G. Low-coordinated edge sites on ultrathin palladium nanosheets boost carbon dioxide electroreduction performance. *Angew. Chem. Int. Ed.* **2018,** 57, 11544-11548.

18. Dongdong X, Ying L, Shulin Z, Yanan L, Min H, Jianchun Bao. Novel surfactant-directed synthesis of ultra-thin palladium nanosheets as efficient electrocatalysts for glycerol oxidation. *Chem. Commun.* **2017,** 53, 1642–1645.

19. Jiang J, Ding W, Li W, Wei Z. Freestanding single-atom-layer Pd-based catalysts: oriented splitting of energy bands for unique stability and activity. *Chemistry.* **2020,** 6, 431-447.

20. Xiao H, Shaozhou L, Yizhong H, Shixin W, Xiaozhu Z, Shuzhou L, Chee LG, Freddy B, Chad A, Mirkin Hua Zhang. Synthesis of hexagonal close-packed gold nanostructures. *Nat. Commun.* **2011,** 2, 1–6.

21. Jiong Z, Qingming D, Alicja B, Gorantla S, Alexey P, Jurgen E, Mark HR. Free-standing single-atom-thick iron membranes suspended in graphene pores. *Science.* **2014,** 343, 1228–1232.

22. Lu T, Mingzi S, Yin Z, Mingchuan L, Fan L, Menggang L, Qinghua Z, Lin G, Bolong H, Shaojun G. A general synthetic method for high-entropy alloy subnanometer ribbons. *J. Am. Chem. Soc.* **2022,** 144, 23, 10582-10590.

23. Yaoda L, Khang ND, Zhengfei D, Qingyu Y. Metallenes: recent advances and opportunities in energy storage and conversion applications. *ACS Mater. Lett.* **2020,** 2, 9, 1148-1172.

24. Jingfang Y, Qiang W, Dermot O, Luyi S. Preparation of two dimensional layered double hydroxide nanosheets and their applications. *Chem. Soc. Rev.* **2017,** 46, 5950–5974.

25. Vasilios G, Jitendra NT, Christian K, Jason AP, Athanasios BB, Kwang SK, Radek Z. Noncovalent functionalization of graphene and graphene oxide for energy materials, biosensing, catalytic, and biomedical applications. *Chem. Rev.* **2016,** 116, 5464–5519.

26. Qi-Qi Y, Rui-Tong L, Chao H, Yi-Fan H, Lin-Feng G, Bing S, Zhi-Peng H, Lei Z, Chen-Xia H, Ze-Qi Z, Chun-Lin S, Qiang W, Yu-Long T, Hao-Li Z. 2D bismuthene fabricated via acid-intercalated exfoliation showing strong nonlinear near-infrared responses for mode-locking lasers. *Nanoscale.* **2018,** 10, 21106–21115.

27. Xin W, Junjie H, Benqing Z, Youming Z, Jiatao W, Rui H, Liwei L, Jun S, Junle Q. Bandgap-tunable preparation of smooth and large two-dimensional antimonene. *Angew. Chem. Int. Ed.* **2018,** 57, 8668–8673.

28. Vidya K, Atanu S, Yuan Z, Sanjit B, Praveena M, Syed A, Anthony SS, Robert V, Abhishek KS, Pulickel MA. Atomically thin gallium layers from solid-melt exfoliation. *Sci. Adv.* **2018,** 4, e1701373.

29. Xuelu W, Chunyang W, Chunjin C, Huichao D, Kui D. Free-standing monatomic thick two-dimensional gold. *Nano Lett.* **2019,** 19, 4560–4566.

30. Kuang Y, Feng G, Li P, Bi Y, Li Y, Sun X. Single-crystalline ultrathin nickel nanosheets array from in situ topotactic reduction for active and stable electrocatalysis. *Angew. Chem. Int. Ed.* **2015,** 55, 693–697.

31. Zhang Q, Li P, Zhou D, Chang Z, Kuang Y, Sun X. Superaerophobic ultrathin Ni–Mo alloy nanosheet array from in situ topotactic reduction for hydrogen evolution reaction. *Small.* **2017,** 13, 1701648.

32. Fukuda K, Sato J, Saida T, Sugimoto W, Ebina Y, Shibata T, Osada M, Sasaki T. Fabrication of ruthenium metal nanosheets via topotactic metallization of exfoliated ruthenate nanosheets. *Inorg. Chem.* **2013,** 52, 2280–2282.

33. Takimoto D, Sugimoto W, Yuan Q, Takao N, Itoh T, Duy TVT, Ohwaki T, Imai H. Two-dimensional effects on the oxygen reduction reaction and irreversible surface oxidation of metallic Ru nanosheets and nanoparticles. *ACS Appl. Nano Mater.* **2019,** 2, 5743–5751.

34. Tong R, Sun Z, Wang X, Yang L, Zhai J, Wang S, Pan H. Mo incorporated Ni nanosheet as high-efficiency co-catalyst for enhancing the photocatalytic hydrogen production of g-C_3N_4. *Int. J. Hydrogen.* **2020,** 45, 18912–18921.

35. Wang L, Zhang W, Zheng X, Chen Y, Wu W, Qiu J, Zhao X, Zhao X, Dai Y, Zeng J. Incorporating nitrogen atoms into cobalt nanosheets as a strategy to boost catalytic activity toward CO_2 hydrogenation. *Nat. Energy.* **2017**, 2, 869–876.
36. Katsutoshi F, Jun S, Takahiro S, Wataru S, Yasuo E, Tatsuo S, Minoru O, Takayoshi S. Fabrication of ruthenium metal nanosheets via topotactic metallization of exfoliated ruthenate nanosheets. *Inorg. Chem.* **2013**, 52, 2280–2282.
37. Michel C. Nobel Prize in chemistry 1912 to Sabatier: organic chemistry or catalysis? *Catal. Today.* **2013**, 218, 162–171.
38. Hammer B, Norskov JK. Theoretical surface science and catalysis calculations and concepts. *Adv. Catal.* **2000**, 45, 71–129.
39. Minghao X, Sishuang T, Bowen Z, Guihua Yu. Metallene-related materials for electrocatalysis and energy conversion. *Mater. Horiz.* **2023**, 10, 407–431.
40. Wang L, Zhu Y, Wang JQ, Liu F, Huang J, Meng X, Basset JM, Han Y, Xiao FS. Two-dimensional gold nanostructures with high activity for selective oxidation of carbon–hydrogen bonds. *Nat. Commun.* **2015**, 6, 6957.
41. Gu K, Pan X, Wang W, Ma J, Sun Y, Yang H, Shen H, Huang Z, Liu H. In situ growth of Pd nanosheets on g-C_3N_4 nanosheets with well-contacted interface and enhanced catalytic performance for 4-nitrophenol reduction. *Small.* **2018**, 14, 1801812.
42. Xu YF, Yang MZ, Chen HY, Liao JF, Wang XD, Kuang DB. Enhanced solar-driven gaseous CO_2 conversion by $CsPbBr_3$ nanocrystal/Pd nanosheet Schottky-junction photocatalyst. *ACS Appl. Energy Mater.* **2018**, 1, 5083–5089.
43. Zhu Y, Xu Z, Jiang W, Zhong S, Zhao L, Bai S. Engineering on the edge of Pd nanosheet cocatalysts for enhanced photocatalytic reduction of CO_2 to fuels. *J. Mater. Chem. A.* **2017**, 5, 2619–2628.
44. Xu D, Liu X, Lv H, Liu Y, Zhao S, Han M, Bao J, He J, Liu B. Ultrathin palladium nanosheets with selectively controlled surface facets. *Chem. Sci.* **2018**, 9, 4451–4455.
45. Qian A, Han X, Liu Q, Ye L, Pu X, Chen Y, Liu J, Sun H, Zhao J, Ling H, Wang R, Li J, Jia X. Ultrathin Pd metallenes as novel co-catalysts for efficient photocatalytic hydrogen production. *Appl. Surf. Sci.* **2023**, 618, 156597.
46. Hongjie Y, Shaojian J, Wenjie Z, Kai D, Ziqiang W, You X, Hongjing W, Liang W. Hydrogen-intercalation enhanced the anti-CO poisoning ability of palladium-copper metallene for direct formate fuel cells. *Mater. Today Phys.* **2023**, 38, 101216.
47. Ross MB, Luna PD, Li Y, Dinh CT, Kim D, Yang P, Sargent EH. Designing materials for electrochemical carbon dioxide recycling. *Nat. Catal.* **2019**, 2, 648–658.

CHAPTER 2

Fundamentals of 2D Metals

Pintu Barman, Apurba Das, and Anindita Deka

2.1 INTRODUCTION

2D metals are an atomically thin class of materials, consisting of one or two atomic layers and encompass a wide range of materials from conducting graphene to semiconducting transition metal dichalcogenides (TMDs) and insulating hexagonal boron nitride (h-BN) along with single layers, phosphorene and silicene, multilayer black phosphorus, etc. [1–3]. The properties of 2D materials appear to be promising for the creation of next-generation ultrathin semiconductor devices. Their remarkable physical, chemical, electrical, optical, and mechanical properties make them fascinating materials for use in technological progress. This makes the study of the properties of 2D materials important in the field of nanoscience. The first groundbreaking report of the synthesized graphene film was published in 2004 by Novoselov and group, which sparked tremendous interest in the research community for which, later on, he was awarded the Nobel Prize in Physics (2010) [4]. Graphene is made up of carbon atoms organized in a hexagonal lattice which possess some unique features such as amazing heat conductivity, great electrical conductivity, and exceptional mechanical strength. These unique properties of 2D materials led to their utilization in a wide range of applications that include electronics, sensing, catalysis, etc. Following the discovery of graphene, other 2D materials became the subject of significant investigation and study. Scientists began to explore different methods for the creation and modification of ultrathin layers of materials. This led to the emergence of an expanding family of 2D materials, each with unique features and possible uses. To fully understand the behaviour of such materials, it is, therefore, crucial to examine the fundamentals of 2D materials in detail so as to comprehend their significance in materials science. A small subset of these 2D classes is the 2D metals, and a few fundamental ideas about their structure and certain significant properties have been discussed in this chapter.

16 DOI: 10.1201/9781032645001-2

Many features of 2D metals make them an interesting topic of study; one property to note is the thickness at the atomic scale. In contrast to conventional 3D metals, which comprise many layers of metal atoms, 2D metals are made up of one or more layers of metal atoms. Interestingly, in 2D metals, the majority of the atoms are located on the surface and they have a high surface-to-volume ratio, which renders them extremely reactive. Given the critical role that surfaces play in both chemical reactivity and catalytic activity, 2D metals are excellent choices for catalytic applications. Heterostructures of 2D metals with enhanced characteristics can be obtained by hybridizing two or more 2D materials. For example, the transition from indirect to direct bandgap semiconducting 2D materials can be obtained by simply adjusting the number of layers. Regulated electrical and optoelectronic conditions are needed to build high-performance field effect transistors (FET), and this may be accomplished using optimized direct bandgap 2D metals [5]. Electrons in 2D metals are bound to move only in two dimensions. Owing to this constraint energy levels become quantized, resulting in discrete energy states and the quantum size effect, both of which are advantageous for nanoelectronic devices. A notable property of 2D metals is their extraordinary electrical conductivity. High electrical conductivity can be achieved in a 2D lattice due to their optimized atomic arrangements which allow efficient electron transport with a minimal perturbation. Despite having only one or two atomic layers, 2D metals have incredible mechanical flexibility. 2D metals are perfect for applications that involve flexible and wearable electronics because of their electrical conductivity, which is independent of the twisting, stretching, and rolling of the metals. A few 2D metals exhibit exceptional mechanical strength such as single-layer hexagonal boron nitride (h-BN). They have exceptional mechanical stress tolerance as well as tensile strength. This property may be utilized in the creation of robust and low-weight materials. Additionally, 2D metals show unique optical properties. Their ability to interact with light in many ways, such as surface plasmon resonance, makes them valuable for the application of optoelectronic and sensor-based devices.

In this chapter, we focus on multiple aspects of 2D metals. In the beginning, a classification of the 2D metals based on the layered structure is provided. It is followed by a detailed discussion of the state-of-the-art production methodologies. A discussion on the pros and cons of each method is also provided and a comprehensive analysis of the characterization techniques following the fabrication of 2D films is provided at the end. It is strongly believed that the chapter shall serve to be a bridge for the readers to understand the challenges and create a roadmap to tackle them for successful applications of such materials in various devices.

2.2 CLASSIFICATION OF 2D METALS

The structural and compositional understanding of 2D metals is important in order to utilize them in a variety of applications. Based on the layered structure, 2D metals can be categorized into the following classes:

a) **Layered van der Waals (vdW) Solids**: Materials with a layered structure and weak van der Waals forces that keep the layers together are known as layered van der Waals solids. The intriguing thing about these materials is that they have unusual electrical and physical characteristics. Several interesting properties are seen in layered vdW solids owing to the poor bonding between the layers. For instance, they are easily exfoliated or can be peeled off layer by layer to form 2D materials because of their flexibility. Graphene, a carbon atom single layer, is one well-known example. These types of materials show poor out-of-plane van der Waal or hydrogen bonding and substantial in-plane covalent or ionic bonding. Transition metal dichalcogenides (TMDs) are one of the most studied layered 2D metals and particularly MoS_2, $MoSe_2$, and WS_2 are exciting materials that fall in this category. They consist of multilayer structures made up of transition metal atoms (usually Mo, Ti, Bi, V, Nb, Ta, W, Zr, Hf) bonded with chalcogen atoms (usually tellurium, selenium, or sulphur) [6]. Every TMD is composed of three stacked layers per monolayer, giving them all a hexagonal honeycomb structure, where the transition metals are found to be in trigonal prismatic coordinates and octahedral coordinates (for d0, d3, and some d1 metals) within each layer, sharing edges with their six neighbours [7]. The van der Waals surface can serve as a buffer layer to aid in the formation of other materials through epitaxial crystal growth [8]. These materials are used in many applications, such as optics, electronics, etc. Owing to their distinct characteristics, they are used to investigate basic science and to create ultrathin electronic gadgets.

b) **Layered Ionic Solids**: The layered atomic arrangement is the distinguishing feature of layered ionic solids. Layers of stacked anions and cations are present in such a way that the layers may be considered as sheets of ions surrounded by ionic bonds. Ionic bonds, the product of electrostatic attraction between oppositely charged ions, are what keep these layers together. The material is structurally stable because of high electrostatic forces. These kinds of solids are made up of charged polyhedral layers that are held together by electrostatic forces between hydroxide or halide layers. Graphite intercalation compounds, which are composed of layers of carbon atoms (graphene) with intercalated ions, and transition metal dichalcogenides, which comprise layers of transition metal cations positioned between chalcogen anions, are two well-known examples of layered ionic solids. Anisotropic materials, or those with varying electrical and physical characteristics depending on the direction, are frequently found in layered ionic solids. Because of the layered structure, there is anisotropy. By changing the composition of the layers or by introducing various ions between the layers, the properties of layered ionic solids may be tuned, which is important for tailoring materials in a specific application. In lithium-ion batteries, layered ionic materials are used because their layered structure enables lithium ions to intercalate during cycles of charge and discharge. One of the layered structures, graphite can be employed as a lubricant due to its sliding ability between two layers. Various electronic gadgets and sensors employ materials that are layered and possess anisotropic electrical characteristics.

c) **Surface-Assisted Nonlayered Solids**: Materials without a layered or two-dimensional structure, such as graphene or transition metal dichalcogenides, are known as surface-assisted nonlayered solids. These materials interact strongly with surfaces, which can change their characteristics. When these nonlayered materials interact with a surface their properties may change due to the effect of chemical bonding, electron transfer, or other surface interaction phenomena. A few of their characteristics, like electrical conductivity or better catalytic activity, may be improved upon interaction with the surface, or they may display distinct chemical reactivity. Chemical vapour deposition and epitaxial growth are two artificial methods used to create this type of atomically thin-layered material on a substrate. A detailed explanation of various synthesis methods has been introduced in the later section of this chapter. Similar to graphene, silicene is an ideal example of surface-assisted nonlayered solids [9].

2.3 OVERVIEW OF 2D METALS

A crucial aspect of 2D materials is that they can be isolated as freestanding sheets that are one atom thick and do not need the existence of a substrate. The interest in 2D metals among researchers exponentiated after the revolutionary discovery of graphene which opened up a new horizon for investigating the fascinating field of other 2D-layered materials. Graphene's zero bandgap has restricted some of its applications, and that has motivated researchers to look for an alternative. Different from graphene's semimetallic properties, transition metal dichalcogenides are currently some of the most researched 2D materials. Their crystalline structure, the quantity of stacking layers, and the concentration of defects allow them to exhibit unique, varied, and adjustable layer-dependent features. In addition to carbon-based graphene, graphene oxide (GO), and reduced graphene oxide (rGO) scientists have explored and examined a wide range of 2D nanomaterials, including transition metal oxides (TMOs), transition metal dichalcogenides (TMDs), layered double hydroxides (LDHs), black phosphorous, hexagonal boron nitride (h-BN), 2D clay materials, 2D polymers, graphitic carbon nitride, 2D perovskites, and many more [11–21]. The structure of various types of 2D nanomaterials is shown in Figure 2.1. Amidst the several categories of nanomaterials, 2D metals have displayed enthralling characteristics and anisotropy in a structure that finds applications in solar cells, catalysis, bioimaging, sensing, surface-enhanced Raman scattering (SERS) to name a few [22–26]. The 2D transition metal oxides are least explored by the researchers as compared to zero- and three-dimensional TMOs. The TMOs are characterized by a very strong bond that acts between the electrons of transition metals and oxygen atoms [27]. Furthermore, TMOs have been explored for biomedical applications such as targeted drug delivery, MRI, etc. The bulk counterpart of transition metal dichalcogenides shows an indirect bandgap that can be altered by producing 2D TMDs which overcome the zero bandgap limitations of graphene [28].

Layered double hydroxide (LDH) and Laponite clay minerals are an important class of 2D clay materials. The LDH structure is a cationic brucite-like structure composed of positively charged divalent metallic layers such as Zn^{2+}, Fe^{2+}, Ca^{2+}, etc. that have been balanced by anions and water molecules present in the interlayer space. The surface wettability of the LDH layers may be changed by introducing anionic impurities to the interlayer and studies to this effect have been reported by Wang et al. for the Mg–Al surface [29]. A range of material science applications, including wastewater treatment, pharmaceuticals, antacids, and agricultural goods are made possible by the functional diversity that results

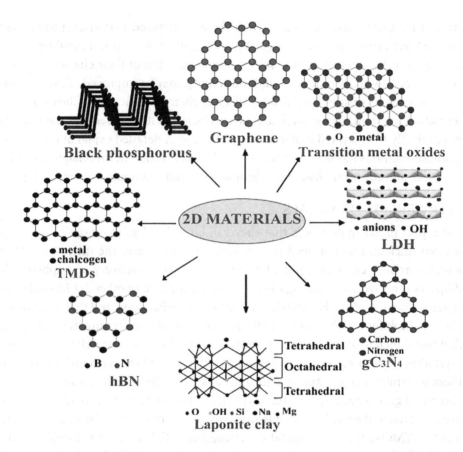

FIGURE 2.1 Schematic illustration of atomic structures of various 2D materials. Adapted with permission from reference [10], Copyright (2018), Elsevier.

from compositional heterogeneity in the layers and interlayer anions of LDH. The ratio of tetrahedral to octahedral sheets in the crystalline structure is denoted by Laponites 2:1 clay mineral. Here, two tetrahedral layers of silicon and sodium are used to sandwich an octahedral layer of magnesium atom. This stacking results in the unique properties of the minerals. Another interesting 2D material is the black phosphorus (BP) also known as phosphorene. It is a unique allotrope of phosphorus that combines some outstanding properties of TMDs, graphene, and other 2D materials and can be considered as the most thermodynamically stable of its three allotropes (white, black, and red). Similar to graphite, BP is composed of various layers. Weak van der Waals forces hold the layers together, while the interlayer phosphorous atoms are confined by covalent bonds in a hexagonal lattice network, which renders it simple for them to split. Anisotropy in structural, mechanical, thermal conductivity, carrier transport, and optical detection of BP is used in a variety of device applications, including gas and humidity sensors, field-effect transistors, and optoelectronic devices [30]. Although BP offers many advantages, there are certain limitations as well. For example, when the BP surface is exposed to air or water, chemical reactions between oxygen and water produce rapid surface degradation. These limitations can be addressed by surface modification and doping processes. A mixture of nitrogen and boron atoms organized in a hexagonal lattice pattern is known as hexagonal boron nitride, or h-BN. Although the characteristics of h-BN are very different from those of graphene, it is commonly referred to as "white graphene" because of its structural similarities to graphene. The lubricating and electrical insulating properties of hexagonal boron nitride can be seen together with its great thermal and chemical durability. Because of its distinctive combination of characteristics, hexagonal boron nitride is a useful material for a wide range of scientific and industrial applications, especially those requiring strong electrical insulation and thermal management [31]. The two-dimensional carbon nitride compound graphitic carbon nitride (g-C_3N_4) has structural similarity to graphene. Its constituent atoms are nitrogen and carbon organized in a planar hexagonal lattice. Despite having some structural similarities to graphene, g-C_3N_4 is a unique material with a wide range of uses because of its unique characteristics. Graphitic carbon nitride is a moderate bandgap semiconductor, thermally and chemically stable and has the ability to absorb visible light making them suitable for photocatalytic processes [32]. It is a promising material for the advancement of technology in fields like environmental remediation, energy conversion, and sensing because of its unique combination of features.

2.4 SYNTHESIS METHODS OF 2D MATERIALS

The fabrication of 2D metals is an interesting branch encompassing materials science, engineering, physics, and chemistry. Since one of the dimensions of the 2D material is restricted to contain structures that are only a few atoms thick, the fabrication process involves sophisticated state-of-the-art techniques that involve complex machineries. The deposition process involves multidisciplinary research that undergoes constant innovation over time. It is not possible to discuss all these techniques within the limitations of this book. However, the most commonly used techniques are briefly outlined. A schematic representation of various techniques used for the growth of 2D metals is summarized in Figure 2.2.

2.4.1 Mechanical Exfoliation (ME) Technique

Mechanical exfoliation (ME) is a conceptually simple technique involving the use of scotch tape to exfoliate a single or a few layers of sheets from bulk crystals [33]. Although the technique has been used successfully for 2D materials like graphene, it can also be applied to certain metals that exhibit a layered structure, such as some transition metals. A synopsis of the mechanical exfoliation technique for the generation of 2D metals is presented below:

(a) **Materials Selection**: Mechanical exfoliation is mostly successful when the materials have a layered Van der Walls structure that allows the isolation of the individual layers. In the case of 2D metals, the bulk target materials should have a layered structure. Common examples include transition metals such as Tantalum (Ta) or niobium (Nb).

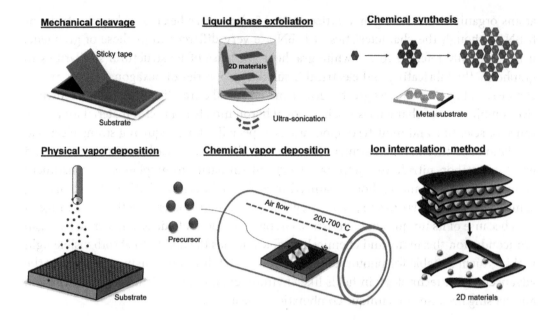

FIGURE 2.2 Schematic representations of various synthesis methods including top-down and bottom-up approaches of 2D materials. Adapted with permission from reference [34], Copyright (2019), Elsevier.

(b) **Sample Preparation:** A high quality of high purity material is chosen as the bulk metal. Preferably, the crystal chosen must be free from defects and impurities to ensure successful exfoliation.

(c) **Exfoliation Process:** The mechanical exfoliation (ME) process involves the following steps:

- Cleavage: A small piece of the bulk metal crystal is attached to an adhesive material, such as scotch tape or similar polymer tape. The tape is then pressed onto the crystal and subsequently peeled off.

- Layer separation: The peeling action can lead to the separation of individual layers of the metal. This process is repeated multiple times to increase the chances of isolating single- or few-layered structures.

- Transfer: Once a thin metal layer is obtained on the tape, it can be transferred onto a substrate for further analysis for device fabrication.

Several new modifications of the exfoliation techniques have been developed and reported to remove the drawbacks of the traditional ME and in one of them, ultra-high vacuum (UHV) and various substrate materials have been used. The general idea is to bring atomically clean surfaces of the bulk crystals and the substrate into contact inside a high vacuum, thereby establishing bonding between the two to facilitate the exfoliation of the single layers. The process is reportedly simple and does not require specialized equipment beyond the ultra-high vacuum and the generated single layers can be used without any post-synthesis cleaning treatment.

As mentioned previously, not all materials can be exfoliated using the ME technique. It is immensely successful if the materials have a layered crystal structure and weak Van der Wall's interlayer forces. In addition, the scalability and reproducibility of the films still remain a big challenge, thus, making them more suitable for research rather than large-scale technological applications.

2.4.2 Physical Vapour Deposition (PVD)

The PVD technique is another most commonly used technique for the deposition of 2D metallic films. In the PVD process, generally, a material is transferred from a source to a substrate through a physical mechanism rather than chemical reactions [35]. In this process, the metal to be deposited is typically in the solid form, often in the form of thin cylindrical metal targets (in case of sputtering) or pellets (in case of pulsed laser deposition (PLD)). The choice of the metal source depends upon the thin film and its compatibility with the PVD technique. The substrate chosen for the deposition of 2D metal films depends on the intended application and its compatibility with the 2D films. However, in general, glass slides and silicon wafers are used as test substrates for preliminary deposition and extraction of the properties of the films. Figure 2.3 shows the synthesis methods of 2D MoS_2 films on Si substrate, their structure and topographical information in the form of SEM micrograph. There is a plethora of PVD processes that are available for deposition of the thin films. Out of them, some of the commonly adopted methods are discussed below:

(a) **Sputtering:** In the process, a highly energetic plasma is used for dislodging the atoms or molecules from the metallic source, which are then deposited onto the substrate. Plasma can be created from various gases, however, in general, for all practical purposes, Ar is used as it is non-reactive and economic. Specific advantages of the technique include a high deposition rate and uniform thickness of the deposited films.

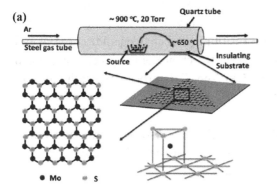
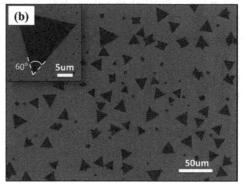

FIGURE 2.3 (a) Schematic illustrations of synthesis of MoS_2 using PVD technique and the demonstration of crystal structure MoS_2. (b) Morphological SEM image of as-grown MoS_2 on Si substrate. Adapted with permission from reference [36], Copyright (2013), American Chemical Society.

Additionally, the reaction parameters (deposition rate, pressure, deposition temperature, thickness) can be precisely controlled and monitored to obtain films with the desired properties. Several variations of sputtering process exist and usually for the deposition of metallic films, direct current (DC) sputtering is used.

(b) **Pulsed Laser Deposition (PLD)**: The process uses a laser to dislodge the atoms or molecules from a target, which are then deposited on a substrate material with the aid of Ar plasma. The method also has a very high deposition rate, however, as compared to sputtering, the films are not very uniform. The technique also provides control over the deposition parameters (such as deposition rate, pressure, temperature, etc.) to obtain films with any desired properties. In situ thickness of the films can also be monitored by integrating devices such as the quartz crystal microbalance (QCM) in the deposition chamber.

(c) **Evaporation**: In this process, the metal source is heated and vaporized. The metal vapour then condenses on the substrate to form the 2D metal film. Various evaporation techniques, such as resistive heating or electron beam evaporation can be used to obtain the desired films.

Thus, in general, PVD is preferred where films with high-quality and well-defined thickness are of utmost necessity. This makes the process very valuable for the production of 2D metal films.

2.4.3 Chemical Vapour Deposition (CVD)

CVD is another versatile process that is followed to obtain 2D structures; the decomposition or reaction of gas, solid, or liquid precursors is carried out inside a controlled atmosphere. It usually involves the controlled deposition of atoms and molecules on a substrate to create a single or few atomic layers of the material [37]. Catalytically active substrates are favourable for the growth of the 2D structures in CVD and the reactions can take place in a wide range of pressures from UHV to atmospheric. However, the use of common substrates such as silicon wafers containing a thin layer of SiO_2 or other materials like copper foils is also seen among the research community. The CVD process has been widely used in the deposition of graphene and transition metal chalcogenides (TMD) and the most important step in any CVD process involves the choice of the precursors, which for instance can be methane (CH_4) for graphene [38]. The CVD process typically involves the following steps:

(a) **Precursor Delivery to the Reaction Chamber**: The precursor gases are introduced into the CVD chamber. This is accomplished using carrier gases like H_2 or Ar to control the flow and concentration of the precursors.

(b) **Heating of the Substrate**: The substrate has to be heated to a specific temperature. The temperature and time are some of the critical parameters that influence the growth of the 2D metals. The choice of temperature usually depends on the material and the desired properties of the deposited layers.

(c) **Chemical Reaction Inside the CVD Chamber**: The precursors decompose on the heated substrate and the atoms rearrange and nucleate on the substrate to form the 2D metal. For instance, in the case of TMDs, the transition metal and chalcogen atoms rearrange to form the desired crystal structure. In the case of graphene, the carbon atoms arrange into the regular hexagonal lattice structure.

(d) **Growth and Nucleation of the 2D Structure**: As long as the precursor gas is supplied, the reaction on the CVD chamber continues and the nucleation of the atoms on the substrate continues to occur. The growth rate, nucleation density, and other factors are controlled through the process parameters.

After the successful deposition of the 2D materials, it needs to be transferred from the substrate where it has been grown to another substrate or device. For this purpose, methods like polymer stamps or chemical etching can be used to achieve the same. All the benefits provided by the CVD process, make it an excellent choice among researchers for deposition of the 2D metals and their commercial applications in advanced electronics, optoelectronic devices, photonics, catalysis, and surface modifications.

2.5 CHARACTERIZATION TECHNIQUES AND APPLICATIONS OF 2D MATERIALS

2.5.1 Characterization Techniques

The characterization of the as-deposited 2D metal films is essential for evaluating the properties of the metal films, such as thickness, structure, composition, electrical, and magnetic properties. Various methods are employed to characterize the deposited films and the choice of the technique depends on the specific properties of interest. Some commonly used characterization techniques for 2D metal films are summarized below.

1. **Atomic Force Microscopy (AFM)**: This technique is used to obtain high-resolution topographic images of 2D metal films. It can be used for measuring film thickness, roughness, and mechanical properties at the nanoscale.

2. **X-ray Diffraction (XRD)**: XRD is employed to determine the crystallographic structure of 2D metal films along with the different phases that exist in the deposited films, crystal orientation, strain, and the lattice parameters of the film.

3. **Scanning Electron Microscopy (SEM)**: SEM is a powerful technique that is used for imaging of the 2D metal films. It is used for generating high-resolution images that provide details about the film's topography, morphology, and the size of the grains along with the defects.

4. **Raman Spectroscopy**: Raman spectroscopy is valuable for understanding and characterizing of 2D materials. It can be used for obtaining information about the bonds as well as the structural information of the deposited films.

5. **X-ray Photoelectron Spectroscopy (XPS)**: XPS is basically a surface-sensitive technique that is used for obtaining information about the elemental composition,

chemical states, and their oxidation states. XPS also provides valuable information about the defect states that can yield valuable information about the spectroscopic properties of the deposited films.

6. **Fourier Transform Infrared Spectroscopy (FTIR)**: FTIR is complementary to Raman spectroscopy and can be used for obtaining information about the vibrational modes of various bonds in the films. They also provide valuable information about the chemical composition and bonding of the various groups constituting the thin film.

7. **Transmission Electron Microscopy (TEM)**: TEM is a sophisticated imaging that yields several critical information about the surface under study. Information related to the structure, thickness, and defect states at the atomic level can be obtained from TEM. Additionally, TEM can also generate diffraction patterns that provide complementary information about the specimens that have already been revealed from the XRD analysis.

8. **Sheet Resistance Measurement (SRM)**: The technique has been used primarily for measuring the conductivity of the fabricated thin films by measuring the sheet resistance. This is accomplished by measuring the current–voltage (V–I) relationship of the thin films using four probe devices that can be used for extraction of the properties at both low and high temperatures.

9. **Hall Effect Measurements**: Hall measurements can provide information about the carrier concentration and mobility of charge carriers in the 2D films. This information is crucial for understanding their electrical behaviour. A complete understanding of the electrical properties is essential to their applications in devices such as sensors, transducers, actuators, etc.

10. **Optical Spectroscopy**: Spectroscopic techniques like UV–Vis absorption spectroscopy and ellipsometry are used for determining the optical properties of the deposited 2D films such as their bandgap, refractive index, extinction coefficient along with thickness of the deposited films.

11. **Photoluminescence Spectroscopy (PL)**: The photoluminescence measurements are used for the investigation of photoluminescence properties of the thin films that provide important information about their electronic band structure, which in turn is useful for their optoelectronics applications.

The choice of characterization techniques depends on the specific properties and the parameters of interest for the 2D thin films. Multiple techniques are often employed for the extraction of the same properties that provide a comprehensive understanding of the film's properties. The preparation of WS_2/Graphen films on Si substrate prepared by the CVD process and their various characterization techniques such as SEM, AFM, and Raman analyses have been shown in Figure 2.4.

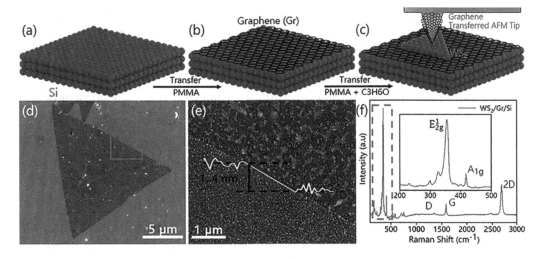

FIGURE 2.4 (a)–(c) Schematic representation of the growth of WS$_2$/graphene on Si substrate via CVD process and their (d) SEM, (e) AFM, and (f) Raman spectrum. Adapted with permission from reference [39], Copyright (2020), American Chemical Society.

2.5.2 Applications

Due to their special qualities, two-dimensional (2D) metals like graphene can be used in a wide range of applications. The following are a few uses for 2D metals:

1. **Electronics**: Due to their superior electrical conductivity, 2D metals like graphene hold great promise as building blocks for electronic gadgets of the future. They may be utilized to create faster and more effective electronics by being included in transistors, interconnects, and other components [40].

2. **Sensors**: Two-dimensional metals are extremely responsive to environmental changes. They can be used as sensors to identify different kinds of gases, chemicals, and biological substances. For instance, gas sensors based on graphene can identify traces of gases for use in medical and environmental monitoring [41].

3. **Energy Storage**: 2D metals have been investigated for use in batteries and supercapacitors, among other energy storage devices [42]. Their excellent electrical conductivity and large surface area present very rapid charging and discharging capabilities in such materials.

4. **Catalysis**: Due to their substantial surface area and adjustable electrical characteristics, 2D metals can be effective catalysts. They are employed in a variety of catalytic processes, including water splitting, fuel cell electrocatalysis, and hydrogen evolution reactions [43].

5. **Films and Coatings**: 2D metal thin films and coatings have good heat conductivity and corrosion resistance. They are used to safe and improve the performance of materials and components in the electronics, automotive, and aerospace sectors.

6. **Flexible Electronics**: 2D metals are a good choice for flexible and wearable electronics due to their mechanical strength and flexibility. It is possible to incorporate graphene and other 2D materials into flexible screens, sensors, and electronic textiles.

7. **Optoelectronics**: Devices like light-emitting diodes (LEDs) and photodetectors can be made with 2D metals [44]. High-performance optoelectronic component development is made possible by their special electrical qualities.

8. **Thermal Management**: Due to their high thermal conductivity, two-dimensional metals are perfect for applications involving thermal management. To effectively disperse heat from electronic devices and power electronics, they can be employed as heat sinks and thermal interface materials.

9. **Water Purification**: Applications for desalination and water purification using 2D metals have been researched. Heavy metals and organic pollutants can be effectively removed from water, thanks to their large surface area and selective permeability.

10. **Medical Devices**: 2D metals are also used in medication delivery systems and medical equipment. They are useful in medicine due to their biocompatibility and capacity to transport and release medicinal chemicals.

11. **Nanoelectromechanical Systems (NEMS)**: NEMS are small-scale mechanical systems with electronic capability. Two-dimensional metals (two-dimensional) can be employed in NEMS [45]. They can be used in resonators and sensors, among other applications.

12. **Aerospace and Defence:** Due to their strength and lightweight, 2D metals find application in several fields. To increase functionality and decrease weight, they can be used in coatings, composites, and componentry.

Research on 2D materials is still in its early stages, but it is expected to yield new uses and strategies for utilizing these materials' special qualities across a range of industries.

2.6 CONCLUSIONS AND FUTURE OUTLOOK

A general overview of the conclusion and future outlook on 2D metals is as follows:

Conclusion

(a) Exceptional Properties: 2D metals have exceptional mechanical strength, flexibility, and great electrical and thermal conductivity.

(b) Versatile Applications: These materials have demonstrated potential in a number of fields, such as energy storage, electronics, catalysis, and sensors.

(c) Challenges: Nevertheless, there are still issues that need to be resolved, including large-scale production, stability, and interaction with current technologies.

(d) Diverse 2D Materials: The family of 2D materials includes materials other than graphene, such as phosphorene and transition metal dichalcogenides (TMDs), each with a special set of properties and possible uses.

Future Outlook:

(a) Electronics: The creation of high-performance transistors and other electronic components is one area in which 2D metals may be extremely important to the advancement of next-generation electronics.

(b) Energy Storage: 2D metals' large surface area and high conductivity make them desirable for use in energy storage devices like supercapacitors and batteries, where they may result in improved efficiency and performance.

(c) Catalysis: Because of their large surface area and capacity to tune their electrical properties – a necessary skill for a variety of chemical reactions – 2D metals may find use in catalysis.

(d) Sensors: 2D metals are a promising option for sensors due to their unique properties, particularly when it comes to detecting small amounts of gases or biomolecules.

(e) Materials Science: To increase the number of potential applications, materials science research is continuously seeking to identify and comprehend novel two-dimensional materials and their characteristics.

(f) Integration Difficulties: Including 2D metals in current manufacturing procedures and technology presents a number of difficulties. Solving this problem is essential to ensuring that these materials are widely used.

(g) Commercialization: Overcoming production obstacles, guaranteeing stability, and proving scalable for industrial uses are necessary for 2D metals to become commercially viable.

It is imperative to acknowledge that the domain of two-dimensional materials is undergoing rapid evolution, with novel advancements consistently emerging over time. It's possible that scientists and researchers are still investigating the possibilities of two-dimensional metals and working to overcome current obstacles in order to fully realize their potential. It is anticipated that a number of intriguing discoveries will transform the field over the next ten years, and that many devices based on two-dimensional metals may find their way into every branch of contemporary science.

REFERENCES

1. G. Li, Y. Li, X. Qian, H. Liu, H. Lin, N. Chen, Y. Li, Construction of tubular molecule aggregations of graphdiyne for highly efficient field emission, *J. Phys. Chem. C* 115 (2011) 2611–2615.
2. D. Pacilé, J.C. Meyer, Ç.Ö. Girit, A. Zettl, The two-dimensional phase of boron nitride: Few-atomic-layer sheets and suspended membranes, *Appl. Phys. Lett.* 92 (2008) 133107.

3. D.J. Late, Temperature dependent phonon shifts in few-layer black phosphorus, *ACS Appl. Mater. Interfaces* 7 (2015) 5857–5862.

4. K.S. Novoselov, A.K. Geim, S.V. Morozov, D.E. Jiang, Y. Zhang, S.V. Dubonos, I.V. Grigorieva, A.A. Firsov, Electric field effect in atomically thin carbon films, *Science* 306 (2004) 666–669.

5. D.J. Late, B. Liu, H.R. Matte, V.P. Dravid, C.N.R. Rao, Hysteresis in single-layer MoS_2 field effect transistors, *ACS Nano* 6 (2012) 5635–5641.

6. D.L. Duong, S.J. Yun, Y.H. Lee, van der Waals layered materials: opportunities and challenges, *ACS Nano* 11 (2017) 11803–11830.

7. S.Z. Butler, S.M. Hollen, L. Cao, Y. Cui, J.A. Gupta, H.R. Gutiérrez, T.F. Heinz, S.S. Hong, J. Huang, A.F. Ismach, E. Johnston-Halperin, Progress, challenges, and opportunities in two-dimensional materials beyond graphene, *ACS Nano* 7 (2013) 2898–2926.

8. E. Wisotzki, A. Klein, W. Jaegermann, Quasi van der Waals epitaxy of ZnSe on the layered chalcogenides InSe and GaSe, *Thin Solid Films* 380 (2000) 263–265.

9. A. Zia, Z.P. Cai, A.B. Naveed, J.S. Chen, K.X. Wang, MXene, silicene and germanene: Preparation and energy storage applications, *Mater. Today Energy* 30 (2022) 101144.

10. A. Jayakumar, A. Surendranath, P.V. Mohanan, 2D materials for next generation healthcare applications, *Int. J. Pharm.* 551 (2018) 309–321.

11. K.S. Novoselov, V.I. Fal'ko, L. Colombo, P.R. Gellert, M.G. Schwab, K. Kim, A roadmap for graphene, *Nature* 490 (2012) 192–200.

12. Y.K. Kim, M.H. Kim, D.H. Min, Biocompatible reduced graphene oxide prepared by using dextran as a multifunctional reducing agent, *Chem. Commun.* 47 (2011) 3195–3197.

13. C. Ataca, H. Sahin, S. Ciraci, Stable, single-layer MX2 transition-metal oxides and dichalcogenides in a honeycomb-like structure, *J. Phys. Chem. C* 116 (2012) 8983–99.

14. M. Chhowalla, H.S. Shin, G. Eda, L.-J. Li, K.P. Loh, H. Zhang, The chemistry of two-dimensional layered transition metal dichalcogenide nanosheets, *Nat. Chem.* 5 (2013) 263–275.

15. G. Mishra, B. Dash, S. Pandey, Layered double hydroxides: A brief review from fundamentals to application as evolving biomaterials, *Appl. Clay Sci.* 153 (2018) 172–186.

16. V. Eswaraiah, Q. Zeng, Y. Long, Z. Liu, Black phosphorus nanosheets: synthesis, characterization and applications, *Small* 12 (2016) 3480–3502.

17. Y. Lin, J.W. Connell, Advances in 2D boron nitride nanostructures: Nanosheets, nanoribbons, nanomeshes, and hybrids with graphene, *Nanoscale* 4 (2012) 6908–6939.

18. W. Dong, Y. Lu, W. Wang, M. Zhang, Y. Jing, A. Wang, A sustainable approach to fabricate new 1D and 2D nanomaterials from natural abundant palygorskite clay for antibacterial and adsorption, *Chem. Eng. J.* 382 (2020) 122984.

19. W. Wang, A.D. Schlüter, Synthetic 2D polymers: A critical perspective and a look into the future, *Macromol. Rapid Commun.* 40 (2019) 1800719.

20. N. Rono, J.K. Kibet, B.S. Martincigh, V.O. Nyamori, A review of the current status of graphitic carbon nitride, *Crit. Rev. Solid State Mater. Sci.* 46 (2021) 189–217.

21. X. Zhao, T. Liu, Y.L. Loo, Advancing 2D perovskites for efficient and stable solar cells: challenges and opportunities, *Adv. Mater.* 34 (2022) 2105849.

22. A.P. Kulkarni, K.M. Noone, K. Munechika, S.R. Guyer, D.S. Ginger, Plasmon-enhanced charge carrier generation in organic photovoltaic films using silver nanoprisms, *Nano Lett.* 10 (2010) 1501–1505.

23. N. Yang, Z. Zhang, B. Chen, Y. Huang, J. Chen, Z. Lai, Y. Chen, M. Sindoro, A.L. Wang, H. Cheng, Z. Fan, Synthesis of ultrathin PdCu alloy nanosheets used as a highly efficient electrocatalyst for formic acid oxidation, *Adv. Mater.* 29 (2017) 1700769.

24. P.K. Jain, K.S. Lee, I.H. El-Sayed, M.A. El-Sayed, Calculated absorption and scattering properties of gold nanoparticles of different size, shape, and composition: applications in biological imaging and biomedicine, *J. Phys. Chem. B* 110 (2006) 7238–7248.

25. E. Martinsson, M.M. Shahjamali, K. Enander, F. Boey, C. Xue, D. Aili, B. Liedberg, Local refractive index sensing based on edge gold-coated silver nanoprisms, *J. Phys. Chem. C* 117 (2013) 23148–23154.
26. H. Hou, P. Wang, J. Zhang, C. Li, Y. Jin, Graphene oxide-supported Ag nanoplates as LSPR tunable and reproducible substrates for SERS applications with optimized sensitivity, *ACS Appl. Mater. Interfaces* 7 (2015) 18038–18045.
27. K. Kalantar-zadeh, J.Z. Ou, T. Daeneke, A. Mitchell, T. Sasaki, M.S. Fuhrer, Two dimensional and layered transition metal oxides, *Appl. Mater. Today* 5 (2016) 73–89.
28. W. Choi, N. Choudhary, G.H. Han, J. Park, D. Akinwande, Y.H. Lee, Recent development of two-dimensional transition metal dichalcogenides and their applications, *Mater. Today* 20 (2017) 116–130.
29. B. Wang, H. Zhang, D.G. Evans, X. Duan, Surface modification of layered double hydroxides and incorporation of hydrophobic organic compounds, *Mater. Chem. Phys.* 92 (2005) 190–196.
30. P. Chen, N. Li, X. Chen, W.J. Ong, X. Zhao, The rising star of 2D black phosphorus beyond graphene: Synthesis, properties and electronic applications, *2D Mater.* 5 (2017) 014002.
31. H. Fang, S.L. Bai, C.P. Wong, "White graphene"–hexagonal boron nitride based polymeric composites and their application in thermal management, *Compos. Commun.* 2 (2016) 19–24.
32. X. Kong, X. Liu, Y. Zheng, P.K. Chu, Y. Zhang, S. Wu, Graphitic carbon nitride-based materials for photocatalytic antibacterial application, *Mater. Sci. Eng. R Rep.* 145 (2021) 100610.
33. E. Gao, S.Z. Lin, Z. Qin, M.J. Buehler, X.Q. Feng, Z. Xu, Mechanical exfoliation of two-dimensional materials, *J. Mech. Phys. Solids* 115 (2018) 248–262.
34. C. Murugan, V. Sharma, R.K. Murugan, G. Malaimegu, A. Sundaramurthy, Two-dimensional cancer theranostic nanomaterials: Synthesis, surface functionalization and applications in photothermal therapy, *J. Control. Release* 299 (2019) 1–20.
35. C. Muratore, A.A. Voevodin, N.R. Glavin, Physical vapor deposition of 2D Van der Waals materials: a review, *Thin Solid Films* 688 (2019) 137500.
36. S. Wu, C. Huang, G. Aivazian, J.S. Ross, D.H Cobden, X. Xu, Vapor–solid growth of high optical quality MoS_2 monolayers with near-unity valley polarization, *ACS Nano* 7 (2013) 2768–2772.
37. J. Zhang, F. Wang, V.B. Shenoy, M. Tang, J. Lou, Towards controlled synthesis of 2D crystals by chemical vapor deposition (CVD), *Mater. Today* 40 (2020) 132–139.
38. Z. Cai, B. Liu, X. Zou, H.M. Cheng, Chemical vapor deposition growth and applications of two-dimensional materials and their heterostructures, *Chem. Rev.* 118 (2018) 6091–6133.
39. J. Tian, X. Yin, J. Li, W. Qi, P. Huang, X. Chen, J. Luo, Tribo-induced interfacial material transfer of an atomic force microscopy probe assisting superlubricity in a WS2/graphene heterojunction, *ACS Appl. Mater. Interfaces* 12 (2019) 4031–4040.
40. J. Park, W.H. Lee, S. Huh, S.H. Sim, S.B. Kim, K. Cho, B.H. Hong, K.S. Kim, Work-function engineering of graphene electrodes by self-assembled monolayers for high-performance organic field-effect transistors, *J. Phys. Chem. Lett.* 2 (2011) 841–845.
41. K.R. Nemade, S.A. Waghuley, Chemiresistive gas sensing by few-layered graphene, *J. Electron. Mater.* 42 (2013) 2857–2866.
42. S. Ding, D. Luan, F.Y.C. Boey, J.S. Chen, X.W. Lou, SnO_2 nanosheets grown on graphene sheets with enhanced lithium storage properties, *Chem. Commun.* 47 (2011) 7155.
43. H. Xu, H. Shang, C. Wang, Y. Du, Recent progress of ultrathin 2D Pd-based nanomaterials for fuel cell electrocatalysis, *Small* 17 (2021) 2005092.
44. Q. Zeng, Z. Liu, Novel optoelectronic devices: Transition-metal-dichalcogenide-based 2D heterostructures, *Adv. Electron. Mater.* 4 (2018) 1700335.
45. P.F. Ferrari, S. Kim, A.M. van der Zande, Nanoelectromechanical systems from two-dimensional materials, *Appl. Phys. Rev.* 10 (2023) 031302.

CHAPTER 3

Synthesis and Characterization of 2D Metals

Hsu-Sheng Tsai, Jing Li, and Zhengguang Shi

3.1 THE PREPARATION METHODS OF 2D METALS

Generally, 2D metals, reported so far, can be prepared by chemical synthesis, physical processes, or physicochemical methods. The chemical synthesis can be categorized into ligand-confined synthesis, space-confined synthesis, template-directed synthesis, and topotactic metallization. The physical processes include various exfoliation methods, electron beam dealloying, and physical vapor deposition. Besides, there are also physicochemical methods such as repeated size reduction and plasma-assisted synthesis. Each of these preparation methods is introduced in this section.

3.1.1 Ligand-Confined Synthesis

The chemical synthesis of 2D metals is generally implemented by hydrothermal, solvothermal, and chemical bath methods. So far, the most popular 2D metal is the Pd metallene mainly synthesized by using the Pd(II) acetylacetonate ($Pd(acac)_2$) as the precursor of metal source [1–4]. Organic chemicals probably with reducing ability, such as dimethylformamide (DMF), oleylamine, and benzene ethanol, are employed as the solvents during synthesis. Ascorbic acid (AA) can be utilized as the reductant to further reduce the $Pd(acac)_2$. It is noted that the addition of CO molecules as the structure-directing agent, which can be acquired from the gaseous CO or the decomposition of metal carbonyls (e.g. $W(CO)_6$, $Mo(CO)_6$, and $Co_2(CO)_8$), is crucial for the anisotropic growth of Pd crystals. The CO molecules adsorbed on the surfaces of Pd crystals can suppress the out-of-plane crystal growth, resulting in the formation of Pd metallene. Besides, the reaction temperature for the synthesis of Pd metallene is between 50°C and 80°C, depending on the different

32   DOI: 10.1201/9781032645001-3

reductants, solvents, and sources of the structure-directing agent. A small amount of acetic acid could be added to the solution for synthesis to slightly reduce the reaction temperature.

Xu et al. reported another approach of the ligand-confined synthesis of Pd metallene using the H_2PdCl_4, AA, and H_2O as the precursor of metal source, reductant, and solvent, respectively, to prevent the use of toxic CO replaced by that of docosylpyridinium bromide (C22-pyB) as a surfactant [5]. During the synthesis process at nearly room temperature, the C22-pyB molecules with long carbon chains first tend to be attracted to the $PdBr_4^{2-}$ or $PdCl_4^{2-}$ ions due to the electrostatic interaction, leading to the formation of lamellar inorganic–organic assemblies. Then, the elemental Pd would be obtained after adding the AA and tiny Pd nanosheets start to form owing to the capping of C22-pyB molecules at the surface. Finally, the tiny Pd nanosheets laterally aggregate into the larger ones, which is caused by the hydrophobic interactions from the carbon chains of C22-pyB. Figure 3.1(a) illustrates this synthesis process.

The Rh metallene could be obtained by the ligand-confined synthesis using the rhodium(II) acetate ($Rh_2(CH_3COO)_4$) as the precursor of metal source in addition to the $Rh(acac)_3$. The former is reduced by the solvent of DMF and the CO gas is infused into the reaction vessel to achieve the lateral growth of Rh crystals at 150°C for 3 hours [6]. Polyvinylpyrrolidone (PVP) is utilized as the surfactant for the latter in the solution of benzyl alcohol and formaldehyde at 180°C for 8 hours; thus the obtained Rh metallene is capped by PVP [7]. On the other hand, the $Co(acac)_3$ is also used as the precursor for the ligand-confined synthesis of Co metallene at 220°C for 48 hours [8]. The DMF mixed with

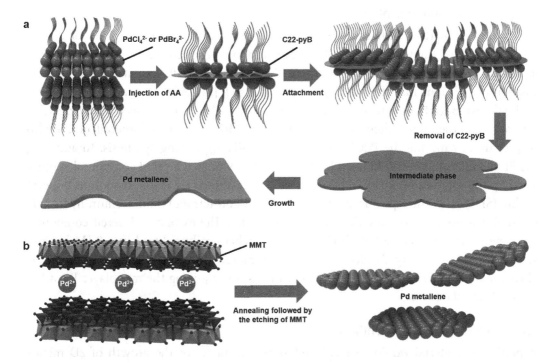

FIGURE 3.1 (a) Illustration of the ligand-confined synthesis of Pd metallene. (b) Illustration of the space-confined synthesis of Pd metallene.

34 ■ 2D Metals

H_2O is used as the solvent as well as reductant and the *n*-butylamine is employed as the surfactant.

Apart from the elemental metallene, it is imaginable that the 2D alloys may also be prepared by the ligand-confined synthesis. The PdMo bimetallene could be synthesized by dissolving $Pd(acac)_2$, AA, and $Mo(CO)_6$ in the oleylamine at 80°C for 12 hours, where AA plays the role of reductant, Mo and CO originated from the decomposition of $Mo(CO)_6$ are the metal source and the structure-directing agent individually [4]. The RuRh bimetallene has been synthesized by Mu et al. using benzoic acid to reduce the $RuCl_3$ and $RhCl_3$ at 150°C for 12 hours [9]. The polyethylene glycol-block-polypropylene glycol-block-polyethylene glycol (PEG-b-PPG-b-PEG) is used as the surfactant to dominate the morphology of RuRh bimetallene during synthesis. The $Pd(acac)_2$ and $Ir_4(CO)_{12}$ provide their metal elements for the formation of PdIr bimetallene at 150°C for 8 hours, where the Pd is obtained by the reduction using NH_4Br and the Ir is released from the thermal decomposition of $Ir_4(CO)_{12}$ [10]. Similarly, the CO molecules confine the out-of-plane growth of PdIr crystals, leading to the atomically 2D bimetallene. Li et al. replaced a reductant by an additive of tri-n-octylphosphine oxide (TOPO) which can extract the metal elements from the precursors during the synthesis of PdFe bimetallene at 60°C [11]. The hydrogen atoms may stabilize the surface of metallene, thus the formaldehyde has been used as the solvent to provide the hydrogen atoms for the synthesis of hydrogen-stabilized RhPd bimetallene (RhPd-H) [12]. Moreover, the ternary 2D alloys, PtPdM (M = Fe, Co, Ni), have also been successfully prepared by Lai et al. for the application of electrocatalysis [13].

3.1.2 Space-Confined Synthesis

This synthesis method utilizes layered compounds to confine one direction of crystal growth, realizing the formation of 2D metals or alloys. The layered sodium-montmorillonite (Na-MMT) has been used as a frame dispersed in the $PdCl_2$ solution for the intercalation of Pd^{2+} ions into the interlayer space of MMT [14]. After annealing the product under the Ar/H_2 (9:1) ambiance at 500°C for 2 hours, the MMT is removed by the HF solution to obtain the Pd metallene as shown in Figure 3.1(b). The PdCo bimetallene can also be prepared by mixing the $PdCl_2$ solution with $Co(NO_3)_2$ during synthesis. In addition, the Pt metallene has been obtained by reducing the exfoliated layered compound synthesized by the solution of K_2PtCl_4, hexamethylenetetramine (HMT), and sodium dodecyl sulfate (SDS) [15]. The aqueous solution of ethanol and tetrabutylammonium hydroxide (TBAOH) is used to exfoliate the layered compound. The exfoliated layered compound could be chemically reduced by $NaBH_4$ or electrochemically reduced in a 0.1M Na_2SO_4 solution at −1.0 V (Ag/AgCl). However, the chemical reduction results in a porous Pt metallene. It is noted that the addition of SDS can ensure acquiring the monolayer Pt rather than the multilayer one.

3.1.3 Template-Directed Synthesis

A particular material would be employed as the template for the growth of 2D metals during this synthesis process. Zhao et al. have monitored the Fe metallene suspended in the pores of graphene by using a low-voltage spherical aberration-corrected transmission electron microscope (LVACTEM) [16]. The graphene monolayer conventionally grown by the chemical vapor deposition (CVD) is transferred from the substrate of Ni-coated Mo foil onto a standard lacey carbon TEM grid. The spontaneous formation of free-standing Fe metallene in the pores of graphene is attributed to the decomposition of $FeCl_3$ which is used to etch the Ni during the transfer process of graphene. The graphene oxide (GO) prepared by the modified Hummers method [17] is utilized as the template as well for the synthesis of Au metallene [18]. The as-prepared GO is injected into an organic solution containing the $HAuCl_4$ and 1-amino-9-octadecene at 55°C for 16 hours, obtaining the Au metallene on GO as shown in Figure 3.2(a). Additionally, the 2D high-entropy alloys (HEAs) have been synthesized by using the Ag nanowires as templates made from the ethylene glycol solution containing the $AgNO_3$, PVP, and $FeCl_3$ at 140°C for 5 hours [19]. The as-prepared Ag nanowires are added into the solution with several metal acetylacetonates at 200°C to initially obtain the Ag/HEAs core/shell nanowires. Then the dealloying process is implemented by using concentrated HNO_3 to remove excess Ag, leading to the obtainment of 2D HEAs.

3.1.4 Topotactic Metallization

This preparation of 2D metals is accomplished by the chemical reduction of layered metal compounds or the solid-state reactions. The 2D Bi (bismuthene) can be acquired by using an aqueous solution of $NaBH_4$ to reduce the $BiCl_3$ dissolved in 2-ethoxyethanol ($C_4H_{10}O_2$) [20]. Liu et al. have synthesized a jelly-like $RhCl_3$-$K_3Co(CN)_6$ cyanogel and injected formaldehyde solution into the cyanogel for reduction to obtain Rh metallene [21]. A layered ruthenate ($H_{0.2}RuO_2 \cdot 0.5H_2O$) synthesized by Fukuda et al. has been deoxidized into the Ru metallene after annealing under the N_2/H_2 ambiance at 200°C [22]. In addition, the Ni foam has been used as the substrate for the topotactic metallization to prepare the 2D NiFe nanomesh [23]. After the pretreatment of Ni foam, it is immersed into an aqueous solution

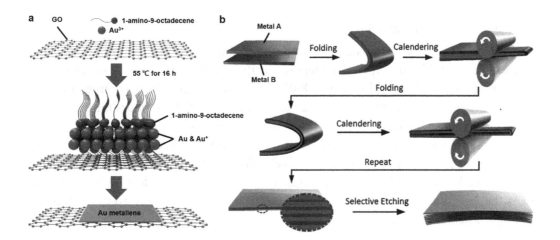

FIGURE 3.2 (a) Illustration of the template-directed synthesis of Au metallene. (b) Illustration of the repeated size reduction for preparation of metallenes. Adapted with permission from [45]. Copyright (2016) John Wiley & Sons.

containing the Zn-, Fe-, and Ni- acetate salts at 60°C for 5 hours to form the NiFeZn alloy nanosheets. Finally, the 2D NiFe nanomesh on Ni foam is obtained by using the KOH solution to etch the Zn element.

The topotactic metallization for the preparation of 2D metals can also be achieved by the electrochemical reduction of layered metal compounds. In general, a specific precursor of a metal compound should be synthesized for the following electrochemical reduction to obtain the metallene. For example, the Bi-based compound precursors include the Bi-based metal-organic layers (Bi-MOLs), Bi_2O_3, scheelite bismuth vanadate (S-BiVO$_4$), and $(BiO)_2CO_3$ [24–27]. After the precursor is ready, it is dispersed into a mixture containing the Nafion solution to prepare a slurry. Then the slurry is coated on carbon paper to form a thin film as the working electrode. Finally, cyclic voltammetry (CV) is implemented to electrochemically reduce the Bi-based compound precursor into the bismuthene.

3.1.5 Exfoliation

Exfoliation methods such as liquid-phase exfoliation, mechanical exfoliation, and solid-melt exfoliation have been applied to prepare the 2D metals. The liquid-phase exfoliation utilizes the intercalation of molecules or ions and/or ultrasonication to peel the layered metal crystals whose interlayer interaction is van der Waals force [28–32]. An ice bath with N_2 bubbling can be used during the liquid-phase exfoliation to prevent the oxidation of metals [30]. The van der Waals layered metals could also be exfoliated by the mechanical force which is the shear stress produced by using a mortar or kitchen blender [33–35]. While the solid-melt exfoliation may only be used for the low melting point metals, especially for the Ga [36]. A Ga droplet on the glass substrate is initially heated to 50°C for the formation of uniform melt. After the temperature is decreased to 30°C, a proper substrate is directly contacted with the Ga droplet. A temperature lower than 30°C at the substrate surface causes the solidification of Ga, thus the Ga metallene is exfoliated and transferred onto the substrate.

3.1.6 Electron Beam Dealloying

This approach may only be implemented under the TEM analysis because of the electron beam source itself. The $Au_{25}Ag_{75}$ alloy prepared by the repetitive arc-melting of Au and Ag wires is first homogenized by heating to 900°C for 100 hours [37]. After compressing and cutting it into the cuboids, they are further annealed at 600°C for 4 hours for the crystal recovery. Then the cuboids of $Au_{25}Ag_{75}$ alloy are thinned by grinding followed by the Ar ion milling for the preparation of TEM sample. The energy for the dealloying of Au (~407 keV) is much higher than that of Ag (~202 keV), implying that the Ag atoms are much easier to knock out than the Au atoms under the irradiation of an electron beam at 300 kV. As the Ag atoms are knocked out, the Au atoms tend to diffuse toward the vacancies, forming a hexagonal Au metallene.

3.1.7 Physical Vapor Deposition

It is well known that ultrahigh vacuum (UHV) growth techniques such as molecular beam epitaxy (MBE) and pulsed laser deposition (PLD) can directly prepare various 2D metals on the substrates. There are several 2D metallenes including 2D Ga, Sb, Bi, and Hf, which could be directly grown on specific substrates by using MBE [38–41]. The 2D Sn (stanene) can be epitaxially grown on a Si(111) substrate by using a buffer layer of $PbTe/Bi_2Te_3$ for the lattice match [42]. The Pd(111) substrate is used for the growth of 2D Pb (plumbene) and the layer structure of $Pb/Pd_{(1-x)}Pb_x/Pd(111)$ would be formed after annealing. In fact, the $Pd_{(1-x)}Pb_x$ layer is first formed on Pd(111) during epitaxy and the Pb atoms are segregated from the $Pd_{(1-x)}Pb_x$ layer onto the surface to crystallize into the plumbene after annealing [43]. Furthermore, the bismuthene can also be deposited on a SiO_2/Si substrate by using PLD [44].

3.1.8 Repeated Size Reduction

The concept of this preparation method originates from a traditional Chinese food, so-called "thousand-layer pie" [45,46]. First, a metal material with a mechanical property similar to that of another one, which would be prepared as metallene, is selected. These two metal foils with the same dimension are pressed face to face by mechanical rolling. Then this bilayer foil is folded and calendared again, leading to a quadruple-layer foil. The thickness of each layer could be reduced to around 0.1 nm after repeated folding and calendaring. Eventually, metallene can be obtained by the chemical etching of the other metal. This preparation process is illustrated in Figure 3.2(b) [45].

3.1.9 Plasma-Assisted Synthesis

This process utilizes the N_2 plasma to introduce nitrogen into a binary compound to induce the selective reaction during the post-annealing. One element in the binary compound would spontaneously react with nitrogen to form a nitride layer. Simultaneously, the other element in the binary compound is expelled to construct a 2D material on the nitride layer. Taking the synthesis of 2D Sb (antimonene) as an example, the bulk InSb is chosen as the binary compound substrate. The InN would be formed, and the Sb atoms are squeezed out to rearrange into the antimonene at the surface of InN after the process [47].

3.2 THE CHARACTERIZATIONS OF 2D METALS

The characterizations of as-prepared 2D metals are necessary to identify their dimensions, morphologies, chemical compositions, crystal structures, physical and chemical properties, and so on. Analysis techniques for characterizations of 2D metals will be introduced by using examples from recent studies in this section.

3.2.1 Dimension Measurement

The dimension of as-prepared 2D metals could be commonly measured by the atomic force microscope (AFM), scanning electron microscope (SEM), and TEM. These techniques can directly monitor the 2D metals to estimate their scales, while their thickness can be measured by the AFM under tapping mode or the cross-sectional TEM image. The SEM is always employed to inspect the morphology of materials. The thickness of 2D metals is generally less than 10 nm which is too thin to be observed by the SEM.

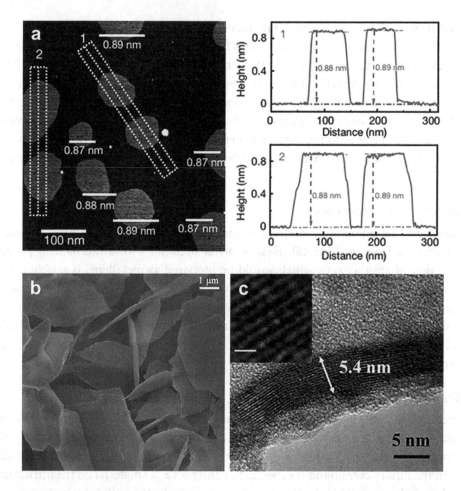

FIGURE 3.3 (a) The AFM image and corresponding height profiles of PdMo bimetallene. Adapted with permission from [4]. Copyright (2019) Springer Nature. (b) The SEM image of the sheet-like Co(OH)$_2$ intermediate. Adapted with permission from [8]. Copyright (2016) Springer Nature. (c) The cross-sectional TEM image of the 5.4 nm-thick multilayer bismuthene. Inset: The high-resolution TEM image of multilayer bismuthene (scale bar = 1 nm). Adapted with permission from [44]. Copyright (2019) John Wiley & Sons. Distributed under a Creative Commons Attribution License 4.0 (CC BY) https://creativecommons.org/licenses/by/4.0/

Figure 3.3(a) shows the AFM image and corresponding height profiles of PdMo bimetallene with an average thickness of ~0.9 nm prepared by the ligand-confined synthesis [4]. The morphology of Co(OH)$_2$ intermediate produced during the ligand-confined synthesis of Co metallene is observed by the SEM as shown in Figure 3.3(b), indicating that the *n*-butylamine can promote the lateral growth of Co crystals [8]. The cross-sectional TEM image of Figure 3.3(c) exhibits the 5.4 nm-thick multilayer bismuthene grown by the PLD [44].

3.2.2 Elemental Analysis

The chemical composition and/or electron configuration of 2D metals and alloys could be determined by the techniques of elemental analysis including energy dispersive spectroscopy (EDS), electron energy loss spectroscopy (EELS), X-ray photoelectron spectroscopy

(XPS), X-ray absorption spectroscopy (XAS), inductively coupled plasma atomic emission spectroscopy (ICP-AES), and inductively coupled plasma mass spectrometry (ICP-MS). The EDS, generally attached to SEM and TEM, can detect the energy of characteristic X-rays emitted from the atoms to distinguish the elements in materials and estimate their percentage. Although it is more difficult to detect the light elements by using the EDS, the EDS analysis of metallenes is feasible since most of the metal elements possess higher atomic numbers. The TEM image and EDS mapping of Figure 3.4(a) show the 2D HEA subnanometer ribbon, uniformly containing the PtPdIrRuAg elements, synthesized by Tao et al [19]. The EELS, usually attached to TEM, determines the elements in materials by detecting the energy loss of electrons after the collision with atoms. Fan et al. utilized the EELS spectrum to confirm the hydrogenation of RhPd bimetallene as shown in Figure 3.4(b) [12].

The XPS utilizes the X-ray to knock the core electrons out from the atoms for the detection of binding energy to discriminate the elements and analyze their oxidation states according to the chemical shift. The Co 2p XPS spectrum of Figure 3.4(c) indicates the characteristic peaks at 778.3 eV and 794.2 eV corresponding to the $2p_{3/2}$ and $2p_{1/2}$ states of Co metallene obtained by the liquid-phase exfoliation, respectively [32]. While the electrons in atoms could be excited by the absorption of synchrotron X-ray to the unoccupied levels above the Fermi level. Based on this principle, the XAS spectrum composed of the

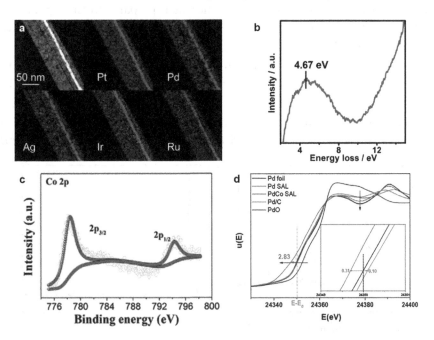

FIGURE 3.4 (a) The TEM image and EDS mapping of 2D HEA subnanometer ribbon. Adapted with permission from [19]. Copyright (2022) American Chemical Society. (b) The EELS spectrum of 2D RhPd-H. Adapted with permission from [12]. Copyright (2020) American Chemical Society. (c) The Co 2p XPS spectrum of Co metallene. Adapted with permission from [32]. Copyright (2021) American Chemical Society. (d) The Pd K-edge XANES spectra of Pd-based nanomaterials. Adapted with permission [14]. Copyright (2020) Elsevier.

extended X-ray absorption fine structure (EXAFS) and X-ray absorption near-edge structure (XANES) can be acquired for studying the electronic structures of materials. The Pd K-edge XANES spectra of Figure 3.4(d) exhibit the negative shifts in the Pd (0.31 eV) and PdCo (2.83 eV) single-atom-layers (SAL) compared with that of bulk Pd, implying that the lowered unoccupied d-orbitals might improve the chemical activity and stability of metallene simultaneously [14].

The inductively coupled plasma could be applied to create the excited atoms and ions in materials emitting the electromagnetic radiation with specific wavelengths for the spectroscopic analysis of ICP-AES. It can also be used to ionize the materials to produce the atomic and molecular ions for the mass spectrometric analysis of ICP-MS. These techniques of elemental analysis with high resolution can detect the trace elements, which is suitable to characterize the 2D alloys [12,13,19].

3.2.3 Crystal Structure Identification

Beyond the dimension measurement and elemental analysis, the crystallinity and phases of 2D metals should be accurately identified. Initially, X-ray diffraction (XRD) and Raman spectroscopy are employed to evaluate the crystallinity of materials and confirm the crystal structure of phases. The crystal structure identification of 2D metals would be finally implemented by the TEM providing the atomic-level images and diffraction patterns for the structure analysis. The XRD pattern of Rh metallene prepared by the topotactic metallization indicates the diffraction peaks located at 40.7°, 47.7°, 69.8°, and 84.4° individually corresponding to the (111), (200), (220), and (311) crystal planes of Rh with the face-centered cubic (FCC) structure as shown in Figure 3.5(a) [21]. Moreover, the diffraction peaks slightly shift to lower 2θ values, implying that the interlayer distance of 2D metals increases with the decrease of thickness owing to the weakening of interlayer interaction. The Raman spectrum of bismuthene prepared by the topotactic metallization as shown in Figure 3.5(b) exhibits the two characteristic peaks at 74 cm^{-1} and 99.7 cm^{-1} separately

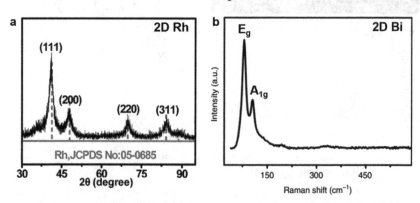

FIGURE 3.5 (a) The XRD pattern of Rh metallene. Adapted with permission from [21]. Copyright (2018) Royal Society of Chemistry. (b) The Raman spectrum of bismuthene. Adapted with permission from [20]. Copyright (2020) Springer Nature. Copyright The Authors, some rights reserved; exclusive licensee Springer Nature. Distributed under a Creative Commons Attribution License 4.0 (CC BY) https://creativecommons.org/licenses/by/4.0/

corresponding to the E_g and A_{1g} modes of metallic Bi, which represent the primary vibration modes in the Bi lattice [20]. The Raman peak positions with slight upshift compared with those of bulk Bi suggest that the obtained Bi is much thinner.

The TEM analysis based on the atomic-level images and diffraction patterns is very powerful for the crystal structure identification of materials. For instance, the crystal structure of multilayer antimonene first prepared by the plasma-assisted synthesis has been precisely identified by the TEM analysis [47]. During the synthesis, the nitrogen produced by the low-temperature plasma is diffused into the InSb to form an N-rich layer at the surface. Then the nitrogen preferentially reacts with indium to form the InN and the Sb atoms are squeezed out toward the surface to construct the multilayer antimonene. As can be seen in Figure 3.6(a), the cross-sectional view of the multilayer antimonene/InN/InSb layer structure corresponding to the atomic model can be observed in the TEM image. The diffraction pattern of multilayer antimonene could be directly obtained under the TEM observation or indirectly obtained from the atomic-level images processed by the fast Fourier transform (FFT). In Figure 3.6(b), the length of the dashed line, which contains a double reciprocal vector, is equal to 8.09 nm^{-1}, so that the length of the reciprocal vector is

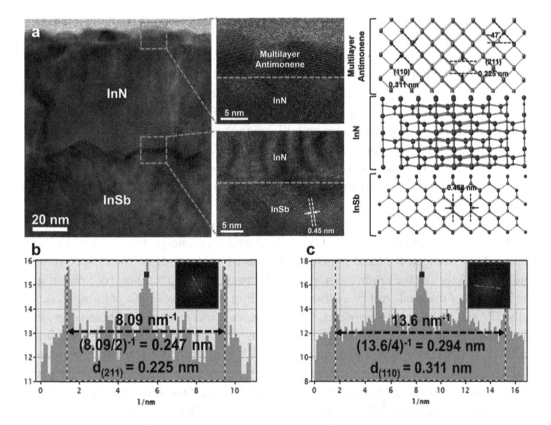

FIGURE 3.6 (a) The cross-sectional TEM image of multilayer antimonene/InN/InSb layer structure and its theoretical atomic model. (b) The distance between (211) reciprocal points of multilayer antimonene. (c) The distance between (110) reciprocal points of multilayer antimonene. Adapted with permission from [47]. Copyright (2016) Royal Society of Chemistry.

42 ▪ 2D Metals

4.045 nm^{-1}. The inverse of 4.045 nm^{-1} is equal to 0.247 nm corresponding to the (211) inter-planar distance (0.225 nm) of multilayer antimonene. On the other hand, the length (13.6 nm^{-1}) of the dashed line as shown in Figure 3.6(c) contains a quadruple reciprocal vector. Hence the length of the reciprocal vector is 3.4 nm^{-1}. The inverse of 3.4 nm^{-1} is equal to 0.294 nm corresponding to the (110) interplanar distance (0.311 nm) of multilayer antimonene. These calculations, derived from the experimental results, agree with the theory and the error rate is less than 9%, indicating that they are believable. In addition, the angle between the two lines representing the (211) and (110) plane groups in the diffraction pattern is nearly equal to 47°C, which is consistent with that between the (211) and (110) plane groups in the real lattice as can be seen in the theoretical atomic model of Figure 3.6(a). As the three lattice parameters obtained by the experiments are simultaneously identical to the theoretical values, suggesting that the as-synthesized material is indeed multilayer antimonene, since the coincident consistency is unique for each material.

3.2.4 Characterizations of Other Properties

There are still some analysis techniques, such as the Fourier-transform infrared spectroscopy (FTIR), ultraviolet photoelectron spectroscopy (UPS), photoluminescence spectroscopy (PL), ultraviolet-visible-near infrared (UV-vis-NIR) spectroscopy, and transient absorption (TA) pump-probe spectroscopy, which may be applied to characterize the physical or chemical properties of 2D metals. The FTIR is generally utilized to confirm the various functional groups of materials according to their different vibration energies. While the FTIR has also been used to recognize bulk and 2D Bi. The characteristic peaks of bismuthene located at 1387 cm^{-1} and 1645 cm^{-1} in the FTIR spectrum tend to disappear as the bismuthene gradually becomes bulk Bi due to the much weaker lattice vibration in the bulk Bi [28]. The UPS, which is the XPS replacing its X-ray source with ultraviolet light, can estimate the work function of metals. Zhang et al. have determined the work function of antimonene on poly[bis(4-phenyl)(2,4,6-trimethylphenyl)amine] (PTAA) as 5.28 eV by using the UPS [30].

The bandgap opening occurs in some metals as the monolayer is reached because of the quantum confinement effect. The PL, UV-vis-NIR, and TA pump-probe spectroscopies are employed to investigate the band structures of materials with bandgap. The PL spectrum is obtained by detecting the luminescence emitted through the electron-hole recombination. The bandgap of materials can be estimated according to the wavelength of luminescence. In the UV-vis-NIR absorption spectra, the absorbance is increased as the photon energy is higher than the bandgap of materials. The bandgap of materials can also be derived from the UV-vis-NIR diffuse reflectance spectra. The TA pump-probe spectroscopy is mainly used to deeply comprehend the carrier relaxation processes in materials, such as the exciton–exciton annihilation, the trapping of excitons by the surface states, and the interband carrier-phonon scattering [48]. Furthermore, the carrier lifetimes for various relaxation processes could be estimated from the TA pump-probe spectra. The bandgap (2.33 eV) of antimonene is confirmed by Zhang et al. using the PL spectroscopy [30]. Yang et al. have applied UV-vis-NIR spectroscopy to determine the bandgap of bismuthene as 0.61 eV [28]. This bismuthene is further characterized by using TA pump-probe spectroscopy to study its carrier dynamics [28]. The carrier lifetime of this bismuthene could be initially divided into two components including one (16 µs) at the microsecond scale and the other (494 ns) at sub-microsecond scale under a probe wavelength of 600 nm. The shorter one is associated with the carrier relaxation to the trapping states and the longer one corresponds to the electron-hole recombination. Then the femtosecond broadband pump-probe technique is implemented to analyze the carrier trapping process, revealing that it contains the intraband, interband, and trap state-mediated carrier relaxations with the lifetimes of 2.8 ps, 689 ps, and >7 ns, respectively.

3.3 SUMMARY

Among the preparation methods of 2D metals, the production yield of chemical synthesis from laboratories is too low to be applied to the industries. Hence it should be greatly improved by the precise control of experimental environments and conditions. The physical processes for the preparation can generally achieve a much higher production yield than that of chemical synthesis. However, the larger-scale 2D metals used for thin film applications cannot be obtained by exfoliation methods. The electron beam dealloying can only be implemented under the TEM observation, which is limited to research rather than real applications. Physical vapor deposition can prepare more commercial products for thin-film applications despite the higher cost. The repeated size reduction may accomplish higher production yield than that of chemical synthesis, but the products are also powders. The plasma-assisted synthesis can prepare the 2D thin metal films at a lower cost than that of physical vapor deposition. Nevertheless, the substrate used for this method must be fixed. Therefore, a suitable preparation method for 2D metals should be chosen according to the following applications.

Up to now, the preparation of 2D metal powders is more popular than that of 2D thin metal films. Nowadays, analysis techniques can satisfy the characterizations of 2D metals, especially for powder materials. There are some difficulties in the nondestructive analysis of 2D metal thin films. In general, people used to characterize the as-prepared materials by the nondestructive methods (e.g. XRD and Raman spectroscopy) before the destructive analysis (e.g. TEM). Nonetheless, the XRD patterns and Raman spectra usually cannot be obtained as the 2D metal films are thinner than 10 nm. Consequently, the detection limits of these nondestructive analyzes should be improved for the efficient characterization of 2D metal thin films in the future.

REFERENCES

1. H. Yu, T. Zhou, Z. Wang, Y. Xu, X. Li, L. Wang, H. Wang, Defect-rich porous palladium metallene for enhanced alkaline oxygen reduction electrocatalysis, *Angew. Chem. Int. Ed.* 60, **2021**, 12027–12031.
2. L. Wang, Z. Zeng, W. Gao, T. Maxson, D. Raciti, M. Giroux, X. Pan, C. Wang, J. Greeley, Tunable intrinsic strain in two-dimensional transition metal electrocatalysts, *Science* 363, **2019**, 870–874.
3. W. Zhu, L. Zhang, P. Yang, C. Hu, Z. Luo, X. Chang, Z. J. Zhao, J. Gong, Low-coordinated edge sites on ultrathin palladium nanosheets boost carbon dioxide electroreduction performance, *Angew. Chem. Int. Ed.* 57, **2018**, 11544–11548.

4. M. Luo, Z. Zhao, Y. Zhang, Y. Sun, Y. Xing, F. Lv, Y. Yang, X. Zhang, S. Hwang, Y. Qin, J. Y. Ma, F. Lin, D. Su, G. Lu, S. Guo, PdMo bimetallene for oxygen reduction catalysis, *Nature* 574, **2019**, 81–85.

5. D. Xu, Y. Liu, S. Zhao, Y. Lu, M. Han, J. Bao, Novel surfactant-directed synthesis of ultra-thin palladium nanosheets as efficient electrocatalysts for glycerol oxidation, *Chem. Commun.* 53, **2017**, 1642–1645.

6. L. Zhao, C. Xu, H. Su, J. Liang, S. Lin, L. Gu, X. Wang, M. Chen, N. Zheng, Single-crystalline rhodium nanosheets with atomic thickness, *Adv. Sci.* 2, **2015**, 1500100.

7. H. Duan, N. Yan, R. Yu, C. R. Chang, G. Zhou, H. S. Hu, H. Rong, Z. Niu, J. Mao, H. Asakura, T. Tanaka, P. J. Dyson, J. Li, Y. Li, Ultrathin rhodium nanosheets, *Nat. Commun.* 5, **2014**, 3093.

8. S. Gao, Y. Lin, X. Jiao, Y. Sun, Q. Luo, W. Zhang, D. Li, J. Yang, Y. Xie, Partially oxidized atomic cobalt layers for carbon dioxide electroreduction to liquid fuel, *Nature* 529, **2016**, 68–71.

9. X. Mu, J. Gu, F. Feng, Z. Xiao, C. Chen, S. Liu, S. Mu, RuRh bimetallene nanoring as high-efficiency pH-universal catalyst for hydrogen evolution reaction, *Adv. Sci.* 8, **2021**, 2002341.

10. F. Lv, B. Huang, J. Feng, W. Zhang, K. Wang, N. Li, J. Zhou, P. Zhou, W. Yang, Y. Du, D. Su, S. Guo, A highly efficient atomically thin curved PdIr bimetallene electrocatalyst, *Natl. Sci. Rev.* 8, **2021**, nwab019.

11. X. Li, P. Shen, Y. Luo, Y. Li, Y. Guo, H. Zhang, K. Chu, PdFe single-atom alloy metallene for N_2 electroreduction, *Angew. Chem. Int. Ed.* 61, **2022**, e202205923.

12. J. Fan, J. Wu, X. Cui, L. Gu, Q. Zhang, F. Meng, B. H. Lei, D. J. Singh, W. Zheng, Hydrogen stabilized RhPdH 2D bimetallene nanosheets for efficient alkaline hydrogen evolution, *J. Am. Chem. Soc.* 142, **2020**, 3645–3651.

13. J. Lai, Fei Lin, Yonghua Tang, Peng Zhou, Yuguang Chao, Yelong Zhang, and Shaojun Guo, Efficient bifunctional polyalcohol oxidation and oxygen reduction electrocatalysts enabled by ultrathin PtPdM (M = Ni, Fe, Co) nanosheets, *Adv. Energy Mater.* 9, **2019**, 1800684.

14. J. Jiang, W. Ding, W. Li, Z. Wei, Freestanding single-atom-layer Pd-based catalysts: Oriented splitting of energy bands for unique stability and activity, *Chem* 6, **2020**, 431–447.

15. A. Funatsu, H. Tateishi, K. Hatakeyama, Y. Fukunaga, T. Taniguchi, M. Koinuma, H. Matsuura, Y. Matsumoto, Synthesis of monolayer platinum nanosheets, *Chem. Commun.* 50, **2014**, 8503–8506.

16. J. Zhao, Q. Deng, A. Bachmatiuk, G. Sandeep, A. Popov, J. Eckert, M. H. Rümmeli, Free-standing single-atom-thick iron membranes suspended in graphene pores, *Science* 343, **2014**, 1228–1232.

17. X. Zhou, X. Huang, X. Qi, S. Wu, C. Xue, F. Boey, Q. Yan, P. Chen, H. Zhang, In situ synthesis of metal nanoparticles on single-layer graphene oxide and reduced graphene oxide surfaces. *J. Phys. Chem. C* 113, **2009**, 10842–10846.

18. X. Huang, S. Li, Y. Huang, S. Wu, X. Zhou, S. Li, C. L. Gan, F. Boey, C. A. Mirkin, H. Zhang, Synthesis of hexagonal close-packed gold nanostructures, *Nat. Commun.* 2, **2011**, 292.

19. L. Tao, M. Sun, Y. Zhou, M. Luo, F. Lv, M. Li, Q. Zhang, L. Gu, B. Huang, S. Guo, A general synthetic method for high-entropy alloy subnanometer ribbons, *J. Am. Chem. Soc.* 144, **2022**, 10582–10590.

20. F. Yang, A. O. Elnabawy, R. Schimmenti, P. Song, J. Wang, Z. Peng, S. Yao, R. Deng, S. Song, Y. Lin, M. Mavrikakis, W. Xu, Bismuthene for highly efficient carbon dioxide electroreduction reaction, *Nat. Cummun.* 11, **2020**, 1088.

21. H. M. Liu, S. H. Han, Y. Zhao, Y. Y. Zhu, X. L. Tian, J. H. Zeng, J. X. Jiang, B. Y. Xia, Y. Chen, Surfactant-free atomically ultrathin rhodium nanosheet nanoassemblies for efficient nitrogen electroreduction, *J. Mater. Chem. A* 6, **2018**, 3211–3217.

22. K. Fukuda, J. Sato, T. Saida, W. Sugimoto, Y. Ebina, T. Shibata, M. Osada, T. Sasaki, Fabrication of ruthenium metal nanosheets via topotactic metallization of exfoliated ruthenate nanosheets, *Inorg. Chem.* 52, **2013**, 2280–2282.

23. Y. Sun, T. Jiang, J. Duan, L. Jiang, X. Hu, H. Zhao, J. Zhu, S. Chen, X. Wang, Two-dimensional nanomesh arrays as bifunctional catalysts for N_2 electrolysis, *ACS Catal.* 10, **2020**, 11371–11379.

24. C. Cao, D. D. Ma, J. F. Gu, X. Xie, G. Zeng, X. Li, S. G. Han, Q. L. Zhu, X. T. Wu, Q. Xu, Metal-organic layers leading to atomically thin bismuthene for efficient carbon dioxide electroreduction to liquid fuel, *Angew. Chem. Int. Ed.* 59, **2020**, 15014–15020.

25. M. Zhang, W. Wei, S. Zhou, D. D. Ma, A. Cao, X. T. Wu, Q. L. Zhu, Engineering a conductive network of atomically thin bismuthene with rich defects enables CO_2 reduction to formate with industry-compatible current densities and stability, *Energy Environ. Sci.* 14, **2021**, 4998–5008.

26. W. Ma, J. Bu, Z. Liu, C. Yan, Y. Yao, N. Chang, H. Zhang, T. Wang, J. Zhang, Monoclinic scheelite bismuth vanadate derived bismuthene nanosheets with rapid kinetics for electrochemically reducing carbon dioxide to formate, *Adv. Funct. Mater.* 31, **2021**, 2006704.

27. C. J. Peng, X. T. Wu, G. Zeng, Q. L. Zhu, *In situ* bismuth nanosheet assembly for highly selective electrocatalytic CO_2 reduction to formate, *Chem. Asian J.* 16, **2021**, 1539–1544.

28. Q. Q. Yang, R. T. Liu, C. Huang, Y. F. Huang, L. F. Gao, B. Sun, Z. P. Huang, L. Zhang, C. X. Hu, Z. Q. Zhang, C. L. Sun, Q. Wang, Y. L. Tang, H. L. Zhang, 2D bismuthene fabricated via acid-intercalated exfoliation showing strong nonlinear near-infrared responses for mode-locking lasers, *Nanoscale* 10, **2018**, 21106–21115.

29. W. Zhang, Y. Hu, L. Ma, G. Zhu, P. Zhao, X. Xue, R. Chen, S. Yang, J. Ma, J. Liu, Z. Jin, Liquid-phase exfoliated ultrathin Bi nanosheets: Uncovering the origins of enhanced electrocatalytic CO_2 reduction on two-dimensional metal nanostructure, *Nano Energy* 53, **2018**, 808–816.

30. F. Zhang, J. He, Y. Xiang, K. Zheng, B. Xue, S. Ye, X. Peng, Y. Hao, J. Lian, P. Zeng, J. Qu, J. Song, Semimetal-semiconductor transitions for monolayer antimonene nanosheets and their application in perovskite solar cells, *Adv. Mater.* 30, **2018**, 1803244.

31. Y. Li, J. Chen, J. Huang, Y. Hou, L. Lei, W. Lin, Y. Lian, Z. Xiang, H. H. Yang, Z. Wen, Interfacial engineering of Ru–S–Sb/antimonene electrocatalysts for highly efficient electrolytic hydrogen generation in neutral electrolyte, *Chem. Commun.* 55, **2019**, 10884–10887.

32. J. Yin, Z. Yin, J. Jin, M. Sun, B. Huang, H. Lin, Z. Ma, M. Muzzio, M. Shen, C. Yu, H. Zhang, Y. Peng, P. Xi, C. H. Yan, S. Sun, A new hexagonal cobalt nanosheet catalyst for selective CO_2 conversion to ethanal, *J. Am. Chem. Soc.* 143, **2021**, 15335–15343.

33. X. Wang, J. He, B. Zhou, Y. Zhang, J. Wu, R. Hu, L. Liu, J. Song, J. Qu, Bandgap-tunable preparation of smooth and large two-dimensional antimonene, *Angew. Chem. Int. Ed.* 57, **2018**, 8668–8673.

34. R. Gusmão, Z. Sofer, D. Bouša, M. Pumera, Pnictogen (As, Sb, Bi) nanosheets for electrochemical applications are produced by shear exfoliation using kitchen blenders, *Angew. Chem. Int. Ed.* 56, **2017**, 14417–14422.

35. C. C. Mayorga-Martinez, R. Gusmão, Z. Sofer, M. Pumera, Pnictogen-based enzymatic phenol biosensors: Phosphorene, arsenene, antimonene, and bismuthene, *Angew. Chem. Int. Ed.* 58, **2019**, 134–138.

36. V. Kochat, A. Samanta, Y. Zhang, S. Bhowmick, P. Manimunda, S. A. S. Asif, A. S. Stender, R. Vajtai, A. K. Singh, C. S. Tiwary, P. M. Ajayan, Atomically thin gallium layers from solid-melt exfoliation, *Sci. Adv.* 4, **2018**, e1701373.

37. X. Wang, C. Wang, C. Chen, H. Duan, K. Du, Free-standing monatomic thick two-dimensional gold, *Nano Lett.* 19, **2019**, 4560–4566.

38. M. L. Tao, Y. B. Tu, K. Sun, Y. L. Wang, Z. B. Xie, L. Liu, M. X. Shi, J. Z. Wang, Gallenene epitaxially grown on Si(1 1 1), *2D Mater.* 5, **2018**, 035009.

39. M. Fortin-Deschênes, O. Waller, T. O. Menteş, A. Locatelli, S. Mukherjee, F. Genuzio, P. L. Levesque, A. Hébert, R. Martel, O. Moutanabbir, Synthesis of antimonene on germanium, *Nano Lett.* 17, **2017**, 4970–4975.
40. F. Reis, G. Li, L. Dudy, M. Bauernfeind, S. Glass, W. Hanke, R. Thomale, J. Schäfer, R. Claessen, Bismuthene on a SiC substrate: A candidate for a high-temperature quantum spin Hall material, *Science* 357, **2017**, 287–290.
41. L. Li, Y. Wang, S. Xie, X. B. Li, Y. Q. Wang, R. Wu, H. Sun, S. Zhang, H. J. Gao, Two-dimensional transition metal honeycomb realized: Hf on Ir(111), *Nano Lett.* 13, **2013**, 4671–4674.
42. M. Liao, Y. Zang, Z. Guan, H. Li, Y. Gong, K. Zhu, X. P. Hu, D. Zhang, Y. Xu, Y. Y. Wang, K. He, X. C. Ma, S. C. Zhang, Q. K. Xue, Superconductivity in few-layer stanene, *Nat. Phys.* 14, **2018**, 344–348.
43. J. Yuhara, B. He, N. Matsunami, M. Nakatake, G. Le Lay, Graphene's latest cousin: Plumbene epitaxial growth on a "Nano WaterCube", *Adv. Mater.* 31, **2019**, 1901017.
44. Z. Yang, Z. Wu, Y. Lyu, J. Hao, Centimeter-scale growth of two-dimensional layered high-mobility bismuth films by pulsed laser deposition, *Info Mat* 1, **2019**, 98–107.
45. H. Liu, H. Tang, M. Fang, W. Si, Q. Zhang, Z. Huang, L. Gu, W. Pan, J. Yao, C. Nan, H. Wu, 2D metals by repeated size reduction, *Adv. Mater.* 28, **2016**, 8170–8176.
46. J. Gu, B. Li, Z. Du, C. Zhang, D. Zhang, S. Yang, Multi-atomic layers of metallic aluminum for ultralong life lithium storage with high volumetric capacity, *Adv. Funct. Mater.* 27, **2017**, 1700840.
47. H. S. Tsai, C. W. Chen, C. H. Hsiao, H. Ouyang, J. H. Liang, The advent of multilayer antimonene nanoribbons with room temperature orange light emission, *Chem. Commun.* 52, **2016**, 8409–8412.
48. H. S. Tsai, Y. H. Huang, P. C. Tsai, Y. J. Chen, H. Ahn, S. Y. Lin, Y. J. Lu, Ultrafast exciton dynamics in scalable monolayer MoS_2 synthesized by metal sulfurization, *ACS Omega* 5, **2020**, 10725–10730.

CHAPTER **4**

Structure, Stability, and Properties of 2D Metallenes

Akanksha A. Sangolkar, Ramakrishna Kadiyam, and Ravinder Pawar

4.1 INTRODUCTION

The exclusive covalent tetrahedral arrangement of C atoms in diamond endows the carbon framework with a distinct electronic feature compared to graphite consisting of 2D C sheets held together by van der Waals (vdW) interactions. Likewise, the properties exhibited by carbon skeletons in low-dimensional morphologies including fullerenes, nanotubes, carbon dots, nanodiamonds, graphene, nanothreads, etc. are entirely different from those of diamond and graphite. Therefore, the arrangement of atoms in a chemical structure as well as the size plays a crucial role in imparting novel properties to a material [1].

An innovative discovery of an atomically thin layer of C (graphene) by Novoselov and Giem in 2004 has enlightened the field of 2D materials [2]. Confinement of electrons in a 2D framework bestows exquisite properties to a material like enhanced structural stability in graphene and direct energy gap in transition metal dichalcogenides (TMDCs). These 2D materials hold promise for the development of next-generation functional and technological devices [3, 4]. Therefore, work has triggered to explore other novel 2D materials and to date, a variety of materials including TMDCs, silicene, graphitic carbon nitride (g-CN), boron nitride, and so forth have been extensively investigated [4].

4.2 THE TWO-DIMENSIONAL (2D) METALLIC LAYERS

4.2.1 The 2D Metals Remained Unexplored for Long

Despite many advances made on 2D materials, rarely any investigation has been made on 2D metals. The question then arises "Why did 2D metals remain unexplored in the

DOI: 10.1201/9781032645001-4

category of 2D materials for a long time?" The answer lies in the intrinsic behavior of metals which is distinct from other elements in the periodic table.

The C atoms in graphene are *sp*2-hybridized and bonded with alternating single and double covalent bonds in a 2D framework. Owing to the directional nature of the hybridized orbitals of C, graphene particularly obtains a honeycomb (HC) structure. These atomically thin graphene sheets stack on each other with weak vdW force of attraction to form a 3D bulk graphite structure (Figure 4.1). Therefore, the 3D counterpart of graphene is composed of layered graphene sheets with strong intralayer covalent bonding and weak interlayer vdW interaction. This facilitates easier exfoliation of graphene sheets from graphite [2]. Similar exfoliation techniques are applicable to obtain the monolayer sheets of other materials like TMDC.

Unlike graphite, the metal atoms are packed together in a nonlayered structure with metallic bonding in their 3D metal crystals. Models for the nonprimitive cell of 3D structures with layered and nonlayered crystal lattices are depicted in Figure 4.1. Therefore, simple exfoliation techniques are not suitable for the experimental fabrication of 2D metals [5–7]. Further, the nondirectional nature of metallic bonding favors the formation of chemical bonding between metal atoms in any direction. Therefore, metals can form bonds with metal atoms in 3D space without any specifically defined orientation. Even a small number of metal atoms prefers 3D clustering over 2D planar morphology. Although studies have revealed that 2D metal clusters can be stabilized up to a certain (four to six) number of atoms, nevertheless the stability of 2D metal sheets is often challenging. This imposes a major practical hurdle to generating 2D metal layers using chemical vapor deposition techniques. Even though 2D metal structures were created, the stability of metals in atomically thin 2D structures was ambiguous and they were believed to be thermodynamically unstable. Therefore, only a few reports previously focused on 2D metals, and the field of 2D metals was in its infancy.

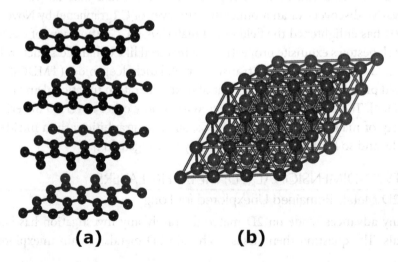

FIGURE 4.1 Supercell for the 3D crystal structures: (a) 3D graphite with layered structure. (b) 3D Cu with nonlayered metallic crystal.

4.2.2 A Breakthrough in the Experimental Fabrication of Thin Iron Membrane

It is worth mentioning that the 2D patches of iron were fabricated and characterized *in situ* employing transmission electron microscopy [8]. Since monoatomically thin metal membranes are less stable than 3D clusters, Rümmeli and co-workers adopted a novel synthetic approach of rigid substrate-based immobilization of metal. Herein, an attempt was made to stabilize the freestanding atomically thin layer of Fe inside the rigid stabilizing nanopore of graphene and ultimately success was achieved. This groundbreaking discovery not only provided the first experimental evidence, but also enhanced expectations of achieving freestanding monoatomic 2D metal layers [8]. This study has thus inspired the scientific research community for theoretical and experimental advances on metal layers and currently 2D metals are the focus of many investigations for their emerging applications. Although challenging, the development of sophisticated analytical techniques and new synthetic approaches has enabled the fabrication and characterization of delicate frameworks that cannot be accessed by classical experimental methods.

4.2.3 Emergence of Novel Methodologies to Obtain Freestanding Metallic Monolayers

The atomically thin metallic monolayers are regarded as less stable structures than their 3D counterparts. However, the sophisticated instrumentations made these 2D metallic structures experimentally accessible following novel synthetic procedures and well-developed analytical techniques facilitated their characterization. Thus, analogous to the Fe membrane, freestanding layers of Sn, Cr, Au were also grown inside the nanopore of graphene [9–11]. Besides, monoatomically thin membranes of Au were obtained by mechanical thinning and electron beam irradiation techniques of the bulk materials [12, 13]. The experimentally observed structures of atomically thin Au membranes characterized with electron microscopy along with theoretically simulated structures are displayed in Figure 4.2. Moreover, the patches of Mo were obtained by selective ionization of Se atoms from $MoSe_2$ monolayers [14]. These experimental achievements in the fabrication and characterization of atomically thin freestanding metal sheets have ensured their existence and formation.

4.3 THE MOST STABLE 2D MORPHOLOGY OF ELEMENTAL METALS

Carbon atoms are arranged in a planar honeycomb (HC) lattice in graphene [3]. Likewise, B and N form planar HC lattice in boron nitride monolayer while Si atoms prefer buckled HC arrangement in silicene. The characterization of the Fe membrane embedded inside the graphene pores reveals the square (SQ) planar arrangement in the 2D lattice [8]. However, patches of Sn, Cr, and Au inside the graphene pore were found to align themselves in a close-packed arrangement resembling the HX lattice [9–11]. What is the most preferred lattice for arranging metal atoms in their atomically thin dimensions with and without template support?

Metal layers were mostly studied in their planar HC, SQ, HX, buckled SQ (bSQ), and HC (bHC) 2D lattices as shown in Figure 4.3 [5, 6, 15, 16]. These 2D crystal lattices were

50 ■ 2D Metals

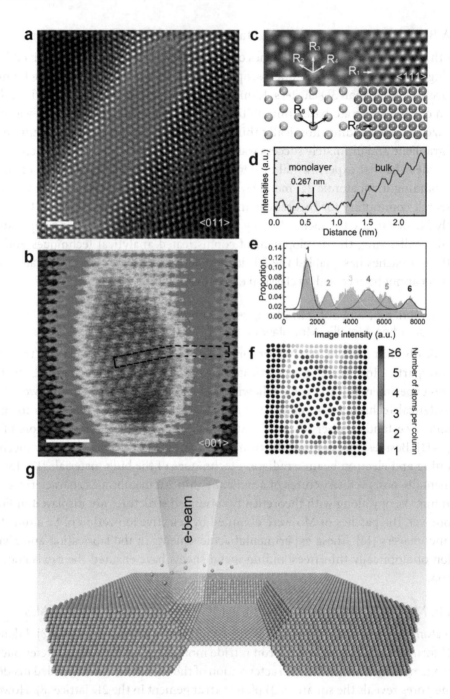

FIGURE 4.2 The electron micrographs, experimentally realized and theoretically simulated structures of atomically thin Au membranes. Adapted with permission from [12]. Copyright (2019), American Chemical Society.

mostly evaluated to investigate the stability of 2D layers with different coordination numbers (CNs) of metal atoms and planar *vs* buckled structures. Theoretical results reveal that the planar hexagonal close-packed arrangement with hexacoordinate metals is the

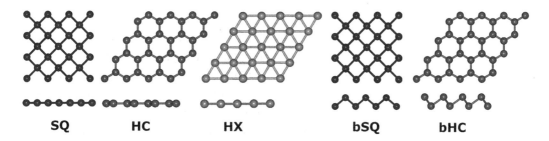

FIGURE 4.3 Two-dimensional crystal lattice of metals.

most stable structure of Cu, Ag, and Au monolayer [17–19]. 2D Cu, Ag, and Au were also found to attain similar arrangements inside the nanopore of the graphene skeleton [20]. Similarly, Ag and Au clusters favor achieving a similar atomic arrangement on the surface TMDC monolayer [21]. These results are consistent with the structure of experimentally obtained metal patches inside the pores of graphene [9–11]. However, the elemental metals in the periodic table also show stability in 2D HC, SQ, bHC, and bSQ crystal lattices [6]. Most often, elemental metals are preferentially stabilized in their planar HX close-packed arrangement therefore, studies usually investigate 2D HX lattices for atomically thin metal layer.

Moreover, atomic size, lattice constant, and number of valence electrons are other factors that control the structure of 2D layers. These factors actually control the extent of orbital overlap between atoms in the 2D lattice and are mostly predicted by the periodicity of the elements in the periodic table [22].

One of the approaches to obtain Au monolayer involves mechanical trimming or electron beam ionization technique for removal of metal atoms [12, 13]. Therefore, it can be expected that the surface obtained will depend upon the facet of the crystal used to remove the metal atoms. Keeping this in mind, the properties of ultrathin metal slabs consisting of one to four layers along the (100), (110), and (001)/(111) crystal planes were scrutinized for elemental metals in the periodic table [23]. Also, the H_2S dissociation mechanism was thoroughly investigated on a two-atom thick Lithium slab along the Li(110) plane of the 3D bulk [24].

4.4 STABILITY OF 2D METAL LAYERS

Although the iron layer was obtained inside the graphene pore, the periodic table is full of elemental metals ranging from soft alkali metals like Li and Na to inert metals like Au and Pt. Then what are the prospects for the stability of their 2D metal layers? Experimental determination of stability in 2D morphology of all metals in the periodic table is not only expensive but also a tedious task. Therefore, theoretical simulations employing the density functional theory (DFT) methods can be a cost-effective and efficient strategy to assess the stability of other metal layers in the periodic table.

52 ■ 2D Metals

4.4.1 Prospects for the Stability of Metals as 2D Layers

Inspired by the experimental fabrication of Fe patches, Koskinen and co-workers have probed into the stability of most of the elemental metals in the periodic table. DFT results show that most metals can be stabilized into a 2D framework by overcoming layer bending. The equilibrium bond distance is shortest for the metallic elements located in the center of the d-block series in the periodic table. Therefore, the two metal atoms can be strongly bound in their 2D layers and thus exhibit the highest cohesive energy and bulk moduli. Basic material properties such as interatomic distances, cohesive energies, bulk moduli of 2D structures show strong correlations with their corresponding bulk counterparts. This enables the properties of 2D layers to be quickly determined from the values of their corresponding 3D metallic crystal structures [5].

Furthermore, Ono devoted attention to systematically analyzing the dynamic stability of elemental metals in the periodic table considering atomically thin HX, bHC, and bSQ morphologies. Interestingly, most of the elemental metals were found to be dynamically stable in 2D layers and more particularly in the HX and bHC crystal lattice. Among the studied metals, V, Nb, and Ta show dynamical stability in their trilayered SQ lattice, while no dynamic stability was observed for Cs, Ba, Ga, and Tl metals. The dynamical stability trends observed for 2D metals show a strong correlation with 3D metal crystals, which is consistent with the parameters appraised by Koskinen. In general, the elemental metals that are dynamically stable in the 2D HX lattice are found to have a 3D Face-Centered Cubic (FCC) or Hexagonal Close-Packed (HCP) crystal lattice. Likewise, the metals with dynamical stability in 2D bHC structure have a Body-Centered Cubic (BCC) arrangement in the 3D bulk. If the 2D bSQ lattice is dynamically stable then the metal has an HCP crystal structure in the 3D lattice. Moreover, metal atoms that show dynamical stability in the 2D bHC and bSQ lattice are stable in their 3D HCP or FCC lattices. This infers that the stable structure of the 2D metallic layer is interrelated with the 3D structures and can be expected to be the intrinsic behavior of the elemental metal [6].

The prospects for the stability of substrate-stabilized finite metal patches would be entirely different from that of an atomically thin 2D metal sheet. This is because the nature of the interaction of the metal atoms with the stabilizing support will also govern the formation of template-stabilized metal patches. Therefore, Koskinen and co-workers have performed a comprehensive study combining DFT and liquid drop models to understand the stability of synthetically accessible metal patches inside the nanopore of graphene. Here, the stability of 2D *vs* 3D metal clusters in the gas phase has been systematically assessed. Thereafter, the interaction of metal atoms with the graphene skeleton, carbide formation energy, the induced strain in 2D metals due to the lattice mismatch between the metal and graphene unit cell were carefully analyzed. Based on the overall calculated values, the relative stability of 2D metal clusters over 3D clusters inside the nanopore of graphene was predicted. The summary of the best elemental metals forming stable 2D patches inside graphene pores is shown in Figure 4.4. Most metals show potential to get stabilized as 2D patches inside the pores of the graphene template. Among the 45 elemental metals studied in the periodic table, Cu has been found to be an excellent candidate for the formation of

Structure, Stability, and Properties of 2D Metallenes ■ 53

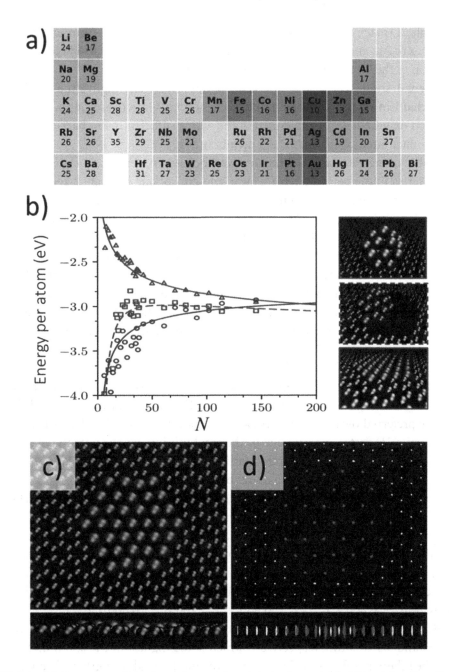

FIGURE 4.4 (a) Summary of the best elemental metals to form stable 2D patches (smaller magnitude indicates higher stability). (b) Variation in energy per atom for N-atoms 3D Cu cluster for on-top adsorption (triangles), 3D Cu clusters on graphene pore edge (squares) and 2D Cu patches inside the graphene pore (circles). (c) Snapshot of 2D Cu_{37} patch in graphene obtained during the molecular dynamics simulations at T = 300 K. (d) Snapshot of the trajectories 2D Cu_{37} patched graphene system obtained after 36.9 ps of the molecular dynamics simulations at T = 300 K. Adapted with permission [7]. Copyright The Authors, some rights reserved; exclusive licensee [The Royal Society of Chemistry]. Distributed under a Creative Commons Attribution License 4.0 (CC BY).

54 ■ 2D Metals

the best stable 2D patch consisting up to 125 Cu atoms with a size of ~8nm^2 inside the pore of graphene. Zn, Ag, and Au are other prominent metal atoms that also hold promise for the formation of the best stable patches inside the graphene pore [7].

Experimental studies have also sparked an interest in envisaging the stability and growth mechanism of metals inside graphene nanopore. Growth studies of Au reveal its easy diffusion on graphene and gold membrane with a sufficiently low barrier of less than 0.5 eV to produce atomically thin 2D Au membrane [25].

The stability of metals hitherto discussed was focused on elemental belonging to the *s*-, *p*-, *d*-block series in the periodic table. Pawar and co-workers have analyzed the stability of lanthanide series elements in their 2D HX lattice with the aid of DFT calculations. All the 2D layers of lanthanide series except Yb and Lu were found to be dynamically stable in their HX lattice without the inclusion of spin-orbit coupling (SOC) during the calculations. SOC is more prominent for heavy elemental metals as *f*-block series elements; therefore, phonon dispersion relation was again evaluated with the inclusion of SOC. In subsequent consideration of SOC, the imaginary frequency appears in the phonon dispersion relations of most 2D HX lanthanides. Interestingly, the calculations showed Ce, Ho, Er, and Tm elements to be dynamically stable structures even in the presence of SOC [16].

4.4.2 How Much Stable Are the 2D Metal Layers?

Although theoretical reports demonstrate stability and experimental evidence proves the formation of atomically thin 2D metal layers, the 3D structure of metals is still thermodynamically preferred over the 2D framework. In other words, a 3D cluster is the global minimum while 2D metals occupy one of the local minima on the potential energy surface (PES). Therefore, atomically thin freestanding layers of metals are metastable forms of metal structures that prefer clustering under extreme conditions [5–7].

The extent of stability of a material can be quantified by its mechanical properties and the maximum temperature until the layer withstands structural deformation. The bulk modulus and elastic constants of 2D layers have been quantified for elemental metals in the periodic table to determine mechanical properties. Noteworthy, the bulk moduli (B) and elastic constant values are found to be highest for the 2D HX lattice and therefore, have higher mechanical strength. The bulk moduli of the 2D lattices are correlated with other 2D lattices as: $B_{SQ} = 0.860 \times B_{HX}$, $B_{SQ} = 0.860 \times B_{HX}$, and $B_{HC} = 0.666 \times B_{SQ}$ [5].

The calculated bulk modulus (B) and elastic constant C_{11}, C_{12}, and C_{66} for 2D HX Li are 5.9, 9.2, 2.6, and 9.2 GPa nm, respectively, which is the highest in the alkali metal series. Owing to the inherent softness of alkali metals, the mechanical strength of 2D alkali metal layers is also the lowest. B, C_{11}, C_{12}, and C_{66} value in the 2D HX Be are 46.3, 76.8, 17.4, 76.3 Gpa nm, respectively, which is remarkably high and shows mechanical strength similar to that of *d*-block series elements. The mechanical strength of the 2D layers increases from the *3d*- to the *5d*-series of the periodic table. Further, elemental metals located in the middle of the transition series show the highest mechanical strength in their 2D layer. Among the transition series elements, the calculated B, C_{11}, C_{12}, and C_{66} for the *4d*-series element Ru is 115.1, 122.3, 107.5, and 125.9 Gpa nm, respectively. The highest calculated B, C_{11}, C_{12}, and C_{66} is obtained for the *5d*-series element as 164.7, 234.7, 88.5, and 235.9 Gpa nm, respectively [5]. The bulk modulus of atomically thin 2D layers of elemental metals follows the same trend observed for ultraincompressible crystals. In 3D ultraincompressible crystal structures, Li in the alkali metals (14 Gpa), Be in the alkaline earth metal (122 Gpa), Ru in the $4d$-series (308 Gpa), and Os in the $5d$-series (402 Gpa) possess the highest bulk modulus values [26].

Ab initio molecular dynamics (AIMD) simulations were also performed at different temperatures to assess the thermal stability of the metal layers. Interestingly, the Au monolayer was observed to retain the 2D framework even after 10 ps of simulations at 1400 K and holes begin to appear after 4 ps of AIMD simulations at 1600 K [19]. The Ag monolayer maintains stability at temperatures up to 800 K while the Cu monolayer is stable till 1200 K [17, 18]. The melting temperature of these coinage metal layers is reported to be higher in other DFT-based simulations which may be attributed to the size of the supercell and computational method employed for the simulations [27].

4.4.3 2D Metals as Atomically Thin Flat Liquids

It is worth mentioning that the position of metal atoms in an atomically freestanding layer is not fixed, rather the atoms move due to the lack of covalent bonding. This behavior of diffusion of metallic atoms inside the monolayer is contradictory to other 2D materials with rigid structures like graphene and TMDCs that exhibit atomic vibrations in the equilibrium position due to covalent chemical bonding. Therefore, atomically thin freestanding 2D layers of metals are recognized to exist in a 2D liquid phase at high temperatures. Fluidity associated with 3D materials was known; however, the motion of metal atoms in atomically thin freestanding 2D layers demonstrates a fascinating nanoscale phenomenon of fluxional behaviour in two-dimensionality [27, 28]. A patch of Au_{49} shows self-healing of small pores generated during the diffusion whereas a patch with a lower number of atoms (Au_{40}, Au_{47}, and Au_{48}) creates pores that become larger without self-healing. The Au_{39} patch created in smaller sized pores shows stability in a solid phase; however, the liquid phase could not be encountered due to the close arrangement of atoms. Accordingly, the number of metal atoms and the size of a graphene pore is critical for the stability and behavior of the embedded metal patches [28].

4.5 ORIGIN OF THE STABILITY OF 2D METAL LAYERS

As it is known 3D metallic structures are thermodynamically stable while their 2D layers are metastable on PES. The answer to the query "What makes 2D metal layers stable?" is of immense interest. Cohesive energy of a structure is the amount of energy released when N metal atoms combine to form a material. Thus, the cohesive energy per atom (E_{coh}) of a 2D metal layer can be theoretically calculated using the following equation:

$$E_{coh} = E_M - \frac{E}{N} \tag{4.1}$$

where, E_M is energy of the free metal atom, $\dfrac{E}{N}$ denotes energy of each individual atom in the unit cell of a 2D metal layer. The amount of energy released (E_{coh}) to form the 3D bulk

56 ■ 2D Metals

is higher than that of the 2D layer. This explains why the thermodynamic stability of the 3D structure is attributed to the formation of a large number of chemical bonds in the 3D space [29].

The calculated cohesive energy per atom (E_{coh}) for 3D bulk crystal of Cu, Ag, and Au is 3.48, 2.49, and 3.11 eV/atom, respectively. The same (E_{coh}) for 2D Cu, Ag, and Au monolayer is 2.76, 2.01, and 2.82 eV/atom, respectively. Akin to the coinage metal layers, the cohesive energy of other elemental metals including the elements of the lanthanide series lies in the same range. The cohesive energies of bulk crystals lie between 3.21 and 4.45 eV/atom while for 2D HX structures of the lanthanide series it ranges within 2.58–3.20 eV/atom. These computed cohesive energies are substantially lower than the cohesive energies of 7.85, and 3.98 eV/atom for graphene and silicene, respectively. This is attributed to the metallic nature of the bonding in the metallic structures in contrast to the covalency in the chemical bonding in graphene and silicene. Nevertheless, the cohesive energies are large enough to facilitate the formation of 2D metallic monolayers.

Confinement of metal atoms to an atomically thin 2D membrane allows bond formation only along xy-plane and restricts bonding along the z-direction. Therefore, the CN of metal atoms in the 2D layer is much lower than their 3D counterparts. This results in a stronger in-plane bonding interaction between the metal atoms and substantially enhances the bond strength per atom. The bond strength per atom is the energy released during the formation of each bond in the 2D layer and can be calculated as:

$$Bond\,strength\,per\,atom = \frac{E_{coh}}{N} \tag{4.2}$$

Each individual atom forms 12 bonds in the 3D FCC crystal structure of Cu, Ag, and Au while it forms six bonds in the 2D HX lattice. The calculated bond strength per atom for each bond in the 3D crystal structures of Cu, Ag, and Au is 0.58, 0.21, and 0.52 eV/bond, respectively. Notably, enhanced bond strength per atom of 0.92, 0.33, 0.94 eV/bond are obtained for the 2D monolayers of Cu, Ag, and Au, respectively. Also, the bond strength per atom for the 2D HX lattice of the lanthanides series elements lies within 0.44–0.53 eV/bond, while for the 3D crystals it lies between 0.27 and 0.37 eV/bond. This suggests that the M–M bond is remarkably stronger in the 2D framework than in the 3D crystal lattice. Thus, reducing the dimensionality of the 3D metallic structure to atomically thin membrane reduces the cohesive energy, yet, the bond strength per atom is remarkably enhanced, which attributes to the stability of the 2D layers [29].

4.6 PROPERTIES OF 2D METAL LAYERS

4.6.1 Electronic Properties

3D metals are known electronic conductors where the energy bands overlap the Fermi level in their electronic band structures. Even after reducing the thickness of metal clusters to atomically thin membrane, the electronic energy bands still overlap the Fermi level in most 2D metal layers. Therefore, it is often found that reducing the thickness of the metallic slabs does not eliminate the conducting nature in the freestanding metallic monolayers

[16, 23]. However, the 2D bHC structure of Bi has been found to be a semiconductor with a low band gap of 0.36 eV. The direct energy gap between the conduction and valence bands is obtained at the Γ high symmetry point of the Brillouin zone of 2D bHC Bi [23].

The intriguing electron dispersion relations were observed in monolayers of alkali metal elements (Na, K, Rb, Cs). These elements exhibit a Dirac cone at the K high symmetry point of the Brillouin zone level in their electronic band structure with the GGA-PBE level of theory. This feature is analogous to Dirac cones reported in graphene skeleton. Despite being a member of the alkali metal series, the Dirac cone is not observed in the monolayer of Li. The electronic band structure of 2D bHC Li and Na are shown in Figure 4.5. Increasing the accuracy of the calculations to the HSE06 level reveals a very narrow band gap in 2D Na (25 meV) and K (71 meV). However, the presence of the Dirac cone in the 2D lattice structure of Rb and Cs was still retained with the HSE06 level of theory. Close analysis of the structure and band structure demonstrates that Dirac cones were observed for 2D lattices that were generated along the (111) plane of the 3D crystal structures of alkali metal elements. These structures resemble the 2D bHC lattices of alkali metals [23]. The 2D HC graphene and 2D bHC lattice of the higher elements belonging to the same group are known for their characteristic Dirac cone in their electronic band structure. All these elements manifesting Dirac cone have either 2D HC or bHC crystal lattice and possess an unpaired electron in the p_z-orbital which facilitates electron delocalization in the 2D lattice. In the case of alkali metals, an unpaired electron is present in the s-orbital of each atom in the 2D bHC lattice. This illustrates that the combination of 2D HC or bHC with an unpaired electron in each atomic orbital is critical for exhibiting a Dirac cone in their electronic band structure. These results further suggest that atomically thin freestanding metallic layers show periodicity of properties in the periodic table [23].

The electronic band structure can be tuned *via* a template-assisted growth of freestanding atomically thin monolayers. For instance, embedding metal patches inside the graphene skeleton alters the electronic band structure of the metal patched graphene layer.

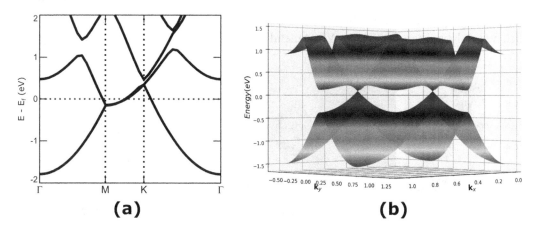

FIGURE 4.5 (a) 2D electronic band structure of the bHC Li layer (GGA-PBE). (b) 3D electronic band structure of bHC Na (GGA-PBE). Adapted with permission [23]. Copyright (2022), John Wiley & Sons Inc.

The variation in band structure depends mainly upon the number and nature of elemental metals confined inside the pore of graphene. The Au$_8$ patch created inside a small graphene pore displays a narrow direct bandgap of 230 meV at the K high symmetry point of the Brillouin zone. The studied Cu and Au patches inside the same pore of graphene involve the overlap of energy bands with the Fermi level and are electronic conductors [20]. Likewise, the energy band diagram can also be modulated by taking semiconducting rigid support to stabilize atomically thin metallic structures [21].

The electron localization function (ELF) is a well-accepted descriptor for chemical bond analysis and provides valuable information on electron delocalization in the molecular as well as periodic systems. The charge accumulation between two metals in atomically thin freestanding layers is larger when compared with their 3D bulk counterparts. This suggests that the electrons are more localized between the metal atoms in the monolayer structure. The ELF calculated for the bulk and 2D Na are shown in Figure 4.6. The ELF emphasizes the enhanced strength of M–M bonds in 2D metal monolayers when compared with their 3D bulk crystals [16, 23].

Work function (WF) is an important property that provides better electronic perception of materials and plays a vital role in understanding important surface phenomena. WF is the amount of energy essential to remove an electron from the Fermi level of the materials to its surface and can be estimated employing the following equation:

$$WF = E_{vac} - E_F \tag{4.3}$$

where, E_{vac} denotes the electrostatic potential of an electron at the vacuum level and E_F indicates the electrostatic potential of an electron at the Fermi level of the material.

The WF of a metal slab depends upon the facet of the exposed surface and the nature of the elemental metals. It is obvious that the WF of metal slabs decreases along the group in the alkali series because the electron becomes loosely bound to the nucleus with the increase in the atomic size. It is worth mentioning that the WF of a monoatomically thin freestanding layer of the transition series elements is anomalously higher than that of the thick metal slabs. The variation in WF of metals belonging to 5d-series is depicted in Figure

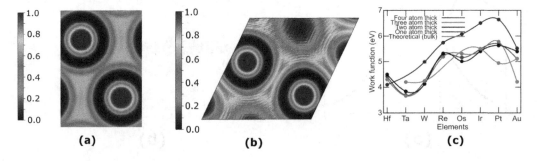

FIGURE 4.6 Electronic properties of metals: (a) 2D display of ELF in bulk Na. (b) 2D display of ELF in 2D bHC Na. (c) Variation in thickness-dependent work function along (001)/(111) facet for 5d-transition metal layers. Adapted with permission [23]. Copyright (2022), John Wiley & Sons Inc.

4.6. The increase in WF of the monolayer compared to the thick metal slabs becomes more pronounced from the *3d*- to *5d*-transition series of the periodic table [23].

As mentioned earlier, the reduction in out-of-plane coordination of metal atoms in atomically thin layers leads to the strengthening of the chemical bonding between metal atoms due to increased electron localization. Therefore, electrons cannot be easily ejected from its Fermi level. Thus the WF of atomically thin membranes is remarkably higher when compared to their thick metal slabs. On the contrary, electrons in thicker metal slabs show repulsive interactions and thereby assist in ejecting out electrons from the metal surface and have lower WF [23].

4.6.2 Magnetic Properties

Two-dimensionality endows outstanding properties to metallic nanostructures; therefore, attention was also paid to the appraise of magnetism in two-dimensional metal layers. DFT-based calculations were performed to determine the magnetic moment in HX, SQ, and HC monolayer of 45 elemental metals in the periodic table. The outcomes reveal that 18 metals out of 45 elements studied show magnetic behavior in their atomically thin layers. However, only 5 of these elemental metals are magnetic in nature in their 3D bulk crystal structures. Therefore, enhanced magnetism has been observed when reducing the thickness of metallic structures to monoatomically thin 2D layers. Reduction in the CN of metals and narrowing of the energy band provides magnetism in 2D metal layers. The 2D HC lattice is more likely to be magnetic than the 2D SQ lattice. Mostly, these elemental layers exhibit ferromagnetic behavior. The calculated magnetic moments are larger for a 2D metallic lattice with low symmetry while high symmetry reduces the magnetic behavior [15].

4.6.3 Catalytic Properties

Metallic structures are known to have intrinsic catalytic behavior, and reducing the CN of metals increases the reactive surface area in 2D structures. This is further beneficial for promoting chemical reactions on the 2D metal surface. The monoatomic Cu(111) layer selectively promotes the epoxidation of propylene oxide using molecular O_2 via an Eley-Rideal mechanism [30]. DFT calculations also reveal that the same Cu(111) monolayer shows selective hydrogenation of carbon dioxide (CO_2) to methanol [31]. Similarly, the atomically thin freestanding Pt(111) layer is selective for oxygen reduction reactions (ORR) [32]. Thus, reducing the thickness of metal structures to ultrathin membrane enhances the catalytic performance of metals [30–32].

4.7 STRUCTURE, STABILITY, AND PROPERTIES OF METAL PATCHES INSIDE THE MOLECULAR RINGS

Hitherto, we mostly discussed the structure, stability, and properties of periodic atomically thin freestanding layers of metals or periodic template-assisted stable patches of metals. However, rigid molecular skeleton with a large empty space can also be used to stabilize the metallic structures. For example, atomically thin Cu and Be nanostructures were realized to be stable inside the hollow space of a [6]-cycloparaphenylene ([6]-CPP) molecular

60 ■ 2D Metals

ring. Obviously, the number of atoms that can be accommodated inside the pore depends on the size of the metal atoms as well as the ring. These kinds of systems render different kinds of charges on the metal atoms depending upon their chemical environment [33, 34]. Akin to the flat layers, the encapsulation of Cu_9 inside [6]-CPP enhances its catalytic behavior [34]. The confinement of beryllene inside [6]-CPP improves the photophysical properties of the nanostructure from the UV to the visible range of the electromagnetic spectrum [33].

4.8 CONCLUSION

Metallic nanostructures can be stabilized as an atomically thin two-dimensional (2D) framework despite the nondirectional nature of metallic bonding in their 3D crystals. 2D metals are metastable structures while 3D bulk counterparts occupy a global minimum on the potential energy surface. The larger cohesive energy of the 3D bulk crystals attributes to the stability of 3D bulk while the enhanced bond strength per atom strengthens the chemical bonding between metal atoms in two-dimensionality. The most favorable arrangement of the metal atoms in the monolayer is the close-packed 2D HX lattice. Atomically thin freestanding layers of metallic elements are recognized as 2D liquids due to the high fluxionality in these layers. The structure and properties of 2D metal layers are correlated with 3D bulk counterparts. The reduction in out-of-plane coordination of the metal atoms in an atomically thin layer remarkably alters the properties of metals. The work function of a monoatomically thick layer is substantially high when compared with a thick metal slab. A large number of metals are found to be magnetic in atomically thin freestanding monolayers, thus reducing the thickness-induced inherent magnetism in 2D metals. A Dirac cone is observed at the K point of 2D bHC lattice structures of alkali metals layers (Na, K, Rb, and Cs). An opening of 0.36 eV energy gap between the valence and conduction bands is achieved in bHC Bi monolayer. Further, the confinement of Au_8 patch inside the small pore of graphene opens an energy gap of 230 meV at the K point. Furthermore, atomically thin 2D layers of metals enhance catalytic performance and photophysical properties.

REFERENCES

1. V. Georgakilas, J.A. Perman, J. Tucek, R. Zboril, Broad family of carbon nanoallotropes: classification, chemistry, and applications of fullerenes, carbon dots, nanotubes, graphene, nanodiamonds, and combined superstructures, *Chem. Rev.* 115 (2015) 4744–4822.
2. K.S. Novoselov, A.K. Geim, S.V. Morozov, D. Jiang, Y. Zhang, S.V. Dubonos, I.V. Grigorieva, A.A. Firsov, Electric field effect in atomically thin carbon films, *Science.* 306 (2004) 666–669.
3. A.K. Geim, K.S. Novoselov, The rise of graphene, *Nature Mater.* 6 (2007) 183–191.
4. G.R. Bhimanapati, Z. Lin, V. Meunier, Y. Jung, J. Cha, S. Das, D. Xiao, Y. Son, M.S. Strano, V.R. Cooper, L. Liang, S.G. Louie, E. Ringe, W. Zhou, S.S. Kim, R.R. Naik, B.G. Sumpter, H. Terrones, F. Xia, Y. Wang, J. Zhu, D. Akinwande, N. Alem, J.A. Schuller, R.E. Schaak, M. Terrones, J.A. Robinson, Recent advances in two-dimensional materials beyond graphene, *ACS Nano.* 9 (2015) 11509–11539.
5. J. Nevalaita, P. Koskinen, Atlas for the properties of elemental two-dimensional metals, *Phys. Rev. B.* 97 (2018) 035411.
6. S. Ono, Dynamical stability of two-dimensional metals in the periodic table, *Phys. Rev. B.* 102 (2020) 165424.

7. J. Nevalaita, P. Koskinen, Stability limits of elemental 2D metals in graphene pores, *Nanoscale.* 11 (2019) 22019–22024.
8. J. Zhao, Q. Deng, A. Bachmatiuk, G. Sandeep, A. Popov, J. Eckert, M.H. Rümmeli, Free-standing single-atom-thick iron membranes suspended in graphene pores, *Science.* 343 (2014) 1228–1232.
9. X. Yang, H.Q. Ta, W. Li, R.G. Mendes, Y. Liu, Q. Shi, S. Ullah, A. Bachmatiuk, J. Luo, L. Liu, J.-H. Choi, M.H. Rummeli, In-situ observations of novel single-atom thick 2D tin membranes embedded in graphene, *Nano Res.* 14 (2021) 747–753.
10. H.Q. Ta, Q.X. Yang, S. Liu, A. Bachmatiuk, R.G. Mendes, T. Gemming, Y. Liu, L. Liu, K. Tokarska, R.B. Patel, J.-H. Choi, M.H. Rümmeli, In situ formation of free-standing single-atom-thick antiferromagnetic chromium membranes, *Nano Lett.* 20 (2020) 4354–4361.
11. L. Zhao, H.Q. Ta, R.G. Mendes, A. Bachmatiuk, M.H. Rummeli, In situ observations of free-standing single-atom-thick gold nanoribbons suspended in graphene, *Adv. Mater. Interfaces.* 7 (2020) 2000436.
12. X. Wang, C. Wang, C. Chen, H. Duan, K. Du, Free-standing monatomic thick two-dimensional gold, *Nano Lett.* 19 (2019) 4560–4566.
13. Q. Zhu, Y. Hong, G. Cao, Y. Zhang, X. Zhang, K. Du, Z. Zhang, T. Zhu, J. Wang, Free-standing two-dimensional gold membranes produced by extreme mechanical thinning, *ACS Nano.* 14 (2020) 17091–17099.
14. X. Zhao, J. Dan, J. Chen, Z. Ding, W. Zhou, K. Loh, S. Pennycook, Atom-by-atom fabrication of monolayer molybdenum membranes, *Adv. Mater.* 30 (2018) 1707281.
15. Y. Ren, L. Hu, Y. Shao, Y. Hu, L. Huang, X. Shi, Magnetism of elemental two-dimensional metals, *J. Mater. Chem. C.* 9 (2021) 4554–4561.
16. A.A. Sangolkar, S. Jha, R. Pawar, Density functional theory-based calculations for 2D hexagonal lanthanide metals, *Adv. Theory Simul.* 5 (2022) 2200057.
17. L.-M. Yang, T. Frauenheim, E. Ganz, Properties of the free-standing two-dimensional copper monolayer, *J. Nanomater.* 2016 (2016) 8429510.
18. L.-M. Yang, T. Frauenheim, E. Ganz, The new dimension of silver, *Phys. Chem. Chem. Phys.* 17 (2015) 19695–19699.
19. L.-M. Yang, M. Dornfeld, T. Frauenheim, E. Ganz, Glitter in a 2D monolayer, *Phys. Chem. Chem. Phys.* 17 (2015) 26036–26042.
20. A.A. Sangolkar, R. Pawar, Structure, stability, properties, and application of atomically thin coinage metal flatland in graphene pore: a density functional theory calculation, *Physica. Status Solidi (b).* 259 (2022) 2100489.
21. A.A. Sangolkar, Pooja, R. Pawar, Structure, stability, and electronic and optical properties of TMDC–coinage metal composites: vertical atomically thin self-assembly of Au clusters, *Phys. Chem. Chem. Phys.* 25 (2023) 4177–4192.
22. P. Hess, Periodicity of two-dimensional bonding of main group elements, *Chem. Phys. Chem.* 23 (2022) e202100880.
23. A.A. Sangolkar, R. Agrawal, R. Pawar, A prospectus for thickness dependent electronic properties of two-dimensional metals using density functional theory calculation, Internat. *J. Quant Chem.* 122 (2022) e26982.
24. A.A. Sangolkar, R. Agrawal, R. Pawar, Dissociative adsorption of H_2S on Li(110) surface using density functional theory calculations and car-parrinello molecular dynamics simulations, *Chem. Phys. Chem.* 23 (2022) e202100658.
25. S. Antikainen, P. Koskinen, Growth of two-dimensional Au patches in graphene pores: a density-functional study, *Comput. Mater. Sci.* 131 (2017) 120–125.
26. R. Jin, X. Yuan, E. Gao, Atomic stiffness for bulk modulus prediction and high-throughput screening of ultraincompressible crystals, *Nat. Commun.* 14 (2023) 4258.
27. L.-M. Yang, A.B. Ganz, M. Dornfeld, E. Ganz, Computational study of Quasi-2D liquid state in free standing platinum, silver, gold, and copper monolayers, *Condensed Matter.* 1 (2016) 1.

62 ▪ 2D Metals

28. P. Koskinen, T. Korhonen, Plenty of motion at the bottom: atomically thin liquid gold membrane, *Nanoscale*. 7 (2015) 10140–10145.
29. A.A. Sangolkar, S. Jha, R. Pawar, Density functional theory-based calculations for 2D hexagonal lanthanide metals, *Adv. Theory Simul.* 5 (2022) 2200057.
30. A.A. Sangolkar, R. Pawar, Enhanced selectivity of the propylene epoxidation reaction on a Cu monolayer surface via Eley-Rideal mechanism, *Chem. Phys. Chem.* 23 (2022) e202200334.
31. S.C. Mandal, K.S. Rawat, P. Garg, B. Pathak, Hexagonal Cu(111) monolayers for selective CO2 hydrogenation to CH3OH: insights from density functional theory, *ACS Appl. Nano Mater.* 2 (2019) 7686–7695.
32. A. Mahata, P. Garg, K.S. Rawat, P. Bhauriyal, B. Pathak, A free-standing platinum monolayer as an efficient and selective catalyst for the oxygen reduction reaction, *J. Mater. Chem. A.* 5 (2017) 5303–5313.
33. A.A. Sangolkar, R.K. Kadiyam, M. Faizan, O. Chedupaka, R. Mucherla, R. Pawar, Electronic and photophysical properties of atomically thin bowl-shaped beryllene encapsulated inside the cavity of [6]Cycloparaphenylene (Ben@[6]CPP), *Phys. Chem. Chem. Phys.* 25 (2023) 23262–23276.
34. Y.-Q. Liu, Z.-Y. Qiu, X. Zhao, W.-W. Wang, J.-S. Dang, Trapped copper in [6]cycloparaphenylene: a fully-exposed Cu7 single cluster for highly active and selective CO electro-reduction, *J. Mater. Chem. A.* 9 (2021) 25922–25926.

CHAPTER 5

Properties and Functionalization of 2D Metals

Harini G. Sampatkumar, G. Banuprakash, and Siddappa A. Patil

5.1 INTRODUCTION

Nanomaterials are being increasingly utilized in various fields due to their unique physico-chemical properties, such as small size, large surface area, high surface activity, and fascinating electronic/magnetic properties. These fields include sustainable chemical industry, medicine, biosensors, cosmetics, and food additives [1]. Due to their intriguing potential for use in nanotechnology, 2D nanomaterials such as graphene, tungsten disulphide (WS_2), MoS_2, boron nitride (BN), carbon nitride (g-C_3N_4), and graphene are gaining importance [2, 3]. Among these 2D nanomaterials, MoS_2 is a promising material because of its numerous notable optical, mechanical, chemical, electrical, catalytic, and sensing properties [4]. MoS_2 is a prevalent type of inorganic compound among the 2D transition metal dichalcogenides (TMDCs), consisting of one atom of molybdenum (Mo) and two atoms of sulphur (S). These materials have been essential throughout the 20th century and possess a wide range of exceptional physicochemical, biological, and mechanical properties that distinguish them from graphene-based nanomaterials. This uniqueness holds great promise for applications in energy storage and conversion, biomedicine and environmental protection, as evidenced by the growing research interest in this area [5].

It exists as multiple polytypIGUREes, just like other TMDCs like molybdenum diselenide ($MoSe_2$), WS_2, and so forth. These polytypes include 1T [6], 2H [7], and 3R [8], which belong to point groups D_{6d}, D_{6h}, and C_{3v}, respectively. The first digit in the polytype name

DOI: 10.1201/9781032645001-5

63

indicates the number of layers in the arrangement, while the alphabet indicates the crystallographic configuration. Specifically, "T" represents trigonal, "H" represents hexagonal, and "R" represents rhombohedral arrangements. Among polytypes, 1T phase coordinates are in an octahedral structure, while 2H and 3R are in a trigonal prismatic structure [9]. The hexagonal unit cell of 2H-MoS$_2$, which is the most thermodynamically stable arrangement, consists of trigonal prismatic MoS$_6$ centres and other metastable polytypes are also possible. The layers of MoS$_2$ are held together by relatively weak Van der Waals forces and each layer is made up of covalently bonded condensed MoS$_6$ centres with alternating sheets of sulphur atoms sandwiching a sheet of Mo atoms (Figure 5.1).

The 1T structure is metallic, while the monolayer of hexagonal MoS$_2$ and the 2H and 3R phases are semiconducting and used as dry lubricants. The 3R phase is particularly useful in nonlinear optical mass sensing in quantum measurements and biomedicine [11] due to its nonlinear optical properties. Additionally, the different phase materials of MoS$_2$ can be advantageous in obtaining high sensitivity and rapid desorption [12] for gas sensors. By implementing certain modifications in the synthesis process, various allotropies and morphologies can be attained, including three-dimensional structures such as flowers, snowflakes, and dandelions; 2D formations like nanosheets, nanostripes, and nanoribbons; one-dimensional (1D) structures such as nanowires and nanorods and zero-dimensional (0D) structures like nanoplatelets.

This chapter provides insight into MoS$_2$'s various properties and its functionalization for various applications.

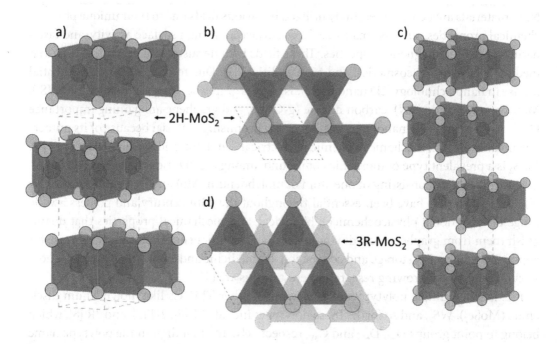

FIGURE 5.1 (a) 2H-MoS$_2$, (b) its c-axis view, (c) 3R-MoS$_2$, and its (d) c-axis view. Adapted with permission from [10]. Copyright (2023) American Chemical Society.

5.2 PROPERTIES OF MOS_2

5.2.1 Electronic Properties

The electronic properties and large surface areas of 2D nanomaterial semiconductors make them highly versatile for various applications [13]. Therefore, it is important to comprehend the electrical properties of 2D MoS_2. In group six dichalcogenide transition metals, the energy gap size in the monolayer is 50% larger than in the bulk mode [14]. The intrinsic structure of bulk MoS_2, where single sheets are stacked and bonded by weak Van der Waals interactions, results in low electrical conductivity and high electrical resistivity. However, due to the direct energy band structure, 2D MoS_2 has a distinct shape and electrical properties that allow for effective electron transmission. The indirect bandgap of 1.2 eV in multilayer MoS_2 is known to grow as the number of layers decreases, ultimately leading to a direct bandgap of 1.8 eV in monolayer MoS_2 [15]. Although the MoS_2 bandgap value is good, it is still far from 1.12 eV direct bandgap of Silicon [16]. The bandgap of MoS_2 is influenced by mechanical strain, resulting in a transition from a direct to an indirect bandgap. This change in bandgap transforms the material from a semiconducting state to a metallic one. This effect is particularly evident in single-layer films, where the energy gap is altered, leading to enhanced photoluminescent features. These characteristics make MoS_2 suitable for various optical and electronic applications. Additionally, the presence of electron-free spaces between bandgaps allows for greater control over the electrical properties, enabling certain electrons to move across these gaps. Doping a monolayer MoS_2 with different elements such as chromium (Cr), copper (Cu), scandium (Sc) (p-type), nickel (Ni), or zinc (Zn) (n-type) can further modify its semiconductor behaviour. For instance, doping with titanium can result in MoS_2 behaving as either a p-type or n-type semiconductor, depending on the doping levels and sites. MoS_2 exhibits p-type behaviour, when the Titanium (Ti) doping is less than 2.04%. The covalent interaction between MoS_2 and Ti is strong in the case of interstitial doping of Ti at 3.57% doping levels. This raises the surface dipole moment, which causes a drop in electron affinity of 0.49 eV, leading to n-type behaviour. The Fermi level merges into the conduction at higher doping levels of 7.69%, where the electron affinity increases and the surface dipole moment decreases, anchoring the Fermi level over the conduction band. With spin polarization equal to 1, MoS_2 transforms into a ferromagnetic half-metal making it promising for spintronics applications. Such unique properties of MoS_2 offer excellent prospects for future use in a variety of fields, such as memory devices, photodetectors, transistors, and the expansion of 2D graphene and graphene-like flexible materials into transparent conducting and semiconducting domains.

5.2.2 Optical Properties

MoS_2 nanomaterials have emerged as potential materials for various applications in optical devices due to their remarkable optical properties. Single-layer MoS_2 is a non-centrosymmetric material, unlike graphene. It has a direct bandgap, which is due to the role of the d-electron orbitals. The relationship between wavelength and bandgap energy is inverse, meaning that photons with larger wavelengths and smaller energy cannot be absorbed. The response of a material when a specific wavelength passes through it is determined by the absorption coefficient and refractive index. A high attenuation to the applied wave is indicated by the high absorption coefficient. In semiconductors, if there is insufficient energy to excite electrons from the valence band to the conduction band, they will show low absorption coefficients for long wavelengths. Conversely, they will show high absorption coefficients for short wavelengths (high energy and frequency spectrum). It has been demonstrated that both multilayer and monolayer MoS_2 have comparatively higher absorption coefficients near the visible spectrum (400–500 nm) as they do not cover the entire visible range (400–700 nm). At 500 nm, there is noticeable decay. For a single layer of MoS_2, the extinction coefficient reaches its peak at 450 nm, indicating that at this specific wavelength, the light is retained by a single layer of MoS_2. Beyond 500 nm, the extinction coefficient decreases significantly, indicating that the single layer of MoS_2 becomes transparent [17]. In the case of multilayer MoS_2, the peak also appears at approximately 400 nm, but the strength of the peak is greater than that of monolayer MoS_2, suggesting that multilayers of MoS_2 retain light better than a single layer at this wavelength. The wide use of MoS_2 in optoelectronics is attributed to its tunable bandgap, which can be altered by changing its size and structure. Different bandgaps result in tunable photoresponsivity, specific detectivity, and response time, providing a wide range of possibilities [18]. The refractive index of MoS_2 varies depending on its layer thickness, with single-layer MoS_2 having a refractive index value between 1 and 2, indicating transparency to visible light for thicknesses less than 250 nm. In contrast, multilayer MoS_2 always has a refractive index greater than 2, with the highest value observed near 500 nm for both monolayer and multilayer MoS_2 [17]. MoS_2 exhibits different photoluminescence (PL) activity due to factors such as bandgap, doping and structure with single-layer MoS_2 having a peak exciton (A) (Figure 5.2). The PL properties of monolayer MoS_2 can be enhanced by adding hydrogen peroxide (H_2O_2) solution [19], while the low PL quantum yield (QY) or the ratio of the number of emitted photons to the number of generated electron-hole pairs of TMDCs ranges from 0.01% to

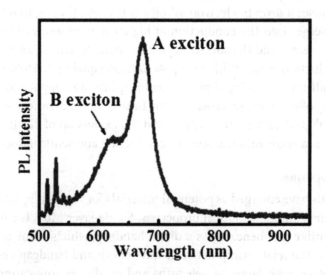

FIGURE 5.2 PL spectra of pristine monolayer of MoS_2. Adapted with permission from [21]. Copyright (2023) Elsevier.

6% and it can be raised to 95% [20] using a chemical treatment of an organic superacid. The observed lifetime of MoS_2 carriers is nearly 10.8 ns, making it a promising material for high-performance lasers and solar cells.

5.2.3 Photoluminescence (PL) Properties

As discussed earlier, the bandgap of MoS_2 undergoes a transition from direct to indirect as it shifts from a monolayer to a bulk configuration, resulting in a significant difference in the optical gap (0.1 eV for bulk to 1.1 eV for single layer) between the two. Bulk MoS_2 exhibits an indirect bandgap and negligible PL [16]. The variation in MoS_2 reflectivity for a single layer depends on the absorption constant. Reflection spectroscopy revealed [21] two peaks, A and B, at 670 nm and 627 nm, respectively (Figure 5.2), with an energy difference due to splitting of the valance band maximum (VBM) at K.

The combination of spin-orbit and interlayer coupling together accounts for the split in the peaks for multilayered MoS_2. Whereas spin-orbit coupling is the cause of the single layer. The bandgap PL in bulk MoS_2 is attributed to a weak phonon-assisted process with the lowest QY. The number of layers in MoS_2 is found to be closely correlated with PL spectra. The clear PL differences between monolayer (1L), bilayer (2L), quadrilayer (4L), and hexalayer (6L) samples have been reported by Andrea Splendiani's group (Figure 5.3) [22]. Thus, as the number of layers of MoS_2 changes from multilayer to monolayer, the band structure changes qualitatively.

5.2.4 Mechanical Properties

As previously mentioned, MoS_2 is a two-dimensional shape and belongs to the transition metal dichalcogenide family. Additionally, it possesses high strength characteristics, making it an appealing compound in the industry [23]. The Young Modulus of MoS_2's sheet is remarkably high, measuring $E = 0.33 \pm 0.07$ TPa, which is equivalent to graphene oxide. According to research, the bulk form of MoS_2 has a Young Modulus of 0.24 TPa. However, further research reveals that a single layer of MoS_2 has a Young Modulus of 0.27 TPa, which is greater than that of the bulk form. This inconsistency in results is attributed to the presence of stacking imperfections in thinner flakes compared to the bulk form of MoS_2. The small pre-tension and high elasticity of MoS_2 make it a promising candidate for composite films and flexible optoelectronic devices in the semiconductor field [24]. The controlled introduction of strain into semiconductors offers significant flexibility in the engineering of devices.

5.2.5 Magnetic Properties

In the context of the nanoscale era and the advancement of spintronics, it became necessary to study the electron spin of promising structures such as MoS_2. Since TMDCs are known to be nonmagnetic, we could potentially use them as tunable semiconductors if magnetism could be added to them [25]. A specific study [26] delved into the magnetic behaviour and characteristics of multilayer MoS_2. The findings revealed that MoS_2 possesses a lengthy spin diffusion length of 235 nm and that an in-plane spin polarization can effectively hinder electron spin relaxation. Additionally, research [27] demonstrated that

68 ■ 2D Metals

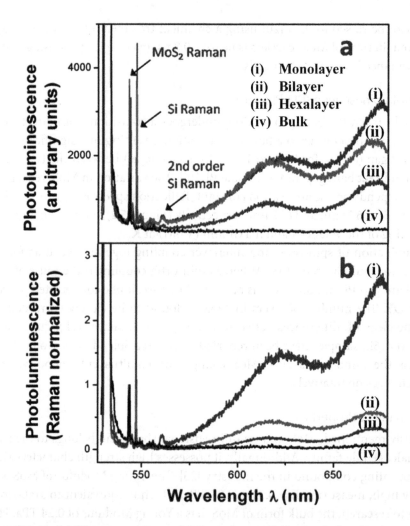

FIGURE 5.3 (a) PL and (b) Raman spectra of different layers of MoS$_2$ and photoluminescence spectra normalized by Raman intensity, respectively. Adapted with permission from [22]. Copyright (2023) American Chemical Society.

when doped with Sc, MoS$_2$ acquires semi-metallic ferromagnetic properties and a unity spin polarization value, making it highly advantageous for spintronics.

5.2.6 Catalytic Properties

MoS$_2$ nanomaterials have attracted significant attention for their diverse applications. It has become particularly notable for its catalytic properties towards the hydrogen evolution reaction (HER) [28], in addition to photoelectrochemical [29] and photoluminescence [30] properties. Consequently, extensive research efforts have been dedicated to enhancing the catalytic properties of 2D MoS$_2$, with the aim of potentially replacing platinum (Pt) as a catalyst. Unlike conventional noble metal catalysts, 2D MoS$_2$ exhibits high chemical stability, cost-effectiveness, and commendable catalytic efficiency, making it a promising nonprecious material. While the catalytic ability of 2D MoS$_2$ may not match that of noble metals because it is resistant to poisoning [31]. One of the benefits of 2D MoS_2 nanosheets is their abundance of exposed edges, which is the foundation for their catalytic activity [32]. Localized to uncommon surface sites, 2D MoS_2 exhibits catalytic activity, whereas bulk MoS_2 is comparatively inert [33]. Many attempts have been made to synthesize MoS_2 nanomaterials with large edge sites and improved catalytic properties, as the identification of active sites is important for assessing a catalyst's performance.

5.3 FUNCTIONALIZATION OF MOS_2

Strategies of functionalization and surface modifications are suggested to enhance the characteristics and achieve the desired performance of MoS_2 for diverse applications. Physical functionalization relies on interactions like electrostatic attraction between materials. Whereas, chemical functionalization is typically achieved by forming new covalent bonds between the modifier and the substrate [34]. These techniques have been thoroughly examined in the sections that follow.

5.3.1 Covalent Methods

New covalent bonds between molecules and nanostructures of MoS_2 are established in chemical functionalization by utilizing defects and vacancies present in the crystal lattice's plane or edges. MoS_2 sheets in two dimensions are frequently illustrated as flawless lattices, however, in actuality, sulphur vacancies are not only prevalent but also essential for catalysis [35]. Depending on the type of surface modifying agent, both the S and Mo atoms in the structure of MoS_2 offer a potential to form covalent bonds [36]. Mo atoms in the MoS_2 lattice will be able to coordinate with target molecules that have sulphur functional groups, such as thiols. The first direct evidence of covalent TMD functionalization is provided by scanning tunnelling microscope studies on the adsorption of dibenzothiophene (DBT) on single-layer MoS_2 nanoclusters (Figure 5.4) [37]. Strong bonding between the organic and inorganic molecules is indicated by the thermal stability and the small Mo-S distance at the occupied corner defects, rather than weak physisorption. Notably, DBT does not modify the edge vacancies that are predominantly formed in the larger clusters under investigation.

The chemically exfoliated MoS_2 was successfully functionalized through the straightforward combination with thiol-terminated polyethylene glycol (PEG) ligands [38]. This process resulted in the formation of two-dimensional materials with exceptional colloidal stability in water, facilitated by the presence of organic molecules containing hydroxyl, carboxyl, and trimethyl ammonium groups. Further, to create the vacancies required for sulphur attachment, Jung et al. used ultrasonication in ethanol/water [39]. They developed a volatile organic compounds (VOCs) sensor functionalized with mercaptoundecanoic acid that displayed a variable resistance response when exposed to VOCs containing oxygen. Given the straightforward nature of this approach, it is plausible to envision the development of a diverse array of sensors with unique surface modifications, capable of identifying specific substances of interest through their distinctive response patterns.

In order to coordinate with the Mo atoms in the structure, Sideri et al. covalently functionalized both the 1T and 2H structures of MoS_2 with dithiolenes via a green root using

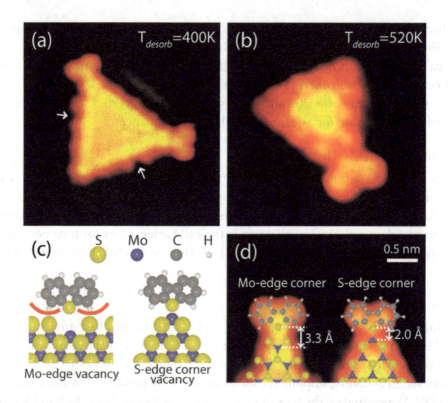

FIGURE 5.4 (a, b) STM images of DBT bound to the corner site of MoS$_2$ at different temperatures with edge sites (indicated by white arrows) remaining vacant. (c) Schematic representation showing possible interaction only at the corner sites. (d) STM of adsorbed (left) and covalently modified (right) nanoflake. Adapted with permission from [37]. Copyright (2023) American Chemical Society.

bis(thiolate) salts as ligands (Figure 5.5(a)) [40]. This innovative synthesis method employs water as a solvent and has demonstrated that incorporating dithiolenes can enhance the electrochemical and photochemical activity of MoS$_2$ nanostructures.

The plasma method was utilized by Seo et al. to enhance the generation of sulphur vacancies within the MoS$_2$ structure. With 3-mercaptoacetic acid as a ligand, 1.8 times greater functionalization compared to conventional methods was observed [41]. Hyperbranched polyglycerol functionalized with lipoic acid and folic acid was synthesized by Xu et al. The compound was then covalently linked to MoS$_2$ through disulphide bonds and Chloroquine and doxorubicin drugs were loaded onto it for targeted delivery to (HeLa-R) cells [42]. The covalent binding of the polymeric compound to MoS$_2$ was facilitated by lipoic acid, which acted as a bidentate sulphur ligand.

Additionally, MoS$_2$'s structure allows for the coordination of sulphur atoms with modifiers that have metal centres. By immersing a CVD-grown MoS$_2$ on the Si/SiO$_2$ substrate in ethanolic solutions of the metal salts for 10 minutes, Liu et al. coordinated transition metals such as gold (Au), nickel (Ni), Cu, Zn, cobalt (Co), manganese (Mn,), and Cr to the structure of MoS$_2$ [43]. It was demonstrated that these complexes greatly enhance MoS$_2$'s electrical and optical characteristics. Notably, they utilized Au–S interactions to deposit individual

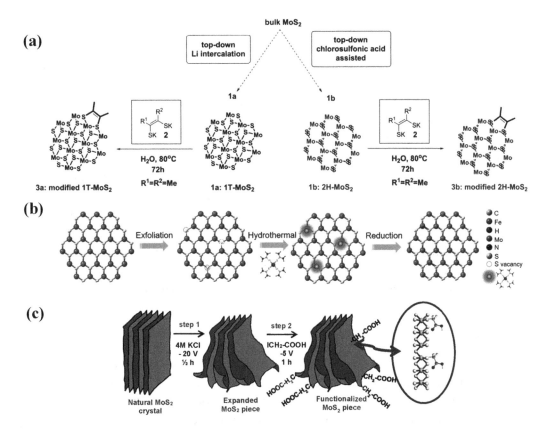

FIGURE 5.5 Schematic illustration of various covalent MoS$_2$ functionalization approaches: (a) coordination of sulphur-containing functional groups to Mo atoms [40]. (b) Coordination of S atoms on the structure of MoS$_2$ to a metal centre containing modifier [44]. (c) Direct formation of C–S bond between MoS$_2$ and a reagent containing a carbon bonded to a leaving group [45]. Adapted with permission from [40], [44], and [45]. Copyright (2023) American Chemical Society.

gold atoms onto the substrate. In order to use transition metals as catalysts in petrochemical-related chemical transformations, Zheng et al. loaded them at the atomic level onto the surface of MoS$_2$ (Figure 5.5(b)) [44]. The chemically exfoliated MoS$_2$ nanosheets were then treated with transition metal-thiourea complexes, and the resulting self-assembled complexes were hydrothermally reduced to metal atoms. By using a cathodic potential in the presence of organoiodides, Paredes et al. developed a novel electrochemical-based functionalization of MoS$_2$ (Figure 5.5(c)) [45]. Applying a –20 V DC potential in a 4 M KCl electrolyte for 30 minutes exfoliated a piece of MoS$_2$ crystal that was used as the working electrode and platinum foil as the counter electrode in a two-electrode configuration. The functionalization was then achieved by changing the electrolyte to an iodoacetic acid solution and applying a –5 V DC potential for 1 hour. This caused nucleophilic addition of sulphur atoms to iodoacetic acid and substitution with iodine. The modified MoS$_2$ platforms were subsequently utilized for catalysing nitroarenes reduction reactions.

72 ■ 2D Metals

5.3.2 Non-Covalent Methods

The main techniques for functionalizing MoS_2 in a non-covalent way typically involve the depositing different mineral salts onto MoS_2, utilizing an ultrasonic method based on Van der Waals interactions to form composites with other nanostructures, and establishing electrostatic interactions with organic molecules. A two-step CVD method was utilized by Mohapatra et al. to cultivate a layer of $\beta\text{-}In_2Se_3$ on the MoS_2 surface, resulting in the creation of a hybrid-layered compound held together by Van der Waals forces [46]. Promising properties were displayed by this hybrid material for optoelectronic uses. To improve MoS_2's biocompatibility, Xie et al. physically altered the surface using an amphiphilic phospholipid [47] and the resultant material successfully delivered as an anti-cancer drug to mice tumours and demonstrated good stability in biological environments. Through electrostatic interactions, Zhang et al. functionalized negatively charged MoS_2 nanoflowers with cationic hydroxyethyl cellulose [48]. An ultrasonic bath was used to sonicate an aqueous mixture of MoS_2 and hydroxyethyl cellulose, which was then heated to 80°C for 4 hours. The resulting carrier was utilized for transdermal atenolol delivery under near IR irradiation.

5.4 CONCLUSIONS

In summary, we highlighted some of the main features of MoS_2 such as its polytypes, unique properties including electronic, optical, photoluminescence, mechanical, magnetic and catalytic properties followed by functionalization of MoS_2. These properties make 2D metal a promising material with a wide range of applications.

REFERENCES

1. M. Amde, J. fu Liu, Z.Q. Tan, D. Bekana, Transformation and bioavailability of metal oxide nanoparticles in aquatic and terrestrial environments. A review, *Environ. Pollut.* 230 (2017) 250–267.
2. T.M. Mohona, A. Gupta, A. Masud, S.C. Chien, L.C. Lin, P.C. Nalam, N. Aich, Aggregation behavior of inorganic 2D nanomaterials beyond graphene: Insights from molecular modeling and modified DLVO theory, *Environ. Sci. Technol.* 53 (2019) 4161–4172.
3. C. Tan, X. Cao, X.J. Wu, Q. He, J. Yang, X. Zhang, J. Chen, W. Zhao, S. Han, G.H. Nam, M. Sindoro, H. Zhang, Recent advances in ultrathin two-dimensional nanomaterials, *Chem. Rev.* 117 (2017) 6225–6331.
4. M.B. Megalamani, Y.N. Patil, S.T. Nandibewoor, Electrochemical sensing of carcinogenic p-dimethylamino antipyrine using sensor comprised of eco-friendly MoS_2 nanosheets encapsulated by PVA capped Mn doped ZnS nanoparticle, *Inorg. Chem. Commun.* 151 (2023) 110617.
5. P. Jia, T. Bu, X. Sun, Y. Liu, J. Liu, Q. Wang, Y. Shui, S. Guo, L. Wang, A sensitive and selective approach for detection of tetracyclines using fluorescent molybdenum disulfide nanoplates, *Food Chem.* 297 (2019) 124969.
6. X. Gan, L.Y.S. Lee, K.Y. Wong, T.W. Lo, K.H. Ho, D.Y. Lei, H. Zhao, 2H/1T Phase transition of multilayer MoS_2 by electrochemical incorporation of S vacancies, *ACS Appl. Energy Mater.* 1 (2018) 4754–4765.
7. S.H. El-Mahalawy, B.L. Evans, J.J. Thomson, The thermal expansion of $2H\text{-}MoS_2$, $2H\text{-}MoSe_2$ and $2H\text{-}WSe_2$ between 20 and 800°C, *J. Appl. Cryst.* 9 (1976) 403–407.

8. X. Li, H. Zhu, Two-dimensional MoS$_2$: Properties, prepafration and applications, *J. Mater.* 1 (2015) 33–44.

9. S.S. Chee, W.J. Lee, Y.R. Jo, M.K. Cho, D.W. Chun, H. Baik, B.J. Kim, M.H. Yoon, K. Lee, M.H. Ham, Atomic vacancy control and elemental substitution in a monolayer molybdenum disulfide for high performance optoelectronic device arrays, *Adv. Funct. Mater.* 30 (2020) 1908147.

10. J. Strachan, A.F. Masters, T. Maschmeyer, 3R-MoS$_2$ in review: History, status, and outlook, *ACS Appl. Energy Mater.* 4 (2021) 7405–7418.

11. S. Manzeli, D. Dumcenco, G. Migliato Marega, A. Kis, Self-sensing, tunable monolayer MoS$_2$ nanoelectromechanical resonators, *Nat. Commun.* 10 (2019) 4831.

12. J. Cao, J. Zhou, J. Chen, W. Wang, Y. Zhang, X. Liu, Effects of phase selection on gas-sensing performance of MoS$_2$ and WS$_2$ substrates, *ACS Omega.* 5 (2020) 28823–28830.

13. H. Xu, M.K. Akbari, S. Zhuiykov, 2D Semiconductor nanomaterials and heterostructures: Controlled synthesis and functional applications, *Nanoscale Res. Lett.* 16 (2021) 94.

14. A. Splendiani, L. Sun, Y. Zhang, T. Li, J. Kim, C.Y. Chim, G. Galli, F. Wang, Emerging photoluminescence in monolayer MoS$_2$, *Nano Lett.* 10 (2010) 1271–1275.

15. P. Johari, V.B. Shenoy, Tuning the electronic properties of semi-conducting transtion metal dichalcogenides by applying mechanical strains, *ACS Nano.* 6 (2012) 5449–5456.

16. Y.P. Venkata Subbaiah, K.J. Saji, A. Tiwari, Atomically thin MoS$_2$: A versatile nongraphene 2D material, *Adv. Funct. Mater.* 26 (2016) 2046–2069.

17. M.S. Ullah, A.H. Bin Yousuf, A.D. Es-Sakhi, M.H. Chowdhury, Analysis of optical and electronic properties of MoS$_2$ for optoelectronics and FET applications, *AIP Conf. Proc.* 1957 (2018) 020001.

18. H.S. Nalwa, A review of molybdenum disulfide (MoS$_2$) based photodetectors: From ultrabroadband, self-powered to flexible devices, *RSC Adv.* 10 (2020) 30529–30602.

19. Y. Cheng, J.Z. Wang, X.X. Wei, D. Guo, B. Wu, L.W. Yu, X.R. Wang, Y. Shi, Tuning photoluminescence performance of monolayer MoS$_2$ via H$_2$O$_2$ aqueous solution, *Chinese Phys. Lett.* 32 (2015) 117801.

20. M. Amani, D.H. Lien, D. Kiriya, J. Xiao, A. Azcatl, J. Noh, S.R. Madhvapathy, R. Addou, K.C. Santosh, M. Dubey, K. Cho, R.M. Wallace, S.C. Lee, J.H. He, J.W. Ager, X. Zhang, E. Yablonovitch, A. Javey, Near-unity photoluminescence quantum yield in MoS$_2$, *Science.* 350 (2015) 1065–1068.

21. M.S. Kim, G. Nam, S. Park, H. Kim, G.H. Han, J. Lee, K.P. Dhakal, J.Y. Leem, Y.H. Lee, J. Kim, Photoluminescence wavelength variation of monolayer MoS$_2$ by oxygen plasma treatment, *Thin Solid Films.* 590 (2015) 318–323.

22. G. Eda, H. Yamaguchi, D. Voiry, T. Fujita, M. Chen, M. Chhowalla, Photoluminescence from chemically exfoliated MoS$_2$, *Nano Lett.* 11 (2011) 5111–5116.

23. R.K. Zahedi, N. Alajlan, H.K. Zahedi, T. Rabczuk, Mechanical properties of all MoS$_2$ monolayer heterostructures: Crack propagation and existing notch study, *Comput. Mater. Contin.* 70 (2022) 4635–4655.

24. X. Zhang, S.Y. Teng, A.C.M. Loy, B.S. How, W.D. Leong, X. Tao, Transition metal dichalcogenides for the application of pollution reduction: A review, *Nanomaterials.* 10 (2020) 1012.

25. Z. Guguchia, A. Kerelsky, D. Edelberg, S. Banerjee, F. Von Rohr, D. Scullion, M. Augustin, M. Scully, D.A. Rhodes, Z. Shermadini, H. Luetkens, A. Shengelaya, C. Baines, E. Morenzoni, A. Amato, J.C. Hone, R. Khasanov, S.J.L. Billinge, E. Santos, A.N. Pasupathy, Y.J. Uemura, Magnetism in semiconducting molybdenum dichalcogenides, *Sci. Adv.* 4 (2018) eaat3672.

26. S. Liang, H. Yang, P. Renucci, B. Tao, P. Laczkowski, S. Mc-Murtry, G. Wang, X. Marie, J.M. George, S. Petit-Watelot, A. Djeffal, S. Mangin, H. Jaffrès, Y. Lu, Electrical spin injection and detection in molybdenum disulfide multilayer channel, *Nat. Commun.* 8 (2017) 14947.

27. Y.C. Tsai, Y. Li, Impact of doping concentration on electronic properties of transition metal-doped monolayer molybdenum disulfide, *IEEE Trans. Electron Devices.* 65 (2018) 733–738.

74 ▪ 2D Metals

28. Y. Shi, J. Wang, C. Wang, T.T. Zhai, W.J. Bao, J.J. Xu, X.H. Xia, H.Y. Chen, Hot electron of Au nanorods activates the electrocatalysis of hydrogen evolution on MoS_2 nanosheets, *J. Am. Chem. Soc.* 137 (2015) 7365–7370..

29. Y. Wang, F. Chen, X. Ye, T. Wu, K. Wu, C. Li, Photoelectrochemical immunosensing of tetrabromobisphenol A based on the enhanced effect of dodecahedral gold nanocrystals/MoS_2 nanosheets, *Sens. Actuat. B Chem.* 245 (2017) 205–212.

30. X. Zhang, W. Guo, Z. Wang, H. Ke, W. Zhao, A. Zhang, C. Huang, N. Jia, A sandwich electrochemiluminescence immunosensor for highly sensitive detection of alpha fetal protein based on MoS_2-PEI-Au nanocomposites and Au@BSA core/shell nanoparticles, *Sens. Actuat. B Chem.* 253 (2017) 470–477.

31. K. Kalantar-Zadeh, J.Z. Ou, Biosensors based on two-dimensional MoS_2, *ACS Sens.* 1 (2016) 5–16.

32. J. Kibsgaard, Z. Chen, B.N. Reinecke, T.F. Jaramillo, Engineering the surface structure of MoS_2 to preferentially expose active edge sites for electrocatalysis, *Nat. Mater.* 11 (2012) 963–969.

33. H.I. Karunadasa, E. Montalvo, Y. Sun, M. Majda, J.R. Long, C.J. Chang, A molecular MoS_2 edge site mimic for catalytic hydrogen generation, *Science.* 335 (2012) 698–702.

34. A. Stergiou, N. Tagmatarchis, Molecular functionalization of two-dimensional MoS_2 nanosheets, *Chem. Eur. J.* 24 (2018) 18246–18257.

35. R.R. Chianelli, M.H. Siadati, M.P. De la Rosa, G. Berhault, J.P. Wilcoxon, R. Bearden, B.L. Abrams, Catalytic properties of single layers of transition metal sulfide catalytic materials, *Catal. Rev. Sci. Eng.* 48 (2006) 1–41.

36. G. Solomon, R. Mazzaro, V. Morandi, I. Concina, A. Vomiero, Microwave-assisted vs. conventional hydrothermal synthesis of MoS_2 nanosheets: Application towards hydrogen evolution reaction, *Crystals.* 10 (2020) 1040.

37. A. Tuxen, J. Kibsgaard, H. Gøbel, E. Lægsgaard, H. Topsøe, J.V. Lauritsen, F. Besenbacher, Size threshold in the dibenzothiophene adsorption on MoS_2 nanoclusters, *ACS Nano.* 4 (2010) 4677–4682.

38. S. Anju, P. V. Mohanan, Biomedical applications of transition metal dichalcogenides (TMDCs), *Synth. Met.* 271 (2021) 116610.

39. M. Fizir, P. Dramou, N.S. Dahiru, W. Ruya, T. Huang, H. He, Halloysite nanotubes in analytical sciences and in drug delivery: A review, *Microchim Acta.* 185 (2018) 389.

40. I.K. Sideri, R. Arenal, N. Tagmatarchis, Covalently functionalized MoS_2 with dithiolenes, *ACS Mater. Lett.* 2 (2020) 832–837.

41. W.S. Seo, D.K. Kim, J.H. Han, K.B. Park, S.C. Ryu, N.K. Min, J.H. Kim, Functionalization of molybdenum disulfide via plasma treatment and 3-mercaptopropionic acid for gas sensors, *Nanomaterials.* 10 (2020) 1860.

42. S. Xu, Y. Zhong, C. Nie, Y. Pan, M. Adeli, R. Haag, Co-delivery of doxorubicin and chloroquine by polyglycerol functionalized MoS_2 nanosheets for efficient multidrug-resistant cancer therapy, *Macromol. Biosci.* 21 (2021) 2100233.

43. H. Liu, D. Grasseschi, A. Dodda, K. Fujisawa, D. Olson, E. Kahn, F. Zhang, T. Zhang, Y. Lei, R.B. Nogueira Branco, A.L. Elías, R.C. Silva, Y.T. Yeh, C.M. Maroneze, L. Seixas, P. Hopkins, S. Das, C.J.S. de Matos, M. Terrones, Spontaneous chemical functionalization via coordination of Au single atoms on monolayer MoS_2, *Sci. Adv.* 6 (2020) eabc9308.

44. J. Zheng, K. Lebedev, S. Wu, C. Huang, T. Ayvall, T.S. Wu, Y. Li, P.L. Ho, Y.L. Soo, A. Kirkland, S.C.E. Tsang, High loading of transition metal single atoms on chalcogenide catalysts, *J. Am. Chem. Soc.* 143 (2021) 7979–7990.

45. S. García-Dalí, J.I. Paredes, S. Villar-Rodil, A. Martínez-Jódar, A. Martínez-Alonso, J.M.D. Tascón, Molecular functionalization of 2H-Phase MoS_2 nanosheets via an electrolytic route for enhanced catalytic performance, *ACS Appl. Mater. Interfaces.* 13 (2021) 33157–33171.

46. P.K. Mohapatra, K. Ranganathan, L. Dezanashvili, L. Houben, A. Ismach, Epitaxial growth of In_2Se_3 on monolayer transition metal dichalcogenide single crystals for high performance photodetectors, *Appl. Mater. Today.* 20 (2020) 100734.
47. M. Xie, N. Yang, J. Cheng, M. Yang, T. Deng, Y. Li, C. Feng, Layered MoS_2 nanosheets modified by biomimetic phospholipids: Enhanced stability and its synergistic treatment of cancer with chemo-photothermal therapy, *Colloids Surf. B.* 187 (2020) 110631.
48. K. Zhang, Y. Zhuang, W. Zhang, Y. Guo, X. Liu, Functionalized MoS_2-nanoparticles for transdermal drug delivery of atenolol, *Drug Deliv.* 27 (2020) 909–916.

CHAPTER 6

2D Metals in Electronics

Reza Rahighi, Seyed Morteza Hosseini-Hosseinabad,
Amirmahmoud Bakhshayesh, and Somayeh Gholipour

6.1 INTRODUCTION

With the discovery of graphene [1], researchers became intrigued on finding other 2D materials exhibiting exotic characteristics. These materials are divided into three electronic groups: insulators [2] (like graphene oxide [3] or 2D crystals of SiTe), semiconductors [4] (e.g., black phosphorus, MoS_2, WTe_2, $TiSe_2$, $TiTe_2$, InSe, In_2Se_3, GaSe, and GaTe), and metals [5] (such as VSe_2, $TaTe_2$, TiS_2, $NiSe_2$, PdS_2, $PdSe_2$, and PtS_2). In 2014, the first 2D metal, phosphorene, was successfully synthesized [6], leading to the creation of a new chapter in materials science. Since then, several other metallenes, such as monometallenes (e.g., Pd [7]), bimetallenes (e.g., PdMo [8]), and trimetallenes (e.g., PtPdNi [9]) have been extensively studied. These materials are intriguing options for a range of applications, such as electronics, energy storage, and catalysis, due to their unique electrical, optical, mechanical, and chemical capabilities [10].

Metallenes, unlike their bulk counterparts, can exhibit remarkable high-temperature superconductivity, comprehending the mechanism of which remains a challenge. Nonetheless, some proposed explanations include the presence of strong electron–electron interactions, specific lattice structures, and the role of quantum fluctuations [11]. Scientists are particularly intrigued by the distinctive electronic and structural characteristics of these materials as they can potentially enable superconductivity at higher temperatures. By simulating metallenes using techniques such as density of states analysis, valuable insights into the distribution and energy levels of electrons within the material emerge, aiding in the understanding of their electronic properties in these extremely thin layers [12]. Furthermore, the structural tunability of metallenes opens up possibilities for tailored mechanical properties to meet specific requirements in various applications, such as flexible electronics, wearable devices, and novel sensors [13]. In addition, these materials

76 DOI: 10.1201/9781032645001-6

exhibit strong reactivity because of the high surface-to-volume ratio endowed from the under-coordinated metal atoms on the surface, which allows for efficient adsorption and catalytic reactions. This makes them highly attractive for applications involving hydrogen evolution and oxygen reduction reactions, like fuel cells [14]. The existence of metal atoms on the surface of metallenes enables the anchoring of functional organic or inorganic molecules, further expanding their applications in sensing, imaging, and drug delivery systems [15].

Although there have been considerable advancements in exploring metallenes and their possible applications, there are still substantial challenges to be overcome. Synthesis methods need to be further developed to produce large-scale, highly uniform metallene samples with controlled composition and morphology [5]. With further advancements in synthesis techniques and a deeper understanding of their structure–function relationships, metallenes hold immense potential for revolutionizing a wide range of technologies.

6.2 WHAT ARE THE BENEFITS OF BEING 2D?

It is anticipated that the unique properties of 2D materials can enable further advancements and innovations toward tremendous transformations in applied science [16, 17]. An illustration of this would be the utilization of 2D materials such as graphene in electronics [18] that has the potential for faster and more advanced gadgets accompanied by superior energy storage capabilities [19–21]. Moreover, in the healthcare field, these materials possess the potential to be utilized in the development of adaptable and wearable sensors capable of monitoring vital signs and identifying diseases at an early stage, which ultimately creates a tangible change in the provision of medical care to patients [22].

The emergence of 2D materials has had a profound impact on the chemical, optical, electrical, and magnetic characteristics of bulk materials, leading to substantial changes and enabling diverse applications. Furthermore, the ability to arrange 2D materials in various sequences to create van der Waals heterostructures has opened up a captivating field of study known as intercalation, wherein the engineering of internal interfaces within layered materials is feasible [23, 24]. As a result of these modifications, the introduction of these novel material architectures has resulted in the emergence of intriguing phenomena, including optical switching [25, 26], mixed ionic and electronic conduction [27, 28], and exotic charge-transport physics [29]. The phenomenon of intercalation in few-layer materials is growing as a burgeoning topic of study aimed at advancing the development of diverse flexible electronic devices, including flexible transducers, circuits, capacitors, field-effect transistors (FETs), light-emitting diodes, and photodetectors [19, 30, 31]. These devices exhibit the capacity to undergo deformation through bending and stretching, rendering them well-suited for a diverse array of electrical applications. The conversion characteristics pertaining to the transition from bulk materials to few-layer materials extend beyond the attributes discussed. Additionally, 2D materials exhibit enhanced surface-to-volume ratios, density and accessibility of active sites, excellent conductivity, and accelerated kinetics, enabling them to engage in surface interactions with intercalants and assume a more pivotal function in facilitating charge transfer between the guest and host for optical and electrical purposes [22, 32, 33].

6.3 HISTORICAL EXPLORATION TREND OF 2D METALS FAMILY

The discovery of graphene in 2004 marked the beginning of the history of 2D metallic nanomaterials, also known as metallenes. Subsequently, a large number of scientists became interested in studying these materials leading to a dramatic growth in the amount of research conducted on 2D materials including transition metal dichalcogenides (TMDs), hexagonal boron nitride, layered double hydroxides, black phosphorus, 2D metal oxides/sulfides, graphitic carbon nitride, MXenes, 2D perovskites, 2D metal–organic frameworks, transition metal carbides, 2D polymers, and 2D metals [34–36]. Soon after the development of graphene [37], researchers began to investigate the possibility of creating metallic counterparts for graphene, which led to the development of metallenes. Metallenes, as a newcomer to the family of 2D materials, are considered a special category of this family, which have garnered significant attention in recent research endeavors owing to their atomic-scale thickness and strong metal–metal bonding, anisotropic structure, and intriguing characteristics.

The study of metallic nanomaterials holds a significant place in the history of modern nanoscience. In 1857, Faraday became the first individual to report the chemical synthesis of gold nanoparticles. Following the development of electron microscopy, several attempts were made to synthesize a variety of metallic nanomaterials with unique morphologies, compositions, structures, and surprising features. Synthesized metals and alloys are typically characterized in three-dimensional (3D) morphology. The reason behind these materials preferential energetics in 3D morphology over 2D morphology is because of their atomic bonding, which is nondirectional. It is important to highlight that the term 2D encompasses more than just monolayer or atomically thin materials. It also contains freestanding ultrathin metallic nanosheets, the features of whose surfaces largely determine their characteristics. Several theoretical studies anticipated the presence of stable 2D metals with a single layer, while only a small number of these metals have been produced successfully using experimental methods. These methods include epitaxial growth on substrates and confined development within the nanopores of 2D templates. In recent years, there have been significant advancements in the development of controlled manufacturing techniques for producing freestanding metallic nanosheets [34, 38, 39]. For instance, certain structures of 2D materials can be synthesized in a hybrid way, or special metal nanostructures with metallic properties can be obtained by controlling their phase, which can have distinct advantages in electronics, and this synthesis method is feasible via advanced techniques of atomic layer deposition (ALD) or plasma-enhanced ALD methodologies [40].

6.4 APPLICATION OF METALLENES IN ELECTRONICS

The use of 2D metallic TMDs as a promising class of 2D materials to create novel electronic phases, magnets, superconductors, and other electronic applications has gained more attention in recent years. Vanadium diselenide (VSe_2) is a typical metallic TMD exhibiting a crystal structure resembling graphite lamellar with weak van der Waals interlayer interactions [41]. The metallic nature of VSe_2 and the occurrence of charge density wave (CDW) phases can be attributed to the significant electron coupling between adjacent

V^{4+}–V^{4+} pairs. The 3d1 odd-electronic configuration of V^{4+} ions offers a substantial amount of spin-related information. This odd-electronic configuration leads to interesting magnetic properties in VSe_2, such as the presence of magnetic ordering and spin fluctuations. Furthermore, the spin-related information provided by the V^{4+} ions can be utilized for potential applications in spintronics and quantum computing [42]. Based on theoretical calculations conducted on a pristine monolayer of VSe_2 in a 2D structure, it has been determined that this material could possess the inherent magnetic ordering, subsequently exhibiting characteristics of unwonted 2D magnetic material. Compared to other 2D conductive materials, VSe_2 nanosheets exhibit higher electrical conductivity than conventional metals [43]. All these scientific research studies have been able to improve their electronic/magnetic properties by tuning the thickness. Numerous scientific investigations have successfully enhanced the electrical and magnetic characteristics of materials by adjusting their thickness, hence opening up novel prospects for the advancement of creative electronic and magnetic devices.

As seen in Figure 6.1 (a), Zhang and co-workers [43] presented a van der Waals epitaxial route for the first time, using a one-step CVD method to make a few-layer metallic 1T-VSe_2 nanosheets on the freshly cleaved (001) facet of a mica substrate. It has been shown that the lattice symmetry of mica and VSe_2 exhibit similar threefold symmetry, which can be used to promote VSe_2 epitaxial growth on mica and is comparable to the traditional heteroepitaxial mode. In this unique structure, to accommodate remaining nonbonding d electrons, the d band typically breaks into two sub-bands, e_g and t_{2g}, as shown schematically in Figure 6.1 (b). 1T-VSe_2 exhibits metallic behavior due to its lower-energy t_{2g} band, which has a filling factor of 1/3.

Furthermore, Figure 6.1 (c) shows how the resistivity of a VSe_2 nanosheet with a thickness of 10 nm varies with temperature. It is evident that the electrical resistance rises with temperature, which is strongly suggestive of the ultrathin VSe_2 nanosheets metallic nature. A typical Hall-bar device constructed on a truncated triangular VSe_2 nanosheet is shown in the inset of Figure 6.1 (c).

The electrical resistivity of the sample exhibits a roughly linear relationship with temperature in the range of 100–300 K, demonstrating the dominating electron–phonon scattering process at this temperature range. Additionally, the blue dotted line in Figure 6.1 (c) depicts the electrical conductivity of VSe_2 nanosheets with a thickness of 10 nm, which was reported as 1.1×10^6 S m^{-1} at 300 K and 3.4×10^6 S m^{-1} near 0 K, while the electrical conductivity of VSe_2 nanosheets with a thickness of 7 nm and 22 nm at 300 K was approximately 5.0×10^6 S m^{-1} and 1.0×10^6 S m^{-1}, respectively. Figure 6.5d illustrates the temperature-dependent electrical resistivity profiles of VSe_2 samples with different thicknesses. The observed anomaly in the resistivity can be ascribed to the CDW transition of the VSe_2 nanosheets. The transition temperature (T_p) of CDW is defined by plotting temperature-dependent differential resistance spectra. Calculations demonstrate that the CDW onset transitions for these nanosheets were below 100 K. It is also possible to conclude that the T_p decreases with decreasing layer thickness.

Recent research indicates that metallic 2D 1T-VSe_2 could be a suitable electrode for supercapacitors. It has a lot of specific surface areas and is a good conductor. Wang et

80 ■ 2D Metals

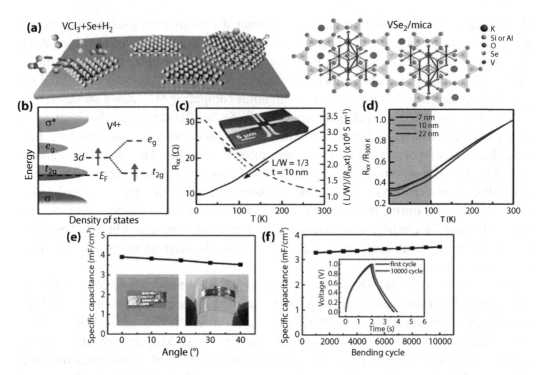

FIGURE 6.1 (a) The left panel: a schematic of the van der Waals epitaxial growth path and the right panel: a schematic representation of the possible occupation of two 60° rotated VSe$_2$ domains on mica. (b) Density of electronic states diagram for 1T-VSe$_2$. (c) Black line and arrow: the relationship between temperature and resistivity; and blue dashed line and arrow: the relationship between temperature and electrical conductivity for the VSe$_2$ device. The inset picture depicts a conventional Hall-bar device constructed using a triangular ultrathin metallic VSe$_2$ nanosheet with a thickness of ~10 nm. (d) The relationship between temperature and resistivity of VSe$_2$ nanosheets with various thicknesses. White and blue regions are attributed to the linear and anomalous regions, respectively. Adapted with permission [43], Copyright (2024), Wiley-VCH. (e) The relationship between specific capacitance and degree of curvature. Capacitance is computed using the CD measurements with a 2 mA/cm² scan rate. (f) The specific capacitance stability of a VSe$_2$ supercapacitor under mechanical bending. The insets depict the CD curves of the supercapacitor after 10,000 bending cycles at a scan rate of 2 mA/cm². Adapted with permission [44], Copyright (2024), Royal Society of Chemistry.

al. [44] synthesized high-quality VSe$_2$ nanosheets by a facile chemical vapor deposition (CVD) technique for use as electrode material in the construction of a flexible solid-state supercapacitor. The supercapacitor that has been manufactured exhibits characteristics of electrical double-layer capacitive behavior and demonstrates a power density that is comparable to supercapacitors based on graphene (Figure 6.1(d)).

Moreover, these flexible supercapacitors exhibit favorable mechanical qualities. Figure 6.1 (e) and (f) depicts the utilization of bending and fatigue cycle tests as a means to evaluate the stability of supercapacitors under deformation. The graph depicted in Figure 6.1 (e) illustrates the relationship between capacitance and bending angle, ranging from 0° to 40°. Based on the results of this experiment, it can be observed that the capacitance remains constant during bending. Also, the capacitance reaches 90% when the bending angle reaches 40°. Figure 6.1 (f) illustrates the relationship between capacitance and the bending cycle. The charge–discharge (CD) tests presented in the inset of Figure 6.1 (f) demonstrate the satisfactory performance of the supercapacitor following 10,000 bending cycles. The deformation stability of the flexible in-plane supercapacitor shows its superior mechanical qualities, making it ideal for wearable devices. The outstanding performance of the flexible VSe$_2$-based in-plane supercapacitor and its advantageous mechanical properties make metallic 2D nanosheets as power sources in advanced flexible electronics very attractive.

Jo and colleagues [45] conducted a study in which they fabricated FETs and photodetector devices utilizing few-layered palladium disulfide (PdS$_2$) films. Figure 6.2 (a) shows a schematic of the PdS$_2$ FET device fabricated, where a few-layered PdS$_2$ thin film was fabricated on a 2D semiconductor sapphire substrate. Subsequently, the film was transferred from the sapphire substrate onto a SiO$_2$/Si substrate with a thickness of 300 nm, employing the wet-transfer technique. They assessed the electron transport characteristics of the fabricated FETs to verify their n-type behavior. Additionally, they measured the capacitance–voltage (C–V) curves of these devices at various frequencies (50, 100, and 200 kHz). The investigation focused on the field-effect mobility of few-layered PdS$_2$ FETs, and the measurements were conducted at room temperature, revealing a field-effect mobility value of 2.85 cm^2 V^{-1} s^{-1}. The devices exhibit normally on n-type transistors in vacuum,

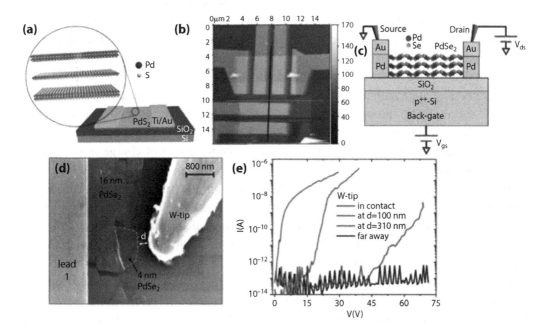

FIGURE 6.2 (a) Schematic of a few-layered PdS$_2$ film FET [45]. (b) AFM images of a PdSe$_2$ flake contacted with Pd/Au leads. (c) Schematic and biasing of the PdSe$_2$ field-effect device. (d) SEM image of the anode W-tip utilized in the FE test and a PdSe$_2$ nanosheet acting as the cathode that is protruding from metal lead 1. (e) Current-voltage curves displaying the evolution from electric contact to the FE regime at growing distances from the PdSe$_2$ nanosheet, and W-tip anode. Adapted with permission [46], Copyright (2024), Wiley-VCH.

with electron mobility of around 20 cm^2 V^{-1} s^{-1}. A transient encounter with air induces a reversible decrease in the channel current and a conversion of the device conductance from n-type to ambipolar.

Bartolomeo et al. [46] used the n-type conductivity that is enhanced by low pressure, the sharp edges of the flakes, and the work function that can be controlled at the layer level to show that 2D palladium diselenide (PdSe$_2$) nanosheets can produce a high field emission (FE) current of up to μA. Figure 6.2 (b) displays an atomic-force microscope (AFM) image of a PdSe$_2$ nanosheet that is contacted by multiple metal leads at a distance of approximately 950 nm from one another. The AFM height measurements demonstrated a uniform nanosheet that is almost 17 nm thick, which is equivalent to approximately 40 atomic layers. These nanosheets have been electrically characterized as the back-gated FET channel. These authors presented a report on the measurement of the FE current originating from exfoliated PdSe$_2$ flakes and depicted the FE current from few-layer PdSe$_2$ for the first time. The electrical measurements were conducted at ambient temperature and within an SEM chamber with regulated pressure.

The specimens were manipulated using piezo-driven tungsten tips (W-tips) that were incrementally displaced by 5 nm. This procedure involves connecting a Keithley 4200 semiconductor analyzer equipment to the W-tips. Figure 6.2 (c) displays a schematic of the three-terminal setup that was used for the electrical tests. The metal leads were used as the source and drain in different combinations, and the Si substrate was used as the back gate of a FET. They conducted the characterization of PdSe$_2$ FETs, with a primary emphasis on the effects of electric stress and air exposure. Furthermore, in order to examine FE originating from few-layer PdSe$_2$ nanosheets, protruding flakes from the metal leads were chosen. As determined by AFM, Figure 6.2 (d) illustrates a flake emerging from lead 1 that has a thickness of approximately 4 nm (about 10 layers). The I–V characteristic between lead 1 and the anode tip in nanosheet-electric contact represented the modestly rectifying p-type behavior suggesting a low Schottky barrier between the W-tip and PdSe$_2$ nanosheet.

The observed phenomenon could be ascribed to the processes of charge trapping and doping caused by adsorbates, along with the variability of the Schottky barrier height at the interfaces. These findings provide valuable insights into the fundamental properties of PdSe$_2$ flakes and their potential applications in FE devices and FETs. With the initiation of the contact condition, the anode W-tip was retracted to predetermined distances from the PdSe$_2$ nanosheet's edge. Then the voltage was swept to 100 V, with the current being monitored. In Figure 6.2 (e), a sequence is depicted, wherein the anode tip was retracted to distances of approximately 100 nm and 310 nm (represented by the dark cyan and red curves, respectively), after the first contact condition (magenta curve). When $d = 100$ nm, the I–V behavior acts like one of the contact situations: Most likely, electrostatic attraction has set up a kind of loose electric contact between the flake and the tip. However, at $d = 310$ nm, a current with an exponential growth pattern becomes detectable above the background noise level at a voltage of 45 V. This voltage is associated with an electric field known as the turn-on electric field, which is equal to V/kd ~ 90 V/μm. Further research in this area could focus on optimizing the device performance by controlling these factors and exploring other possible mechanisms influencing the FE current and device characteristics.

In a study by Chow et al. [47], it was observed that FETs incorporating ultrathin PdSe$_2$ exhibit inherent ambipolar characteristics. Furthermore, the researchers found that these FETs can be adjusted to operate as either hole-transport-dominated devices through molecular doping with F$_4$-TCNQ, or as electron-transport-dominated devices following vacuum annealing. The process of vacuum annealing was conducted with the purpose of enhancing mobility by eliminating surface adsorbates at various annealing temperatures, as depicted in Figure 6.3 (a). The transfer curves exhibit a noticeable trend wherein the threshold voltage gradually moves toward the negative bias (specifically, from +12.1 to −15.9 V). This shift is observed alongside a concurrent rise in drain current subsequent to annealing at temperatures of 400 K and 450 K, respectively. As depicted in Figure 6.3 (b), the elimination of hysteresis in the transfer curve was observed when annealing was conducted at a temperature of 450 K, and an average electron mobility of around 16 cm^2 V^{-1} s^{-1} was achieved. The PdSe$_2$ with a thickness of 9 nm demonstrates the maximum mobility of around 216 cm^2 V^{-1} s^{-1}, along with a favorable on/off ratio of 10^3.

Another application of 2D metals is the utilization of diodes for current rectification. Indeed, a considerable number of heterojunctions comprising pentagonal 2D metals exhibit inherent diode characteristics. Consequently, these heterojunctions have the capability to achieve self-powered photodetection by utilizing the internal electric field [48].

Through the substitution of oxygen atoms for selenium atoms in the pentagonal 2D PdSe$_2$ material, the primary conduction of electrical transport is shifted from n-type to p-type. This substitution enables the successful formation of an in-plane homojunction. As shown in Figure 6.3 (c), the determination of an ideality factor of 1.158 is achieved through the use of the Schottky diode equation, indicating the high quality of the interface in the homojunction diode [49].

The phenomenon of superconductivity in 2D metallic materials has garnered significant interest and witnessed substantial advancements in recent years due to their notable physical characteristics, including a high transition temperature (Tc), a continuous phase

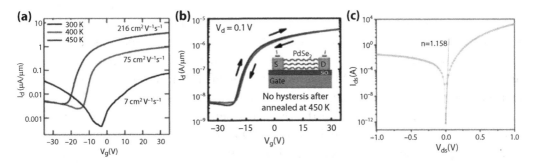

FIGURE 6.3 (a) Transfer curves of PdSe$_2$ FET. The transition from ambipolar (black curve) to electron transport became more pronounced after annealing at temperatures of 400 K (red curve) and 450 K (blue curve). (b) Transfer curve of the 9 nm PdSe$_2$ FET showing the absence of hysteresis after annealed at 450 K. The inset image shows the schematic of PdSe$_2$ FET. Adapted with permission [47], Copyright (2024), Wiley-VCH. (c) Logarithmic plot of the current-voltage curve of the homojunction. The extracted ideality factor of 1.158 indicates that the homojunction is of high quality. Adapted with permission [49], Copyright (2024), American Chemical Society.

84 ■ 2D Metals

transition, and an augmented parallel critical magnetic field (Bc). Considerable efforts have been dedicated to investigating various physical parameters in order to elucidate the underlying mechanisms behind the unanticipated phenomena of superconductivity. These efforts encompass tuning sample thickness, creating diverse heterostructures, and modulating carrier density through electric field and chemical doping, among other approaches [50]. There are many intrinsic superconductors in 2D metallic TMDs such as 2D $1T$-$TiSe_2$, $1T$-TaS_2, $2H$-TaS_2, $2H$-$NbSe_2$, $1T$-MoS_2, $3R$-$TaSe_2$, and so on [51–55].

The advancement of TMDs has significantly expedited investigations in the 2D domains, particularly in relation to layered MoS_2. Significantly, the monolayer of metallic MoS_2 serves as an optimal substrate for the emergence of novel topological electronic states. The $1T'$-MoS_2 nanosheets, which are characterized by their enormous size and high quality demonstrate evident intrinsic superconductivity across all thicknesses, including monolayer. This superconductivity is accompanied by a gradual decrease in the transition temperature from 6.1 to 3.0 K. The centrosymmetric $1T'$-MoS_2 structure exhibits atypical superconducting behavior, characterized by an upper critical field that surpasses the Pauli limit. In a study conducted by Peng et al. in 2019 [56], it was demonstrated that magneto-electrical transport could be achieved by utilizing $1T'$-MoS_2 nanosheets with varying thicknesses.

Figure 6.4 (a) illustrates the relationship between temperature and resistance in a bilayer sample, encompassing a temperature range from ambient temperature to 2 K. The study discovered metallic behavior in a bilayer of $1T'$-MoS_2, where the resistance exhibited a significant decrease from 4.7 K and reached zero at 2.4 K. This observation suggests the occurrence of a superconducting transition. A similar phenomenon of superconductivity was observed in $1T'$-MoS_2 samples of varying thicknesses, notwithstanding the incomplete transition at 2 K in the monolayer. As illustrated in Figure 6.4 (b) and (c), thick $1T'$-MoS_2 shows a superconducting transition temperature of 6.1 K. However, this temperature decreases significantly with decreasing sheet thickness; a monolayer achieved a temperature of 3.0 K at a resistance level of 90% of the normal state. The aforementioned observations provide evidence that superconductivity endures at the atomic level in $1T'$-MoS_2 [56].

In the case of the $2H$-TaS_2 sample, the results of temperature-dependent scanning tunneling spectroscopy showed that the superconductivity of this 2D metal occurs at a temperature below 2.8 K (Figure 6.4 (d)) [57]. The utilization of electrical transport measurement is a direct and efficacious approach to investigating the superconductivity of 2D m-TMDs. Liu et al. [58] conducted a study on the thickness-dependent superconducting property of $2H$-$NbSe_2$, which was synthesized via the CVD method. The researchers achieved this by fabricating Hall-bar devices, as depicted in Figure 6.4 (e). Hunt et al. [59] performed a study on the thickness-dependent superconductivity of $2H$-TaS_2 and $2H$-$NbSe_2$, which were obtained through the exfoliation process. These researchers achieved this by fabricating several multiterminal transport devices, as illustrated in Figure 6.4 (f).

The investigation of superconductivity in 2D m-TMDs with high transition temperatures continues to provide a significant challenge in the fields of scientific research and materials science. Liu et al. documented that the transition temperature of superconductivity in as-grown $NbSe_2$ exhibits a rise from 1.0 to 4.56 K as the thickness progresses from

2D Metals in Electronics ■ 85

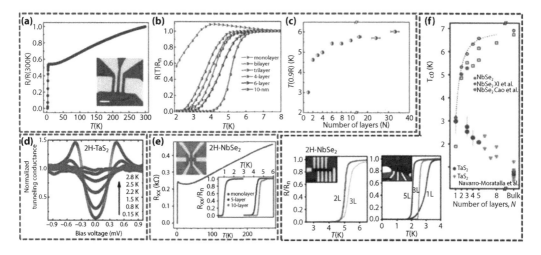

FIGURE 6.4 (a) Normalized resistance of bilayer sample under the temperatures of 300 to 2 K. (Inset: optical picture of a device with a scale bar measuring 10 μm). (b) Normalized resistance for 1T′-MoS$_2$ under varying thickness (under 8 to 2 K temperature dependency). (c) Thickness dependence of superconducting transition temperature, which is defined at 0.9 Rn, where Rn is the normal resistance at 8 K. Adapted with permission [56], Copyright (2024), Wiley-VCH. (d) Scanning tunneling spectroscopy results of 2H-TaS$_2$ nanosheet. Adapted with permission [57], Copyright (2024), American Physical Society. (e) Temperature dependence of longitudinal resistance Rxx as determined by a 2H-NbSe$_2$ device. Adapted with permission [58], Copyright (2024), Nature. (f) Superconductivity of atomically thin 2H-TaS$_2$ and 2H-NbSe$_2$ in zero magnetic field. Adapted with permission [59], Copyright (2024), *Nature*.

monolayer to 10-layer. The phenomenon of superconductivity in 2D-m-TMDs is influenced by some structural characteristics, including the thickness of the layers and phase transitions.

A number of 2D m-TMDs, including 2D 1T-TiSe$_2$, 1T-TaS$_2$, and 1T-TaSe$_2$, demonstrate the occurrence of charge density wave (CDW) phase transitions. These transitions have proven to be advantageous for the development of advanced electronic devices with superior performance characteristics. Notable applications include the utilization of these materials in electrical oscillators [60], non-volatile memory storage devices [61], and very sensitive bolometers operating at room temperature [62]. The variations in the electrical characteristics of 2D m-TMDs are a consequence of the lattice reconstruction-induced phase transitions. The metal-to-insulator transition (MIT) and the insulator-to-metal transition (IMT) in 2D m-TMDs have garnered increasing interest. A number of variables, including thickness, ambient pressure, intercalation, doping, and the use of electric source drain or gate bias, can be used to modify MIT and IMT [63, 64].

A study by Liu et al. [64] showed an integrated 1T TaS$_2$-BN-graphene voltage-controlled oscillator (VCO), which protected TaS$_2$ from oxidation and preserved the CDW phase with h-BN nanosheets (Figure 6.5 (a)). The transistor current density and load resistance match TaS$_2$. Figure 6.5 (b) displays the I–V characteristics of TaS$_2$-BN devices from 78 K to 395 K,

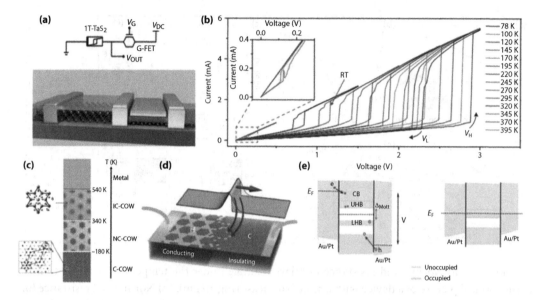

FIGURE 6.5 (a) The top panel displays the equivalent circuit and biasing configuration and the bottom panel displays the integrated oscillator's device structure comprised of a series of G-FETs connected to 1T–TaS$_2$. (b) Current–voltage curve of thin-film 1T-TaS$_2$ at different temperatures from 78 K to 395 K. Adapted with permission [64], Copyright (2024), *Nature*. (c) A graphic representation of the various CDW phases that occur in 1T-TaS$_2$ at various temperatures. (d) An illustration of carrier injection followed by DW formation and carrier trapping. (e) An illustration of the charge injection occurring at the electrodes in the C state, as well as the band structure subsequent to the conversion from the Mott-to-band state. Adapted with permission [65], Copyright (2024), *Nature*.

including the 1T-TaS$_2$ phase transition from nearly commensurate CDW to incommensurate CDW.

At temperatures lower than 320 K, a distinct and sudden increase in current is observed when the voltage surpasses a specific threshold (it is defined with V_H). This phenomenon is commonly known as threshold switching, which involves the transition between states of low and high conductivity. When the voltage is gradually reduced to zero, a reverse threshold switch takes place at a lower voltage V_L, which is associated with a reduced critical electric field. The disparity between these two voltage values gives rise to a hysteresis window. The conversion of IMT and MIT associated with the phase transition of the NC-CDW and IC-CDW phases in 1T-TaS$_2$ served as the 1T TaS$_2$-BN-graphene VCO operational mechanism. These oscillators can function at temperatures up to about 320 K without the need for a complex biasing circuit, making them a form of affordable technology that can be used in communication systems and other systems.

Furthermore, the movement of ions between IMT and MIT represents another potential mechanism. According to Vaskivskyi et al. [65], a stable way to change the resistance of CDW systems was tried, using 50–100 nm 1T-TaS$_2$ nanosheets to change from an insulator to a metal at low temperatures by injecting a short pulse current. Figure 6.5 (c) shows an image of different states of CDW in 1T-TaS$_2$ at various temperatures. As the charge pulse passes through the 1-TaS$_2$ channel, it undergoes a transition from a commensurately ordered polaronic Mott insulating state to a metastable electronic state with textured domain walls (Figure 6.5 (d)). Polarons are transformed into band states as a result of the transitioning process, creating new opportunities for the modification of all electronic states in nonvolatile memory devices.

Figure 6.5 (e) provides a visual representation of how the charge injection occurs at the electrodes in the C state, highlighting the band structure that is associated with the Mott-to-band state conversion process. Understanding this process is crucial for optimizing nonvolatile memory systems and exploring their potential applications in various electronic devices. Figure 6.5 (e) provides a visual representation of how the charge injection occurs at the electrodes in the C state, highlighting the band structure that is associated with the Mott-to-band state conversion process. Understanding this process is crucial for optimizing nonvolatile memory systems and exploring their potential applications in various electronic devices.

REFERENCES

1. K.S. Novoselov, A.K. Geim, S. V Morozov, D. Jiang, Y. Zhang, S.V Dubonos, I.V Grigorieva, A.A. Firsov, Electric field effect in atomically thin carbon films, *Science*. 306 (2004) 666–669.
2. K.Z. Donato, H.L. Tan, V.S. Marangoni, M.V.S. Martins, P.R. Ng, M.C.F. Costa, P. Jain, S.J. Lee, G.K.W. Koon, R.K. Donato, Graphene oxide classification and standardization, *Scientific Reports*. 13 (2023) 6064.
3. E. Bolhasani, F.R. Astaraei, Y. Honarpazhouh, R. Rahighi, S. Yousefzadeh, M. Panahi, Y. Orooji, Delving into role of palladium nanoparticles-decorated graphene oxide sheets on photoelectrochemical enhancement of porous silicon, *Inorganic Chemistry Communications*. 135 (2022) 109081.
4. G. Migliato Marega, Y. Zhao, A. Avsar, Z. Wang, M. Tripathi, A. Radenovic, A. Kis, Logic-in-memory based on an atomically thin semiconductor, *Nature*. 587 (2020) 72–77.
5. A. El Sachat, P. Xiao, D. Donadio, F. Bonell, M. Sledzinska, A. Marty, C. Vergnaud, H. Boukari, M. Jamet, G. Arregui, Effect of crystallinity and thickness on thermal transport in layered $PtSe_2$, *NPJ 2D Materials and Applications*. 6 (2022) 32.
6. S. Li, M. Tian, Q. Gao, M. Wang, T. Li, Q. Hu, X. Li, Y. Wu, Nanometre-thin indium tin oxide for advanced high-performance electronics, *Nature Materials*. 18 (2019) 1091–1097.
7. Z. Wei, Z. Zhao, C. Qiu, S. Huang, Z. Yao, M. Wang, Y. Chen, Y. Lin, X. Zhong, X. Li, Tripodal Pd metallenes mediated by Nb_2C MXenes for boosting alkynes semihydrogenation, *Nature Communications*. 14 (2023) 661.
8. Q. Zhang, S. Chen, Y.B. Yu, J. Hong, Paracetamol degradation and disinfection via electrocatalytic oxidation by using N-doped graphene as anode, *Catalysis Letters*. 152 (2022) 1–18.
9. H. Wang, Y. Li, K. Deng, C. Li, H. Xue, Z. Wang, X. Li, Y. Xu, L. Wang, Trimetallic PtPdNi-truncated octahedral nanocages with a well-defined mesoporous surface for enhanced oxygen reduction electrocatalysis, *ACS Applied Materials & Interfaces*. 11 (2019) 4252–4257.
10. C. Cao, Q. Xu, Q.-L. Zhu, Ultrathin two-dimensional metallenes for heterogeneous catalysis, *Chemical Catalysis*. 2 (2022) 693–723.
11. O. Bubnova, A new metallene arrival, *Nature Nanotechnology*. 13 (2018) 272.
12. M.C. Onbaşli, A. Avşar, S.M. Zanjani, A.M. Cheghabouri, F. Katmis, Theory and modeling of spintronics of nanomagnets, in: *Fundamentals of Low Dimensional Magnets*, CRC Press, 2022, pp. 325–342.
13. J. Zhang, F. Lv, Z. Li, G. Jiang, M. Tan, M. Yuan, Q. Zhang, Y. Cao, H. Zheng, L. Zhang, Cr-doped Pd metallene endows a practical formaldehyde sensor new limit and high selectivity, *Advanced Materials*. 34 (2022) 2105276.

14. V. Jose, V. Do, P. Prabhu, C. Peng, S. Chen, Y. Zhou, Y. Lin, J. Lee, Activating amorphous ru metallenes through co integration for enhanced water electrolysis, *Advanced Energy Materials*. 13 (2023) 2301119.

15. C. Lu, R. Li, Z. Miao, F. Wang, Z. Zha, Emerging metallenes: Synthesis strategies, biological effects and biomedical applications, *Chemical Society Reviews*. 52 (2023) 2833–2865.

16. R. Mas-Balleste, C. Gomez-Navarro, J. Gomez-Herrero, F. Zamora, 2D materials: To graphene and beyond, *Nanoscale*. 3 (2011) 20–30.

17. R. Rahighi, S.M. Hosseini-Hosseinabad, A.S. Zeraati, W. Suwaileh, A. Norouzi, M. Panahi, S. Gholipour, C. Karaman, O. Akhavan, M.A.R. Khollari, Two-dimensional materials in enhancement of membrane-based lithium recovery from metallic-ions-rich wastewaters: A review, *Desalination*. 543 (2022) 116096.

18. R. Rahighi, O. Akhavan, A.S. Zeraati, S.M. Sattari-Esfahlan, All-carbon negative differential resistance nanodevice using a single flake of nanoporous graphene, *ACS Applied Electronic Materials*. 3 (2021) 3418–3427.

19. X. Hu, K. Liu, Y. Cai, S.-Q. Zang, T. Zhai, 2D oxides for electronics and optoelectronics, *Small Science*. 2 (2022) 2200008.

20. N. Briggs, S. Subramanian, Z. Lin, X. Li, X. Zhang, K. Zhang, K. Xiao, D. Geohegan, R. Wallace, L.-Q. Chen, A roadmap for electronic grade 2D materials, *2D Materials*. 6 (2019) 22001.

21. R. Siavash Moakhar, S.M. Hosseini-Hosseinabad, S. Masudy-Panah, A. Seza, M. Jalali, H. Fallah-Arani, F. Dabir, S. Gholipour, Y. Abdi, M. Bagheri-Hariri, Photoelectrochemical water-splitting using CuO-based electrodes for hydrogen production: A review, *Advanced Materials*. 33 (2021) 2007285.

22. M.S. Stark, K.L. Kuntz, S.J. Martens, S.C. Warren, Intercalation of layered materials from bulk to 2D, *Advanced Materials*. 31 (2019) 1808213.

23. L. Oakes, R. Carter, T. Hanken, A.P. Cohn, K. Share, B. Schmidt, C.L. Pint, Interface strain in vertically stacked two-dimensional heterostructured carbon-MoS2 nanosheets controls electrochemical reactivity, *Nature Communications*. 7 (2016) 11796.

24. D.K. Bediako, M. Rezaee, H. Yoo, D.T. Larson, S.Y.F. Zhao, T. Taniguchi, K. Watanabe, T.L. Brower-Thomas, E. Kaxiras, P. Kim, Heterointerface effects in the electrointercalation of van der Waals heterostructures, *Nature*. 558 (2018) 425–429.

25. W. Bao, J. Wan, X. Han, X. Cai, H. Zhu, D. Kim, D. Ma, Y. Xu, J.N. Munday, H.D. Drew, Approaching the limits of transparency and conductivity in graphitic materials through lithium intercalation, *Nature Communications*. 5 (2014) 4224.

26. A.A. Popkova, A.A. Chezhegov, M.G. Rybin, I. V Soboleva, E.D. Obraztsova, V.O. Bessonov, A.A. Fedyanin, Bloch surface wave-assisted ultrafast all-optical switching in graphene, *Advanced Optical Materials*. 10 (2022) 2101937.

27. S.T.M. Tan, A. Gumyusenge, T.J. Quill, G.S. LeCroy, G.E. Bonacchini, I. Denti, A. Salleo, Mixed ionic–electronic conduction, a multifunctional property in organic conductors, *Advanced Materials*. 34 (2022) 2110406.

28. F. Chen, Q. Tang, T. Ma, B. Zhu, L. Wang, C. He, X. Luo, S. Cao, L. Ma, C. Cheng, Structures, properties, and challenges of emerging 2D materials in bioelectronics and biosensors, *InfoMat*. 4 (2022) e12299.

29. Z. Jie, H. Xia, S.-L. Zhong, Q. Feng, S. Li, S. Liang, H. Zhong, Z. Liu, Y. Gao, H. Zhao, The gut microbiome in atherosclerotic cardiovascular disease, *Nature Communications*. 8 (2017) 845.

30. D. Jiang, Z. Liu, Z. Xiao, Z. Qian, Y. Sun, Z. Zeng, R. Wang, Flexible electronics based on 2D transition metal dichalcogenides, *Journal of Materials Chemistry A*. 10 (2022) 89–121.

31. L. Zheng, X. Wang, H. Jiang, M. Xu, W. Huang, Z. Liu, Recent progress of flexible electronics by 2D transition metal dichalcogenides, *Nano Research*. 15 (2022) 1–20.

32. J. Hui, M. Burgess, J. Zhang, J. Rodríguez-López, Layer number dependence of Li+ intercalation on few-layer graphene and electrochemical imaging of its solid–electrolyte interphase evolution, *ACS Nano*. 10 (2016) 4248–4257.
33. W. Zhao, P.H. Tan, J. Liu, A.C. Ferrari, Intercalation of few-layer graphite flakes with $FeCl_3$: Raman determination of Fermi level, layer by layer decoupling, and stability, *Journal of the American Chemical Society*. 133 (2011) 5941–5946.
34. Y. Chen, Z. Fan, Z. Zhang, W. Niu, C. Li, N. Yang, B. Chen, H. Zhang, Two-dimensional metal nanomaterials: Synthesis, properties, and applications, *Chemical Reviews*. 118 (2018) 6409–6455.
35. R. Rudrapati, *Graphene: Fabrication Methods, Properties, and Applications in Modern Industries*, IntechOpen London, UK, 2020.
36. S.M. Hosseini H., R.S.Moakhar, F. Soleimani, S.K. Sadrnezhaad, S. Masudy-Panah, R. Katal, A. Seza, N. Ghane, S. Ramakrishna, One-pot microwave synthesis of hierarchical C-doped CuO Dandelions/g-C_3N_4 nanocomposite with enhanced photostability for photoelectrochemical water splitting, *Applied Surface Science*. 530 (2020) 147271.
37. L. Mahmoudian, A. Rashidi, H. Dehghani, R. Rahighi, Single-step scalable synthesis of three-dimensional highly porous graphene with favorable methane adsorption, *Chemical Engineering Journal*. 304 (2016) 784–792.
38. T. Wang, M. Park, Q. Yu, J. Zhang, Y. Yang, Stability and synthesis of 2D metals and alloys: A review, *Materials Today Advances*. 8 (2020) 100092.
39. F. Zhu, W. Chen, Y. Xu, C. Gao, D. Guan, C. Liu, D. Qian, S.-C. Zhang, J. Jia, Epitaxial growth of two-dimensional stanene, *Nature Materials*. 14 (2015) 1020–1025.
40. M.Z. Ansari, I. Hussain, D. Mohapatra, S.A. Ansari, R. Rahighi, D.K. Nandi, W. Song, S.-H. Kim, Atomic layer deposition—A versatile toolbox for designing/engineering electrodes for advanced supercapacitors, *Advanced Science*. n/a (2023) 2303055.
41. C. Yang, J. Feng, F. Lv, J. Zhou, C. Lin, K. Wang, Y. Zhang, Y. Yang, W. Wang, J. Li, Metallic graphene-like VSe_2 ultrathin nanosheets: Superior potassium-ion storage and their working mechanism, *Advanced Materials*. 30 (2018) 1800036.
42. Y. Ma, Y. Dai, M. Guo, C. Niu, Y. Zhu, B. Huang, Evidence of the existence of magnetism in pristine VX2 monolayers (X = S, Se) and their strain-induced tunable magnetic properties, *ACS Nano*. 6 (2012) 1695–1701.
43. Z. Zhang, J. Niu, P. Yang, Y. Gong, Q. Ji, J. Shi, Q. Fang, S. Jiang, H. Li, X. Zhou, Van der Waals epitaxial growth of 2D metallic vanadium diselenide single crystals and their extra-high electrical conductivity, *Advanced Materials*. 29 (2017) 1702359.
44. C. Wang, X. Wu, Y. Ma, G. Mu, Y. Li, C. Luo, H. Xu, Y. Zhang, J. Yang, X. Tang, Metallic few-layered VSe_2 nanosheets: High two-dimensional conductivity for flexible in-plane solid-state supercapacitors, *Journal of Materials Chemistry A*. 6 (2018) 8299–8306.
45. H.S. Jo, G.H. Oh, S. Kim, Atomically thin PdS_2: Physical characteristics and electronic device applications, *Journal of the Korean Physical Society*. 83 (2023) 1–5.
46. A. Di Bartolomeo, A. Pelella, F. Urban, A. Grillo, L. Iemmo, M. Passacantando, X. Liu, F. Giubileo, Field emission in ultrathin PdSe2 back-gated transistors, *Advanced Electronic Materials*. 6 (2020) 2000094.
47. W.L. Chow, P. Yu, F. Liu, J. Hong, X. Wang, Q. Zeng, C. Hsu, C. Zhu, J. Zhou, X. Wang, High mobility 2D palladium diselenide field-effect transistors with tunable ambipolar characteristics, *Advanced Materials*. 29 (2017) 1602969.
48. Q. Liang, Z. Chen, Q. Zhang, A.T.S. Wee, Pentagonal 2D transition metal dichalcogenides: PdSe2 and beyond, *Advanced Functional Materials*. 32 (2022) 2203555.
49. Q. Liang, Q. Zhang, J. Gou, T. Song, Arramel, H. Chen, M. Yang, S.X. Lim, Q. Wang, R. Zhu, Performance improvement by ozone treatment of 2D PdSe2, *ACS Nano*. 14 (2020) 5668–5677.
50. D. Qiu, C. Gong, S. Wang, M. Zhang, C. Yang, X. Wang, J. Xiong, Recent advances in 2D superconductors, *Advanced Materials*. 33 (2021) 2006124.

51. Y. Deng, Y. Lai, X. Zhao, X. Wang, C. Zhu, K. Huang, C. Zhu, J. Zhou, Q. Zeng, R. Duan, Controlled growth of 3R phase tantalum diselenide and its enhanced superconductivity, *Journal of the American Chemical Society.* 142 (2020) 2948–2955.
52. E. Navarro-Moratalla, J.O. Island, S. Manas-Valero, E. Pinilla-Cienfuegos, A. Castellanos-Gomez, J. Quereda, G. Rubio-Bollinger, L. Chirolli, J.A. Silva-Guillén, N. Agraït, Enhanced superconductivity in atomically thin TaS_2, *Nature Communications.* 7 (2016) 11043.
53. J. Pan, C. Guo, C. Song, X. Lai, H. Li, W. Zhao, H. Zhang, G. Mu, K. Bu, T. Lin, Enhanced superconductivity in restacked TaS_2 nanosheets, *Journal of the American Chemical Society.* 139 (2017) 4623–4626.
54. A. Sanna, C. Pellegrini, E. Liebhaber, K. Rossnagel, K.J. Franke, E.K.U. Gross, Real-space anisotropy of the superconducting gap in the charge-density wave material 2H-NbSe$_2$, *NPJ Quantum Materials.* 7 (2022) 6.
55. D. Wines, K. Choudhary, A.J. Biacchi, K.F. Garrity, F. Tavazza, High-throughput DFT-based discovery of next generation two-dimensional (2D) superconductors, *Nano Letters.* 23 (2023) 969–978.
56. J. Peng, Y. Liu, X. Luo, J. Wu, Y. Lin, Y. Guo, J. Zhao, X. Wu, C. Wu, Y. Xie, High phase purity of large-sized 1T′-MoS$_2$ monolayers with 2D superconductivity, *Advanced Materials.* 31 (2019) 1900568.
57. J.A. Galvis, L. Chirolli, I. Guillamón, S. Vieira, E. Navarro-Moratalla, E. Coronado, H. Suderow, F. Guinea, Zero-bias conductance peak in detached flakes of superconducting 2H-TaS$_2$ probed by scanning tunneling spectroscopy, *Physical Review B.* 89 (2014) 224512.
58. H. Wang, X. Huang, J. Lin, J. Cui, Y. Chen, C. Zhu, F. Liu, Q. Zeng, J. Zhou, P. Yu, High-quality monolayer superconductor NbSe2 grown by chemical vapour deposition, *Nature Communications.* 8 (2017) 394.
59. S.C. De la Barrera, M.R. Sinko, D.P. Gopalan, N. Sivadas, K.L. Seyler, K. Watanabe, T. Taniguchi, A.W. Tsen, X. Xu, D. Xiao, Tuning Ising superconductivity with layer and spin–orbit coupling in two-dimensional transition-metal dichalcogenides, *Nature Communications.* 9 (2018) 1427.
60. A.K. Geremew, S. Rumyantsev, B. Debnath, R.K. Lake, A.A. Balandin, High-frequency current oscillations in charge-density-wave 1T-TaS$_2$ devices: Revisiting the "narrow band noise" concept, *Applied Physics Letters.* 116 (2020) 163101.
61. A. Mraz, R. Venturini, M. Diego, A. Kranjec, D. Svetin, Y. Gerasimenko, V. Sever, I.A. Mihailovic, J. Ravnik, I. Vaskivskyi, Energy efficient manipulation of topologically protected states in non-volatile ultrafast charge configuration memory devices, *ArXiv* Preprint ArXiv:2103.04622. (2021).
62. J. Shi, X. Chen, L. Zhao, Y. Gong, M. Hong, Y. Huan, Z. Zhang, P. Yang, Y. Li, Q. Zhang, Chemical vapor deposition grown wafer-scale 2D tantalum diselenide with robust charge-density-wave order, *Advanced Materials.* 30 (2018) 1804616.
63. B. Zhao, D. Shen, Z. Zhang, P. Lu, M. Hossain, J. Li, B. Li, X. Duan, 2D Metallic transition-metal dichalcogenides: Structures, synthesis, properties, and applications, *Advanced Functional Materials.* 31 (2021) 2105132.
64. G. Liu, B. Debnath, T.R. Pope, T.T. Salguero, R.K. Lake, A.A. Balandin, A charge-density-wave oscillator based on an integrated tantalum disulfide–boron nitride–graphene device operating at room temperature, *Nature Nanotechnology.* 11 (2016) 845–850.
65. I. Vaskivskyi, I.A. Mihailovic, S. Brazovskii, J. Gospodaric, T. Mertelj, D. Svetin, P. Sutar, D. Mihailovic, Fast electronic resistance switching involving hidden charge density wave states, *Nature Communications.* 7 (2016) 11442.

CHAPTER 7

2D Metals (Metallenes) in Spintronics

Puja Kumari and Soumya J. Ray

7.1 INTRODUCTION

During the 1960s, a new type of transistor called the self-aligned planar-gate silicon metal-oxide-semiconductor field-effect transistor (MOSFET) was created and proved to be a significant advancement. This development established a solid basis for the semiconductor integrated circuit industry. The exponential growth of the integrated circuit industry was fueled by Gordon Moore's prediction that transistor count would double every two years, coupled with Robert Dennard's scaling guidelines, resulting in the shrinking of transistor size until 2000 [1]. This period in time is commonly known as the era of geometric scaling. Advancements in technology during the early 2000s brought about a second era of scaling. This was made possible by the introduction of strained silicon and silicon–germanium channels, as well as high – /metal-gate stack and nonplanar fin field-effect transistors (FinFETs). In this timeframe, the speed of electrons and holes in the channel (strain) improved [2], while the thickness of the oxide layer (high - dielectric) decreased [3]. As a result, the integrated circuit industry kept up its pace of doubling the number of transistors every two years. Transistor scaling has led to numerous benefits, including faster and more power-efficient computers, smaller and lighter electronic devices, and lower production costs per transistor. It has also enabled the growth of industries like mobile computing and data centers. In recent years, as transistors continue to shrink, the challenges of maintaining Moore's Law have become more difficult. The laws of physics place limits on how small transistors can be made, and the cost of building ever more complex manufacturing facilities is becoming prohibitive. As transistors become more and more smaller, the distance between the source and drain terminals shrinks. This can lead to increased leakage currents, where electrons tunnel through the insulating layer (gate oxide) in the

DOI: 10.1201/9781032645001-7

transistor's off state. High leakage currents can lead to excessive power consumption and heat generation. At extremely small scales, quantum mechanical effects become significant. For instance, quantum tunneling can cause electrons to pass through barriers that would be insurmountable in classical physics. This can affect the transistor's behavior and create uncertainty in its operation.

The slowing down or potential end of Moore's Law, as the semiconductor industry faces increasing challenges in scaling transistors to smaller sizes, has prompted a search for alternative approaches to sustain and advance computing technology. In this context, spintronics (spin transport electronics) is one of the emerging technologies that hold promise for extending the capabilities of integrated circuits and potentially helping overcome limitations associated with miniaturization of electronic device.

7.2 SPINTRONICS

Spintronics stands out as a swiftly advancing research domain within nanotechnology. Research in spintronics gained momentum following the 1988 discovery of giant magnetoresistance [4], a phenomenon where electrical resistance significantly changes based on the parallel or antiparallel alignment of magnetization in adjacent ferromagnetic layers (Figure 1(a)). Their work demonstrated the practical feasibility of using the relative alignment of magnetic moments to control electrical resistance. Within a decade, this discovery was integrated into hard disk drives (Figure 1(c)), the ubiquitous storage medium, and was subsequently honored with the Nobel Prize in Physics 19 years later.

Spintronics is the science of stimulating and controlling the spin of electrons. It gives a significant methodology for storing data using an electron's spin degree of freedom rather than its charge with ultrafast processing and low power consumption at nano-scale devices. The integral spin properties of the electron can be extremely useful for the growth of complex quantum-computing devices, high-density data storage memories, novel logic devices, high-frequency equipment, magneto-electrics, and so on. As a primary unit of spintronic industry, spin valve or magnetic tunnel junctions (MTJs) have been employed in various applications, including magnetic random-access memory (MRAM) (Figure 7.1(b)). A spin valve is composed of a thin nonmagnetic layer sandwiched between two ferromagnets (FMs), with the sensitivity or efficiency determined by the quality of the interfaces between the FMs and the insulator. The unique hybridization of atomic orbitals of nonmagnetic (NM) and ferromagnets creates a spin-polarized junction and hence are called "spinterfaces" that offer tunneling magnetoresistance (TMR) and spin-injection efficiency at the interfaces. The ability to inject or detect spin-polarized electrons across such an interface is influenced by various factors, including interface roughness, defects, impurities, and disparities in the conductivity of the materials constituting the interface. In recent years, MTJs have been a focus of intensive research aimed at achieving high TMR ratios with low power consumption. During the initial development of MTJs, oxide barriers such as MgO and Al_2O_3 are commonly utilized in combination with traditional ferromagnetic transition-metal electrodes like Ni and Co [6]. The TMR observed in metal oxide-based MTJs is notably affected by the thickness of the tunnel barrier [7]. However, as the system size is scaled down to atomic-scale thickness, achieving precise control over the thickness

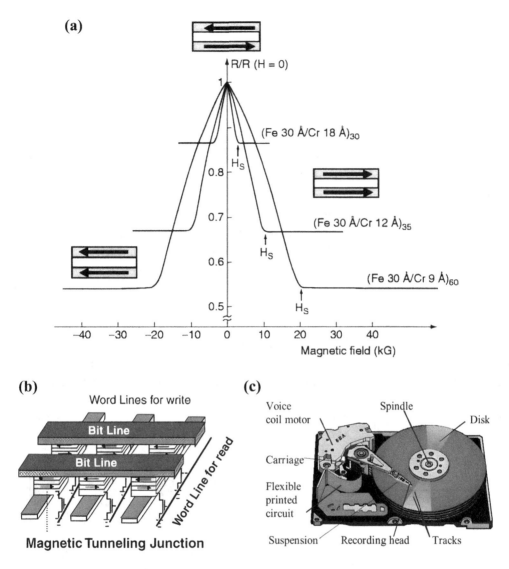

FIGURE 7.1 (a) The variation in electrical resistance in Fe/Cr superlattices at 4.2 K under an external magnetic field H, dependent on the alignment of magnetic moments in adjacent ferromagnetic layers [4] ©1988 American Physical Society. Schematic diagram of: (b) magnetic random-access memory (MRAM) [5] © 2016 Springer Science+Business Media Dordrecht. (c) Hard Disk Drive (HDD).

of the oxide barrier becomes progressively more challenging. This challenge arises due to factors such as the emergence of disorder, the formation of pinholes or point defects, and the formidable task of attaining high crystallinity and smooth interfaces at such ultrathin dimensions. These challenges can be fundamentally mitigated through the utilization of a new class of graphene and graphene-like 2D crystals [8, 9]. These materials offer a renewed opportunity for exploring novel spinterfaces formed by high-quality atomically thin crystals in conjunction with ferromagnetic materials.

7.3 SPIN TRANSPORT IN SPIN VALVE DEVICE

To detect the spin signal, a spin valve device is designed, consisting of a minimum of two ferromagnetic (FM) electrodes and a nonmagnetic layer. When measuring the resistance of a spin valve junction, distinct states are designed based on the relative alignment of the magnetizations of FM metals, namely, parallel and antiparallel configurations. These devices can be classified into two types: (i) Lateral Spin Valve (LSV) and (ii) Vertical Spin Valve (VSV). In the lateral configuration, ferromagnetic (FM) electrodes are placed atop the nonmagnetic layer, allowing spin-polarized current to flow within the plane of the device, establishing a current in-plane (CIP) configuration (Figure 7.2a). Conversely, in the vertical spin valve structure, the nonmagnetic layer is sandwiched between FM electrodes, directing the spin-polarized current to flow perpendicular to the intervening layer, resulting in a current perpendicular-to-plane (CPP) geometry (Figure 7.2c) [11, 12]. In lateral spin valve setups with a CIP state, the spin valve signal can be determined through two distinct methods: local and nonlocal measurements. In both scenarios, the resistance, either nonlocal (RNL) or local (RL), is assessed as a function of an in-plane magnetic field. The nonlocal measurement method is schematically demonstrated in Figure 7.2a, where E_2 and E_3 are FM electrodes while E_1 and E_4 are ideally nonmagnetic metals; however, to simplify the process of device fabrication these are often chosen to be FMs. In this method, an electric current is passed between E_1 and E_2, and the voltage is recorded across electrodes E_3 and E_4. The central electrodes (E_2 and E_3) serve as the spin injector and detector,

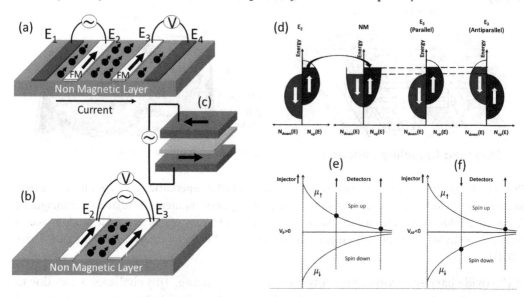

FIGURE 7.2 Diagrams of lateral spin valve devices depict both (a) nonlocal and (b) local geometries designed for magneto transport measurements. (c) Illustrations of a vertical spin valve display the configuration with a non-magnetic spacer positioned between two ferromagnetic (FM) layers. (d) Depiction of the spin-dependent density of states (DOS) illustrating spin injection from E_2 into the non-magnetic layer (NM) [10]. © 2018 Published by Elsevier B.V. The spin signal is detected at E_3, which can be oriented either parallel or antiparallel to E_2. Spin-dependent chemical potential for the (e) parallel and (f) antiparallel spin configurations [10]. © 2018 Published by Elsevier B.V.

respectively, owing to the existence of spin-dependent density of states (DOS) in ferromagnetic (FM) materials at the Fermi level (illustrated in Figure 7.2d). The spins released from E_2 disperse on both sides of the injector. To capture a pure spin current, the voltage is measured across E_3 and E_4, excluding any contribution from the charge current. The nonlocal measurement method effectively decouples charge and spin currents. The voltage recorded across E_3 and E_4 is subsequently converted into resistance by dividing it by the injection current. The nonlocal spin signal is calculated using the provided relation:

$$\Delta R_{NL} = \frac{V_{NL}^P - V_{NL}^{AP}}{I} = \frac{P^2 \lambda_s}{2W\sigma} \exp\left(-L/\lambda_s\right) \tag{7.1}$$

where nonlocal voltages measured across E_3 and E_4 for parallel and antiparallel magnetization states, I is the injection current, P is spin-injection/detection efficiency, λ_s is spin diffusion length, W is width of non-magnetic strip, is conductivity and L is separation between central FM electrodes, demonstrates positive and negative values depending on the relative configuration of magnetization vectors between FM metals.

Upon applying a current between E_1 and E_2, it travels in the leftward direction, as illustrated in Figure 7.2a. The flow of the charge current unidirectionally, however the spin-polarized electrons injected from E_2 diffuse on both sides of the nonmagnetic layer. The accumulation of spin in the nonmagnetic layer is depicted in the density of states (DOS) shown in Figure 7.2d, which also demonstrates the band structures of E_2 and E_3. To simplify the representation of normal ferromagnets, the band structure of a half-metal is employed [13]. Due to spin injection from E_2, the nonmagnetic layer gains a higher population of spin-up electrons. The injected spins travel through the nonmagnetic layer and are detectable at E_3. The nonlocal resistance detected in the spin valve device is determined by the relative magnetization orientations between E_2 and E_3. The alignment of these orientations either in parallel or antiparallel configurations can be done by manipulating the applied magnetic field. To achieve these magnetization configurations, FM electrodes with varying widths and different switching fields of magnetization are employed. When the injector (E_2) and detector (E_3) are aligned in parallel, the resistance is positive because the majority spins in the detector and the spins accumulated in the nonmagnetic layer have the same spin orientation. Conversely, an antiparallel spin alignment leads to a negative resistance since the majority spins in the detector and the spins collected in the nonmagnetic layer have opposing orientations. The spin diffusion towards the detector can be elucidated through the spin-dependent chemical potential difference, as demonstrated in Figures 7.2e and 7.2f.

The relaxation length of spin determines the scale at which the decay of spin density occurs as a result of spin flip scattering. When there is a positive (negative) difference in chemical potential for parallel (antiparallel) alignment between E_2 and E_3, it results in a positive (negative) voltage [14, 15]. In the local measurement technique, measurement entails assessing the resistance across two ferromagnetic (FM) electrodes. Here, a spin-polarized current is injected from one FM metal and passes through the nonmagnetic layer. The other FM film detects the current in this configuration. The local measurement

96 ■ 2D Metals

setup is illustrated in Figure 7.2b. It can be seen that the electrodes E_2 (FM) and E_3 (FM) are used to apply the current, while the voltage is measured across the same electrodes. Detecting the signal of spin valve in the local configuration is a challenging task, mainly due to the presence of charge current that creates significant background signals. However, the nonlocal measurement technique provides an advantage as it removes the charge current between the injector and detector, making the process much more reliable [16]. Vertical magnetic junctions in CPP geometry have their spin valve signal determined by measuring the magnetoresistance ratio (MR) as a variation of magnetic field. The measurement of spin transport for vertical structures, where parallel and antiparallel magnetization states are achieved by sweeping an in-plane magnetic field (B). MR is high (low) when the magnetizations between the two ferromagnetic electrodes, FM-1 and FM-2, are antiparallel (parallel) to each other [17]. *MR* is calculated using the relation:

$$MR = \frac{R_{AP} - R_p}{R_p} = \frac{2P_1P_2}{1 - P_1P_2} \tag{7.2}$$

where R_{AP} and R_P represent the resistance for antiparallel and parallel configurations.

In 1985, Johnson and Silsbee conducted a study on spin signals injection and detection [12]. Their research involved the use of an aluminum bar that had permalloy electrodes deposited on it. These electrodes served as both spin injectors and detectors. This groundbreaking research paved the way for extensive studies on spin injection in metals, including Au, Al, Ag, and Cu. The research scope extended beyond metals to encompass various semiconducting materials such as Si, Ge, GaAs, and many others which were also explored for their spin transport properties.

7.4 2D METALS (METALLENES)

Two-dimensional metallenes refer to a class of two-dimensional materials composed of metal atoms arranged in a planar structure. These materials are often single-atom thick and exhibit unique electronic and physical properties due to their two-dimensional nature. Metallenes can be thought of as the metal counterparts of graphene, which is composed of carbon atoms arranged in a two-dimensional hexagonal lattice. These materials have garnered significant interest in materials science and nanotechnology due to their potential applications in electronics, catalysis, energy storage, and other technological areas. The term "metallene" is derived from the combination of "metal" and the suffix "-ene," denoting a two-dimensional structure. Other members within this family include phosphorene (P-ene), silicene (Si-ene), transition metal-based metallenes, and several others. In the realm of spintronics, these materials offer a remarkable foundation for spintronic investigations due to their distinctive spin-dependent characteristics. These include remarkably prolonged spin relaxation times, extensive spin diffusion lengths, Rashba spin-orbit coupling (SOC), spin-valley coupling, and the quantum spin Hall effect. Additionally, the precise layering of individual 2D metallenes in a predetermined sequence can amalgamate the best attributes of different components into a single, ultimate material. For instance, creating a heterostructure by combining graphene and TMDCs results in exhibiting excellent spin transport performance and larger SOC [18, 19]. In this way, 2D metallenes and their associated heterostructures enable efficient long-range spin transport and effective spin manipulation, thereby realizing magnetic logic gates, MRAM, and various other spintronic devices.

Nevertheless, there are several challenges that still require solutions in the realm of 2D metallenes spintronics. Primarily, the 2D metallenes employed in spintronics are predominantly non-ferromagnetic, or their Curie temperatures (T_c) are often significantly below room temperature. This situation necessitates the injection of spins into the 2D materials through various techniques. However, this introduces a new challenge in enhancing the efficiency of spin polarization during the injection process. Conversely, effectively controlling and preserving spin states remains an ongoing challenge. The capacity to transmit spin information and toggle spin states is yet to be fully realized, rendering certain spin applications unattainable. In recent years, significant efforts have been dedicated to addressing these issues, resulting in substantial progress. This progress includes the continual enhancement of spin-related parameters, the identification and exploration of novel 2D metallenes, and comprehensive investigations into various spin-related phenomena.

7.5 SPIN-INJECTION FROM BULK TO 2D INTERFACE

2D crystals offer nearly ideal solutions for generating atomically thin layers within devices. These crystals naturally form in layers, devoid of dangling bonds or significant defects, thus presenting innovative opportunities for epitaxial growth with ferromagnetic layers [22] (Figure 7.3(a)). Since its first fabrication through mechanical exfoliation from graphite, it has garnered immense interest. Graphene's electronic properties are remarkably unique: it possesses a zero-bandgap electronic structure with a linear band dispersion converging at the Fermi level at the K points in the Brillouin zone, forming what is known as a Dirac cone. Due to the linear energy band dispersion near the Dirac point, graphene shows extraordinarily high carrier mobilities (15,000 $cm^2V^{-1}s^{-1}$) and offering significant potential for applications in nanoelectronics. Consisting of low atomic number element of carbon, graphene has weak spin-orbit interaction, which minimizes unwanted spin relaxation processes, allowing for longer spin lifetimes and improved spin transport [20]. The carrier concentration in graphene can be controlled by applying gate voltages. This tunability enables precise control over the electron density in the channel, which is essential for spintronic devices. In graphene, electron spins can travel relatively long distances without significant scattering or spin relaxation. This property is favorable for maintaining the coherence of spin information in spintronic channels. As a result of these unique properties, graphene proves highly suitable for spintronic applications (Figure 7.3(c–d)), which harness both the spin and charge degrees of freedom of electrons to deliver enhanced operational speed, reduced energy consumption, and expanded functionalities.

Maassen et al. predicted a potential spin-injection efficiency of 60–80% from a ferromagnetic (FM) electrode into graphene [23]. However, experimental data has consistently shown that the actual efficiency remains just a few percent when graphene is directly in contact with FM leads. Although the initial experiment by Hill et al. successfully demonstrated spin injection into graphene using soft magnetic NiFe electrodes [24], subsequent

98 ■ 2D Metals

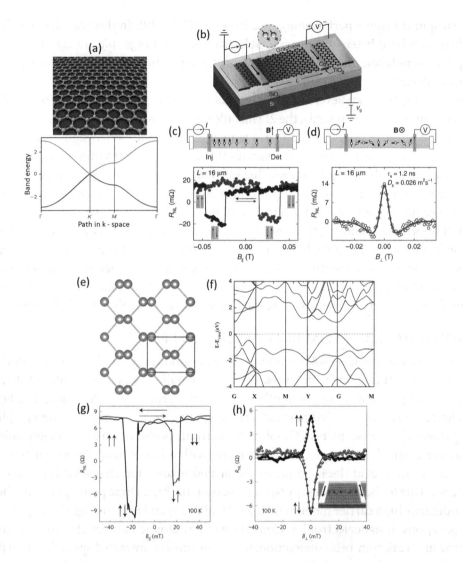

FIGURE 7.3 (a) Single layer of graphene with band structure. (b) A visual diagram illustrating a graphene device with ferromagnetic tunnel contacts designed for spin injection and detection in a nonlocal (NL) geometry [20]. Copyright © 2015, Springer Nature Limited. (c) The figure illustrates a non-local (NL) spin valve signal with a channel length L = 16 μm, featuring an in-plane magnetic field exhibiting high and low values corresponding to the parallel and antiparallel configurations of the ferromagnetic injector and detector electrodes, respectively [20]. Copyright © 2015, Springer Nature Limited. In addition, panel (d) displays the NL Hanle spin precession signal obtained by a perpendicular magnetic field sweep in the parallel configuration of ferromagnetic electrodes for a graphene channel L = 16 μm. The raw data points are fitted with the Hanle equation to extract spin lifetime and spin diffusion constant [20]. Copyright © 2015, Springer Nature Limited. (e, f) Single layer of phosphorene layer with band structure [21]. Copyright © 2015. (g) The graph shows the nonlocal signal plotted against the in-plane magnetic field. The vertical arrows represent the orientations of the relative magnetization of the injector and detector electrodes [9]. Copyright © 2017, Springer Nature Limited. (h) The graph illustrates the nonlocal signal as a function of the perpendicular magnetic field. Measurements were conducted with an injected current of 0.5 μ A at 100 K, and back gate voltage is maintained at 30 V [9]. Copyright © 2017, Springer Nature Limited.

studies have revealed relatively low spin polarization of conduction electrons, as assessed from the magnetoresistance (MR) ratio in CPP spin valves. Despite theoretical expectations of significant spin polarization and outstanding spin-filtering effects at the interfaces between graphene and magnetic metals, the observed low spin-injection efficiency is not unique to graphene [25]; it is a common challenge encountered when attempting spin injection from ferromagnetic (FM) metals to semiconductors. The primary reasons behind this lower-than-expected efficiency include the mismatch in spin conductance between the FM electrode and the semiconductor or graphene, along with the introduction of local electronic states at the interfaces between graphene and FM materials.

To enhance spin-injection efficiency, researchers have proposed employing an insulating oxide, such as Al_2O_3 or MgO, as a tunnel barrier between the ferromagnetic (FM) electrode and graphene [26, 27]. This approach assists in adjusting the interfacial spin-dependent resistivity, leading to an enhancement in spin-injection efficiency. However, the substantial lattice mismatch between graphene and Al_2O_3/MgO poses a challenge, leading to interface strain that could eventually alter the electronic properties of graphene. Therefore, it is crucial to find ways to reduce this lattice mismatch and enhance spin-injection efficiency for graphene and ferromagnetic metal contacts. Certain improvements have been achieved through the adoption of alternative sample growth methods or by introducing an interfacial TiO_2 layer. Employing the latter technique, the measured spin-injection efficiency has reached approximately 30% [28]. Researchers have encountered similar challenges when attempting to grow high-κ dielectric oxides such as HfO_2 or Al_2O_3 on graphene [29, 30]. To tackle these challenges, Wu et al. proposed employing hexagonal boron nitride (h-BN) as the tunnel barrier between graphene and the ferromagnetic (FM) electrode, instead of using oxide [31]. This choice is based on the similar crystal structure and lattice constants shared by h-BN and graphene. Moreover, h-BN serves as a high-κ dielectric, and a high-quality interface with graphene can be established. Transport calculations conducted by Wu et al. have shown that the insulating h-BN tunnel barrier substantially enhances the spin-injection efficiency from the metal lead to graphene [31]. They have successfully demonstrated that utilizing a Ni electrode can achieve highly efficient spin injection in graphene when h-BN is employed as a tunnel barrier. The efficiency of spin injection increases as the thickness of the h-BN layer grows, reaching 100% spin polarization with three atomic layers of h-BN. This enhancement is attributed to the asymmetry in the spin states of graphene induced by h-BN, which obstructs the majority spin transport channel. Previous experimental studies have also shown that h-BN induces a significant improvement in the spin-injection efficiency from metal leads to graphene. Additionally, A. Avsar et al. experimentally reported a spin lifetime of 4 ns for phosphorene (Figure 7.3 (e–h)) [9]. In a recent investigation, spin injection into monolayer phosphorene was studied using the NEGF-DFT quantum transport method [32]. The study employed a 2D magnetic tunneling structure with a monolayer phosphorene situated between ferromagnetic (FM) Ni(111) or Ni(100) leads. Both structures exhibited commendable spin-polarized quantum transport, but the trilayer configuration with Ni(100)/phosphorene/Ni(100) displayed superior characteristics. In this trilayer setup, spin injection and the magnetoresistance (MR) ratio remained nearly constant regardless of the bias voltage. Haoqi Chen et al. investigated

100 ■ 2D Metals

the spin-injection properties of monolayer black phosphorene deposited on a ferromagnetic EuO (111) surface and found that the interaction with the magnetic semiconductor resulted in a spin-injection efficiency of 30% at the Fermi level [33]. Mingyan Chen et al. also examined the spin-injection efficiency through phosphorene with nickel electrodes using first-principles calculations and demonstrated a significant improvement, reaching around 70%, through mechanical deformation [34].

7.6 MAGNETISM IN 2D METALLENES

Certainly, spin injection plays a pivotal and indispensable role in the research and application of spintronics in 2D metallenes. Introducing magnetism into 2D metallenes holds significant promise for the field of spintronics. Pristine graphene, phosphorene or many others, lacking d or f electrons, exhibits strong diamagnetism. However, researchers have been exploring ways to induce magnetic moments in graphene and other 2D metallenes through vacancy, defects, edge-state magnetize and dopant atoms. The emergence of atomically thin permanent magnets could revolutionize spin logic devices, potentially replacing conventional ferromagnetic (FM) thin films with highly scalable, novel spin injector or detector materials. This approach is crucial because conventional FM thin films with thicknesses under 200 nm often undergo substantial changes in their magnetic properties, such as magnetic anisotropy. Even in monolayer or few-layer magnetic thin films, the magnetic ordering temperature is significantly influenced by various topological effects like vacancies, substrate roughness, atomic intermixing, and incomplete crystal structures. Researchers have also encountered challenges with 3D ferromagnetic junctions involving 2D metallenes, including disorder in the electrodes, barrier and electrode-barrier interface issues, interdiffusion affecting FM spin polarization, and challenges related to its interface with the tunnel barrier. To overcome these challenges, utilizing a layer-by-layer assembly of 2D crystals, creating "van der Waals heterostructures," proves advantageous.

One common method to introduce local magnetic moments in 2D metallenes with unpaired electrons is by creating vacancies or adding adatoms (Figure 7.4). In hydrogenated graphene, a strong covalent bond is formed when hydrogen chemisorbs reversibly on graphene, effectively removing one p_z orbital from the π band and creating a sublattice imbalance. According to Lieb's theorem, a single hydrogen adatom can induce a quasi-localized state in graphene with a magnetic moment of 1 μ_B [39]. A recent theoretical study has predicted that zigzag graphene nanoflakes can display robust edge magnetism at room temperature, attributed to the presence of localized edge states. A study conducted by A. V. Krasheninnikov et al. employed density functional theory (DFT) to explore the impact of transition metal (TM) atoms on both pristine and defected graphene, including single and double vacancies [40]. Their findings indicated that TM atoms strongly bind to defected graphene, inducing magnetic properties in the material. In another research effort, Huizhen Zhang and team utilized a first principles approach to investigate the electronic and magnetic characteristics of graphene embedded with an iron (Fe) membrane [41]. Their study revealed that Fe atoms generated substantial magnetic moments, and these magnetic moments varied with increasing strain when the Fe membrane was

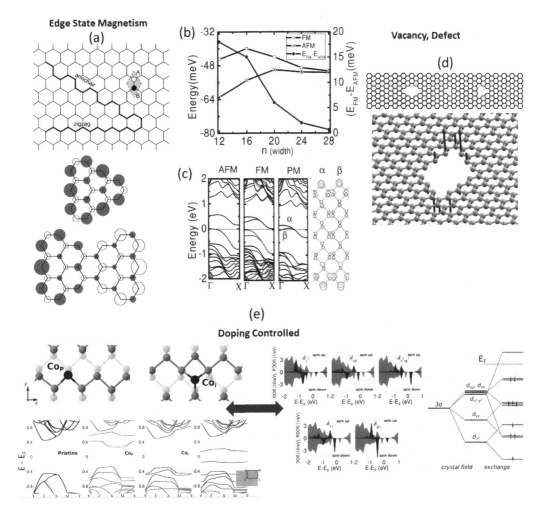

FIGURE 7.4 Different methods to create magnetism in non-magnetic 2D materials through: (a, b, c) Edge state magnetism [35, 36]. Copyright © 2010, IOP Publishing, Copyright © 2014, AIP Publishing LLC. (d) Vacancy, Defect [37], Copyright © 2008, American Physical Society. (e) Dopant atom [38], Copyright © 2015, American Physical Society.

embedded in graphene. However, achieving long-range ferromagnetic order through these methods remains a significant challenge.

Transition metal-doped MoS_2 nanostructures have been successfully prepared through experimental efforts led by J. Wang et al. using the hydrothermal method. Their research revealed that ferromagnetic Mn-doped MoS_2 exhibits a robust temperature-dependent coercivity [42]. The magnetic states observed appear to be influenced by factors such as the type, concentration, and distribution of the doping atoms, as well as the stacking order in the context of 2D metallenes [43].

Limited experimental work has been conducted in the field of magnetic engineering, with most results being theoretical and relying on assumptions. In practical applications, 2D metallenes are frequently supported by substrates, and these substrates can significantly impact the experimental outcomes. Another approach to induce magnetism in 2D crystals involves utilizing the proximity effect in conjunction with a magnetic substrate. When exfoliated graphene is transferred to a ferromagnetic insulator like yttrium iron garnet (YIG) or EuS, a noticeable spin precession can be observed due to a robust exchange field [44]. This phenomenon serves as compelling evidence for the presence of ferromagnetism in graphene. Moreover, magnetic proximity coupling, as demonstrated through first-principles calculations, can significantly enhance valley splitting in WS_2, offering valuable insights into achieving spin polarization [45].

7.7 INTRINSIC 2D MAGNETS

The presence of an atomic magnetic moment, originating from unpaired electrons, is a fundamental requirement for magnetism. In a lattice structure, the magnetic properties of materials are determined by the coupling of adjacent atomic spins through quantum mechanical exchange interactions. The nature of magnetism is influenced by the lattice dimensionality, crystal structure, and the spin dimensionality of the system. For a 2D layer of magnetic moments, there are common configurations of spin dimensionality, such as Ising, XY-model, or Heisenberg kind, as illustrated in Figure 7.5(a–c) [46].

Theoretically, the Mermin-Wagner theorem excludes the existence of continuous symmetry magnetic ordering in 2D dimensions. However, sustainable long-range magnetic ordering can be achieved through anisotropy, which has been experimentally observed in various 2D magnetic crystals. Earlier neutron scattering experiments in layered materials had reported 2D Ising behavior. Recently, the discovery of magnetism down to monolayers in materials like $Cr_2Ge_2Te_6$ (CGT), CrI_3, Fe_3GeTe_2 (FGT), and VSe_2 has revitalized research in MTJs due to their novel physical properties. The deep understanding of magnetic behavior of these materials is mainly explained through Heisenberg Hamiltonian which includes a magnetic anisotropies term:

$$H = \frac{1}{2}\sum_{i,j} J_{ij} S_i S_j + \sum_i A\left(S_i^z\right)^2 - \mu_B \sum_i B S_i^z \tag{7.3}$$

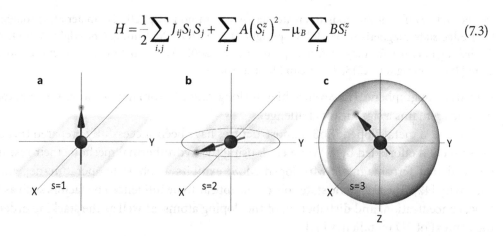

FIGURE 7.5 Spin dimensionality: (a) Ising configuration where spins point either up or down along the z-axis. (b) XY-model where spins are free to orient towards any point on the circle in the XY plane. (c) Heisenberg kind, where spins can orient to any point on the sphere [46]. Copyright © 2020 Author(s).

where S_i represents the spin operator at site i, J_{ij} signifies the exchange interaction between sites i and j, A represents the single-ion anisotropy, g is the Landé g-factor, μ_B denotes the Bohr magneton, and B represents the external magnetic field.

For instance, using CrI_3 as a barrier layer in graphite/CrI_3/graphite tunnel junctions has led to a significant TMR over thousands of percent at low temperatures. FGT, with its metallic behavior at the monolayer limit and a relatively high Curie temperature of 130 K, shows promise for electrode applications in MTJs. MTJs made of FGT/h-BN/FGT exhibited a magnetoresistance of 160% even at low temperatures. The performance of FGT-based MTJ devices, including the switching field and TMR ratio, can be controlled by gate bias. For perpendicular magnetization of FGT driven by spin-orbit torque, a large switching ratio of up to 60% was achieved in FGT/Pt systems. Additionally, researchers observed the antisymmetric magnetoresistance (MR) effect in van der Waals heterostructure FGT/graphite/FGT devices, highlighting the potential for diverse applications in spintronics.

7.8 SPIN FIELD-EFFECT TRANSISTORS

Spin field-effect transistors (FETs) were initially conceptualized by Datta and Das in 1990, promising advantages such as nonvolatile data storage, enhanced speed, and greater energy efficiency compared to traditional transistors. These three-terminal devices regulate the flow of spin-polarized current between source and drain terminals using the gate terminal. In this design, the ferromagnetic source acts as a spin polarizer, aligning the spins of injected carriers. The ferromagnetic drain operates as a spin filter, allowing only electrons with a specific polarization to pass through, while rejecting electrons with different polarizations. A semiconducting channel lies between the ferromagnetic source and drain. Above the channel, a metallic gate terminal, separated by an oxide layer, modulates the strength of the Rashba field through the application of gate voltage, inducing effective magnetic fields due to the Rashba effect. When carriers are injected into the channel, their spins precess about the Rashba field. The gate voltage alters the precessional frequency and controls the angle of spin precession. If this angle is an even multiple of π, the transistor conducts current; if it is an odd multiple of π, current flow is obstructed. The presence of the Dresselhaus interaction induces an effective magnetic field along the channel. However, this does not affect spin precession as the spins are also polarized along the channel. Spin-FETs can be turned OFF by applying a specific voltage at the gate terminal, generating a Rashba-induced magnetic field within the channel. This field changes the spin polarization of electrons moving through the channel, preventing their acceptance by the drain terminal. Consequently, electrons do not flow, rendering the device OFF.

7.9 CHALLENGES AND FUTURE DIRECTIONS WITH 2D MATERIALS IN SPINTRONIC APPLICATIONS

Spintronics holds immense promise, particularly with the advent of 2D crystals, ushering in fresh opportunities. Yet, successful integration into practical devices requires tackling various hurdles. These include merging with magnetic materials, managing spin-orbit coupling, enhancing spin injection and detection efficiency, addressing spin relaxation, ensuring material stability, uniformity, and scalability, refining bandgap engineering,

104 ■ 2D Metals

honing heterostructure engineering, ensuring temperature stability, and achieving standardization and reproducibility. Overcoming these challenges demands interdisciplinary collaboration spanning materials science, physics, chemistry, and engineering to fully harness the potential of 2D crystals in spintronics.

REFERENCES

1. Bondyopadhyay, P. K. "Moore's law governs the silicon revolution." *Proceedings of the IEEE* 86, no. 1 (1998): 78–81.
2. Thompson, S. E., M. Armstrong, C. Auth, M. Alavi, M. Buehler, R. Chau, S. Cea, et al. "A 90-nm logic technology featuring strained-silicon." *IEEE Transactions on Electron Devices* 51, no. 11 (2004): 1790–1797.
3. Chau, R., S. Datta, M. Doczy, B. Doyle, J. Kavalieros, and M. Metz. "High-/spl kappa//metal-gate stack and its MOSFET characteristics." *IEEE Electron Device Letters* 25, no. 6 (2004): 408–410.
4. Baibich, M. N., J. M. Broto, A. Fert, F. N. Van Dau, F. Petroff, P. Etienne, G. Creuzet, A. Friederich, and J. Chazelas. "Giant magnetoresistance of (001) Fe/(001) Cr magnetic superlattices." *Physical Review Letters* 61, no. 21 (1988): 2472.
5. Yoda, H. (2016). MRAM Fundamentals and Devices. In: Xu, Y., Awschalom, D., Nitta, J. (eds) *Handbook of Spintronics*. Springer, Dordrecht.
6. Miyazaki, T., and N. Tezuka. "Giant magnetic tunneling effect in $Fe/Al_2O_3/Fe$ junction." *Journal of Magnetism and Magnetic Materials* 139, no. 3 (1995): L231-L234.
7. Dieny, B., and M. Chshiev. "Perpendicular magnetic anisotropy at transition metal/oxide interfaces and applications." *Reviews of Modern Physics* 89, no. 2 (2017): 025008.
8. Tuček, J., P. Błoński, J. Ugolotti, A. K. Swain, T. Enoki, and R. Zbořil. "Emerging chemical strategies for imprinting magnetism in graphene and related 2D materials for spintronic and biomedical applications." *Chemical Society Reviews* 47, no. 11 (2018): 3899–3990.
9. Avsar, A., J. Y. Tan, M. Kurpas, M. Gmitra, K. Watanabe, T. Taniguchi, J. Fabian, and B. Özyilmaz. "Gate-tunable black phosphorus spin valve with nanosecond spin lifetimes." *Nature Physics* 13, no. 9 (2017): 888–893.
10. Iqbal, M. Z., N. A. Qureshi, and G. Hussain. "Recent advancements in 2D-materials interface based magnetic junctions for spintronics." *Journal of Magnetism and Magnetic Materials* 457 (2018): 110–125.
11. Jedema, F. J., A. T. Filip, and B. J. Van Wees. "Electrical spin injection and accumulation at room temperature in an all-metal mesoscopic spin valve." *Nature* 410, no. 6826 (2001): 345–348.
12. Johnson, M., and R. H. Silsbee. "Interfacial charge-spin coupling: Injection and detection of spin magnetization in metals." *Physical Review Letters* 55, no. 17 (1985): 1790.
13. Ji, Y., A. Hoffmann, J. S. Jiang, J. E. Pearson, and S. D. Bader. "Non-local spin injection in lateral spin valves." *Journal of Physics D: Applied Physics* 40, no. 5 (2007): 1280.
14. Tombros, N., C. Jozsa, M. Popinciuc, H. T. Jonkman, and B. J. Van Wees. "Electronic spin transport and spin precession in single graphene layers at room temperature." *Nature* 448, no. 7153 (2007): 571–574.
15. Han, W., K. M. McCreary, K. Pi, W. H. Wang, Yan Li, Hua Wen, J. R. Chen, and R. K. Kawakami. "Spin transport and relaxation in graphene." *Journal of Magnetism and Magnetic Materials* 324, no. 4 (2012): 369–381.
16. Han, W., R. K. Kawakami, M. Gmitra, and J. Fabian. "Graphene spintronics." *Nature Nanotechnology* 9, no. 10 (2014): 794–807.
17. Julliere, M. "Tunneling between ferromagnetic films." *Physics Letters* A 54, no. 3 (1975): 225–226.

18. Garcia, J. H., M. Vila, A. W. Cummings, and S. Roche. "Spin transport in graphene/transition metal dichalcogenide heterostructures." *Chemical Society Reviews* 47, no. 9 (2018): 3359–3379.
19. Wang, Z., D.-K. Ki, H. Chen, H. Berger, A. H. MacDonald, and A. F. Morpurgo. "Strong interface-induced spin–orbit interaction in graphene on WS2." *Nature Communications* 6, no. 1 (2015): 8339.
20. Kamalakar, M. V., C. Groenveld, A. Dankert, and S. P. Dash. "Long distance spin communication in chemical vapour deposited graphene." *Nature Communications* 6, no. 1 (2015): 6766.
21. Li, X.-B., P. Guo, T.-F. Cao, H. Liu, W.-M. Lau, and L.-M. Liu. "Structures, stabilities and electronic properties of defects in monolayer black phosphorus." *Scientific Reports* 5, no. 1 (2015): 10848.
22. Novoselov, K. S., A. K. Geim, S. V. Morozov, D.-eng Jiang, Y. Zhang, S. V. Dubonos, I. V. Grigorieva, and A. A. Firsov. "Electric field effect in atomically thin carbon films." *Science* 306, no. 5696 (2004): 666–669.
23. Maassen, J., W. Ji, and H. Guo. "Graphene spintronics: The role of ferromagnetic electrodes." *Nano Letters* 11, no. 1 (2011): 151–155.
24. Hill, E. W., A. K. Geim, K. Novoselov, F. Schedin, and P. Blake. "Graphene spin valve devices." *IEEE Transactions on Magnetics* 42, no. 10 (2006): 2694–2696.
25. Han, W., W. H. Wang, K. Pi, K. M. McCreary, W. Bao, Yan Li, F. Miao, C. N. Lau, and R. K. Kawakami. "Electron-hole asymmetry of spin injection and transport in single-layer graphene." *Physical Review Letters* 102, no. 13 (2009): 137205.
26. Popinciuc, M., C. Józsa, P. J. Zomer, N. Tombros, A. Veligura, H. T. Jonkman, and B. J. Van Wees. "Electronic spin transport in graphene field-effect transistors." *Physical Review B* 80, no. 21 (2009): 214427.
27. Józsa, C., M. Popinciuc, N. Tombros, H. T. Jonkman, and B. J. Van Wees. "Electronic spin drift in graphene field-effect transistors." *Physical Review Letters* 100, no. 23 (2008): 236603.
28. Maioli, P., T. Meunier, S. Gleyzes, A. Auffeves, G. Nogues, M. Brune, J. M. Raimond, and S. Haroche. "Nondestructive rydberg atom counting with mesoscopic fields in a cavity." *Physical Review Letters* 94, no. 11 (2005): 113601.
29. Wang, X., S. M. Tabakman, and H. Dai. "Atomic layer deposition of metal oxides on pristine and functionalized graphene." *Journal of the American Chemical Society* 130, no. 26 (2008): 8152–8153.
30. Lee, B., S.-Y. Park, H.-C. Kim, K. Cho, E. M. Vogel, M. J. Kim, R. M. Wallace, and J. Kim. "Conformal Al_2O_3 dielectric layer deposited by atomic layer deposition for graphene-based nanoelectronics." *Applied Physics Letters* 92, no. 20 (2008): 203102.
31. Wu, Q., L. Shen, Z. Bai, M. Zeng, M. Yang, Z. Huang, and Y. P. Feng. "Efficient spin injection into graphene through a tunnel barrier: Overcoming the spin-conductance mismatch." *Physical Review Applied* 2, no. 4 (2014): 044008.
32. Chen, M., Z. Yu, Y. Wang, Y. Xie, J. Wang, and H. Guo. "Nonequilibrium spin injection in monolayer black phosphorus." *Physical Chemistry Chemical Physics* 18, no. 3 (2016): 1601–1606.
33. Chen, H., B. Li, and J. Yang. "Proximity effect induced spin injection in phosphorene on magnetic insulator." *ACS Applied Materials & Interfaces* 9, no. 44 (2017): 38999–39010.
34. Chen, M., Z. Yu, Y. Xie, and Y. Wang. "Spin-polarized quantum transport properties through flexible phosphorene." *Applied Physics Letters* 109, no. 14 (2016): 142409.
35. Yazyev, O. V. "Emergence of magnetism in graphene materials and nanostructures." *Reports on Progress in Physics* 73, no. 5 (2010): 056501.
36. Zhu, Z., C. Li, W. Yu, D. Chang, Q. Sun, and Y. Jia. "Magnetism of zigzag edge phosphorene nanoribbons." *Applied Physics Letters* 105, no. 11 (2014): 113105.
37. Palacios, J. J., J. Fernández-Rossier, and L. Brey. "Vacancy-induced magnetism in graphene and graphene ribbons." *Physical Review B* 77, no. 19 (2008): 195428.

38. Seixas, L., A. Carvalho, and A.H. Castro Neto. "Atomically thin dilute magnetism in Co-doped phosphorene." *Physical Review B* 91, no. 15 (2015): 155138.
39. Šljivančanin, Ž., R. Balog, and L. Hornekær. "Magnetism in graphene induced by hydrogen adsorbates." *Chemical Physics Letters* 541 (2012): 70–74.
40. Krasheninnikov, A. V., P. O. Lehtinen, A. S. Foster, P. Pyykkö, and R. M. Nieminen. "Embedding transition-metal atoms in graphene: structure, bonding, and magnetism." *Physical Review Letters* 102, no. 12 (2009): 126807.
41. Zhang, H., J.-T. Sun, H. Yang, L. Li, H. Fu, S. Meng, and C. Gu. "Tunable magnetic moment and potential half-metal behavior of Fe-nanostructure-embedded graphene perforation." *Carbon* 107 (2016): 268–272.
42. Wang, J., F. Sun, S. Yang, Y. Li, C. Zhao, M. Xu, Y. Zhang, and H. Zeng. "Robust ferromagnetism in Mn-doped MoS2 nanostructures." *Applied Physics Letters* 109, no. 9 (2016): 092401.
43. Kumari, P., S. Majumder, S. Rani, A. K. Nair, K. Kumari, M. Venkata Kamalakar, and S. J. Ray. "High efficiency spin filtering in magnetic phosphorene." *Physical Chemistry Chemical Physics* 22, no. 10 (2020): 5893–5901.
44. Hallal, A., F. Ibrahim, H. Yang, S. Roche, and M. Chshiev. "Tailoring magnetic insulator proximity effects in graphene: first-principles calculations." *2D Materials* 4, no. 2 (2017): 025074.
45. Norden, T., C. Zhao, P. Zhang, R. Sabirianov, A. Petrou, and H. Zeng. "Giant valley splitting in monolayer WS$_2$ by magnetic proximity effect." *Nature Communications* 10, no. 1 (2019): 4163.
46. Dayen, J.-F., S. J. Ray, O. Karis, I. J. Vera-Marun, and M. Venkata Kamalakar. "Two-dimensional van der Waals spinterfaces and magnetic-interfaces." *Applied Physics Reviews* 7, no. 1 (2020): 011303.

CHAPTER 8

2D Metals in Optoelectronics

Swikruti Supriya and Ramakanta Naik

8.1 INTRODUCTION

The search for materials that can push the limits of performance and efficiency is constant in the ever-changing field of optoelectronics. Metals are an important class of materials with substantial scientific value and versatile practical applications; their properties and functionalities are closely correlated with their sizes. 2D metals are one type of material that is currently attracting the interest of researchers and professionals working in this area. Due to the extraordinary physical and chemical properties, as well as numerous potential applications, the family of monolayer or 2D material has accomplished a remarkable growth [1]. Various methods have been developed such as hydrothermal, sol-gel, chemical vapour deposition (CVD) via which the 2D metals are being prepared [2–4]. Several well-defined metallic nanosheets or nanoplates, also referred to as 2D metals and regarded as a significant class of 2D materials, have been synthesized so far [5]. Among the 2D materials, graphene, which is the best-known 2D material, caught the attention due to its various physical and chemical properties. Furthermore, its unique properties made these materials to be utilized in various optoelectronic applications. However, diverse 2D-layered materials have become of importance to researchers for various studies. These different materials are transition metal oxides, MXenes, transition metal dichalcogenides, metal halide perovskites, etc. [6]. The inherent layered designs of the materials that offer strong intralayer chemical interactions yet weak interlayer van der Waals forces, benefit the majority of these present-day fabrication [7]. To prepare these 2D materials, one of the most prominent techniques is to use the solution-based synthesis. Moreover, the bottom-up technique can be utilized to control the size and the morphology of the nanoparticles specifically for some noble metals such as Au, Ag, etc. [8, 9].

The first 2D material after graphene that holds all the attention of the researchers is the transition metal dichalcogenides (TMDs). This was because of the thin, flexible, and transparent nature of the TMDs similar to graphene. However, the most efficient part of

DOI: 10.1201/9781032645001-8

these materials is that they can be made into low-power and ultra-small transistors. TMDs may be deposited onto flexible substrates and withstand the stress and strain compliance of flexible support in addition to having a bandgap in the visible near-infrared range, high carrier mobility, and an on/off ratio similar to accessible silicon [10].

The metal oxides (MOs) in their ultra-thin 2D form has become the prominent material having excellent properties whose crystal structure ranges from cubic to triclinic symmetry. However other low-dimensional material's synthesis procedure highly depends upon the crystal growth mechanism. This makes the difference between the 2D MOs and the other nanostructured materials. These 2D materials can exhibit all the component elements to the surface when the surface-to-volume ratio is nearer to one [11]. Another variety of 2D metals can be metal-halide perovskites which have grabbed the attention due to their outstanding optoelectronic achievement. The power conversion efficiency (PCE) of the dye-sensitized solar cells which was first reported by A. Kojima et al. was found to be 24.8% for methylammonium lead halide perovskite (MAPbI$_3$) [12]. This PCE value is greater than the usual "dye-sensitized" solar cells, and other conventional thin film solar cells which (higher value of PCE) is due to the outstanding light absorption coefficient, low exciton binding energy, etc. [13, 14].

For applications in various electronics and optoelectronic devices another new class of transition metal carbides, carbides, and nitrides are being utilized which is termed as MXenes. Due to the fascinating physical, chemical, optical, and electrical properties, the morphology of these materials can be tuned and can be utilized in various applications, such as photodetectors, Schottky contacts, plasmonic materials, and so on [15]. To synthesize these materials, the most widely used methos is the top-down approach. This includes scrupulous acid etching and subsequent exfoliation [16]. Meanwhile, another material grasped the scrutiny of research with its explicit porosity and structural tuneability, which is known as metal-organic frameworks (MOFs). The 2D MOFs are generally synthesized via hydro/solvo thermal method. This usually gives polycrystalline result [17]. This chapter encloses various types of the 2D materials, their properties, and applications in the field of optoelectronic devices. Finally, it enfolds the application in different optoelectronic devices. The schematic of various 2D materials, properties, and their applications are depicted in the Figure 8.1.

8.2 2D METALS AND THEIR PROPERTIES

2D materials have garnered significant attention in the field of optoelectronics due to their unique properties and potential for various applications. Many 2D materials, such as graphene, TMDs, MOs, metal-halide perovskites, etc. exhibit semiconducting properties. These materials can be utilized in transistors, photodetectors, and light-emitting diodes. TMDs, like MoS$_2$ and WS$_2$, have a direct bandgap that allows for efficient light-matter interactions. The bandgap of 2D materials can be tuned by varying the number of layers, making them suitable for various wavelengths of light which have been discussed further thoroughly [18].

2D Metals in Optoelectronics ■ 109

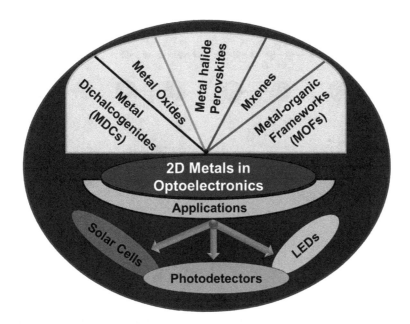

FIGURE 8.1 Schematic presentation of different kinds of 2D metals in the field of optoelectronics.

8.2.1 2D Transition Metal Dichalcogenides (TMDs)

The most considered 2D materials after graphene are the 2D TMDs. These are the atomically thin layers and are of the type MX_2, where M = Mo, W, Re; and X = S, Se, Te. The crystal structure of TMDs, specifically of MoS_2 are isotropic hexagonal phase (2H) and octahedral phase (1T) under ambient conditions. The lattice structure of MTe_2 and ReX_2 are entirely different as compared to MoS_2. These materials manifest a distorted structure which is relative to 1T phase. The deformed 1T phase's monolayer retains its X–M–X structure, with the upper and lower X atoms rotated by 180° to produce an M-centred octahedral structure. Therefore, the X atoms indicate zigzag structure and vary along the atomic position in the perpendicular direction. Two different phases 1T′ phase and Td phase appear due to minute variation in the displacement and stacking [19]. The four phases have been shown in Figure 8.2.

The foremost exciting concern to develop the 2D TMDs is the finite optical bandgap. The optical properties of graphene and the TMDs are different as the monolayer TMDs are generally semiconducting materials, thereby providing wide and comprehensive applications. The bandgap energy range of these materials varies from 1 eV to 2.5 eV. Moreover, when the material is minimized to monolayer from multilayer, the transition of the bandgap from indirect to direct bandgap occurs. The direct bandgap of the monolayer dichalcogenides is, for example, MoS_2 (1.8 eV), WS_2 (2.1 eV), $MoTe_2$ (1.1 eV), WSe_2 (1.7 eV), whereas the bulk materials show indirect bandgap having less energy [20]. The metallic as well as the semiconducting phases are owned by most of the MX_2 materials. However, 2H phase is found to be the most stable phase of the MX_2 materials and by electron beam irradiation or Li-interaction, the 1T phase can also be acquired. So, in case of MoS_2, the 1T phase is 10^7 times conductive than the 2H phase. TMDs have exceptionally strong light-material

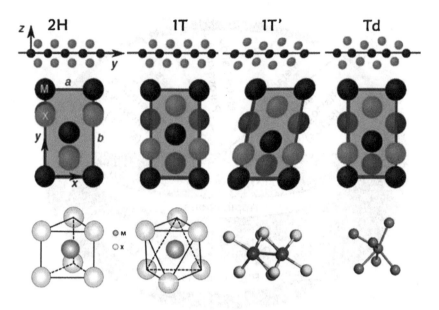

FIGURE 8.2 The four crystalline phases of 2D TMDs. Adapted with permission from reference [19]. Copyright (2017), Wiley.

interactions; at a certain resonance energy, monolayer TMD materials are capable of absorbing 20% of the incident light [21]. Additionally, the nonlinearities of the second-order and the even order are allowed by the TMDs, unlike graphene and even layer TMDs. In recent times, metal–organic–CVD (MOCVD), atomic layer deposition (ALD), pulsed laser deposition (PLD), etc. are being used to produce high-quality TMDs. To form the chemical reaction of the 2D materials, thermal energy in the form of heat substrate, or no thermal energy, like, photon energy or microwave process are utilized. This process mainly relies on the temperature, substrate, atomic glass flux, etc. [22].

8.2.2 2D Metal Oxides

2D MOs, which are composed of metal cations bonded to oxygen anions, offer several advantages for optoelectronic applications. The typical 2D metal oxides are known to be associated with a strong chemical bond in all directions, such as MoO_2, ZnO, GaO_2, etc. withhold from an uncomplicated synthesis of the 2D form. In this context, atoms within individual layers are connected through strong chemical bonds, while the stacking of these layers is facilitated by van der Waals forces. Generally, the 2D oxides possess the wide bandgap. This shows that a large detection range can be achieved by the photodetectors of 2D oxides and can perform in the UV region in a good way. These materials show very strong blue shift compared to the bulk 2D oxides, which has been depicted in Figure 8.3(a) for the material MoO_3 [23]. Similarly, Figure 8.3(b) depicts the absorption of Fe_3O_4 nanosheet, whose absorption edges are 8 cm and 25 cm detected at 463 cm^{-1} and 310 cm^{-1}, respectively with thickness. This corresponds to the light of wavelength 21.6 μm and 32.3 μm. The Tauc plot and the absorbance spectra of the ZnO nanosheet have been illustrated in the Figure 8.3(c), which shows that the cutoff point arises and a tough UV region appears at

2D Metals in Optoelectronics ■ 111

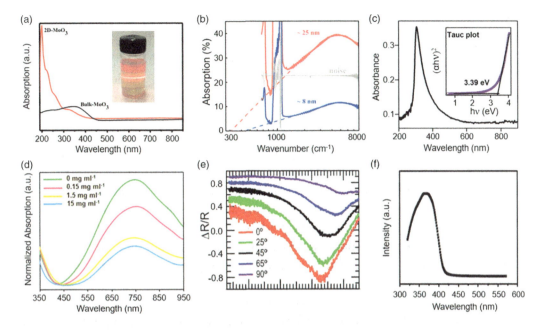

FIGURE 8.3 (a) Absorption spectra of 2D MoO₃ and bulk MoO₃. (b) Infrared absorption spectra of Fe₃O₄ nanosheet synthesized by CVD method. (c) The UV-visible-near-infrared spectra of zinc oxide nanosheets. Produced through liquid-metal exfoliation, with the inset showing a Tauc plot generated using absorption data. (d) Normalized absorption spectra with varied concentration. (e) ΔR/R of MoO₃ under different incident angles. (f) PL spectra of β-Ga₂O₃ nanosheets. Adapted with permission from reference [23]. Copyright (2022), Wiley.

450 nm. Thus, 2D oxides hold promise for the development of highly effective and precise UV photodetectors. Furthermore, M. M. Y. A. Alsaif et al. established that 2D MoO₃ could be potentially used in optical biosensing [24]. Furthermore, its absorption spectra with bovine serum albumin (BSA) have been depicted in Figure 8.3(d). When the concentration of BSA was increased, the 2D MoO₃ material's absorption peak observed to be reduced.

Other 2D metal oxides like β-TeO₂ and β-Ga2O₃, along with MoO₃, exhibit in-plane anisotropy. Figure 8.3 (e) depicts the optical reflectance spectroscopy of MoO₃ nanosheets under various linearly polarized incident light angles. When examining the linear reflectance contrast spectra (ΔR/R), the intensity varies as the incident light transitions from aligning parallel to the *b*-axis to aligning parallel to the *c*-axis. These materials also exhibit properties like fluorescence which can be utilized as the light-emitting devices [25].

These 2D MOs can be synthesized by various methods and their morphology such as 2D nanosheets can be tuned by using various synthesis techniques. However, dangling bonds are seen in formations with reduced dimensionality and no interlayer interactions. This results in improved surface characteristics but also surface instability. The conducting, nonconducting, and semiconducting characteristics of MOs are determined by their fundamental atomic structure [26]. The present-day technical needs call for more surface area due to miniaturization. As a result, the size was reduced to a few nanometres, giving rise to the nanostructured MOs [27].

112 ■ 2D Metals

8.2.3 2D Metal-Halide Perovskites

Depending upon the attachment of the metal-halide octahedra, halide perovskites can be of different dimensions, such as 3D, 2D, 1D, and also 0D. Metal-halide perovskites possess the crystal structure of the formula ABX_3. Here A represents a cation (generally an organic or inorganic compound), B constitutes a metal cation, and X acts for a halide anion (iodide, bromide, or chloride). The crystallographic structures can be tuned by selecting the relevant organic and inorganic components, such as metal-halide octahedra (MX_6). The MX_6 octahedra are attached in the ridged or layered sheets at the corner in case of the 2D structure. Consequently, 2D halide perovskites come in two different varieties. One type is made of 2D nanostructured perovskite crystals and nanostructures that are produced from 3D perovskite materials; these are more analogous to conventional 2D materials like graphene and TMDs. The intrinsic 2D-layered crystal structure is the other kind. Here, the former is given more consideration. However, because of the nonlayered features, specifically in case of the former one, it is very demanding to assemble prime quality 2D perovskites [28].

The exclusive optical properties of the metal-halide perovskites have captivated the attention predominantly. The perovskite family's diversity in chemical composition makes it possible to easily adjust a number of physical characteristics, such as bandgap, electrical conductivity, excitonic behaviour, etc. By minimizing the presence of defect sites within perovskite crystals, it becomes possible to achieve the intended optoelectronic characteristics, effectively diminishing the nonradiative recombination of excited carriers, which will raise the quantum yield of photoluminescence [29]. In particular, when utilized as a light absorber for solar cells, the shape of metal-halide perovskites can be deliberately controlled to modify the optoelectronic properties. The exciton recombination rate in relation to the exciton diffusion length in solar cell devices can have a major impact on the PCE. G. Li et al. proposed a highly effective cross-linking technique based on trimethylaluminum vapour to create $CsPbX_3$ nanocrystals [30]. Metal-halide perovskites offer distinct advantages for various optoelectronic purposes, including photodetectors, solar cells, and laser applications, owing to their high absorption coefficient, robust photoluminescence, and extended carrier diffusion length [31, 32].

Based on theoretical investigations, it is widely acknowledged that the A cation has a minimal direct influence on the properties of perovskites. Instead, the primary factor shaping the band structures of perovskites is typically attributed to the metal–halide–metal bond angles, which can be altered through the tilting of the BX_6 octahedra. In contrast to traditional semiconductors like CdS or MoS_2, hybrid perovskites display a pronounced in-plane crystal structure distortion, which contributes to the increase in bandgap observed in 2D sheets of hybrid perovskites [33]. When compared to conventional 2D covalent semiconductors, 2D perovskites exhibit significantly lower carrier mobility, typically in the range of 6 to 60 cm^2 V–1 s–1, representing a difference of at least one order of magnitude. Because of this, perovskite's carrier lifetime can reach several hundred nanoseconds or even microseconds, and both electron and hole carrier diffusion lengths can reach micrometre scales. This effect leads to the generation of high excitation densities and operates in a distinct manner compared to traditional 2D covalent semiconductors, resulting in the spontaneous creation of free electrons and holes following photo absorption [33, 34].

8.2.4 MXenes

The class of 2D materials such as transition metal carbides/nitrides are called as the MXenes. These new materials have been investigated and found to be synthesized in various processes, and utilized in wide areas. They have the typical chemical formula as $M_{n+1}X_nT_x$. Here, M refers to the early transition metals, X is carbon or nitrogen, T stands for surface terminated group, such as –OH, –Cl, –F, etc. The value of "n" is 1 to 4. Among all the synthesized MXenes, $Ti_3C_2T_x$ is the first one to be prepared [35]. Top-down method is usually utilized to prepare these materials, beginning from the MAX phase. In MAX phase, mainly Ga, In, Al, etc. are employed for A. In a standard MXene synthesis process, the A atoms within the metallic M–A bonds can be selectively removed by etching these bonds using hydrofluoric acid or fluoride salts in acidic solutions, including substances like "hydrochloric acid" and "lithium fluoride." This process results in the retention of covalent M–X bonds, giving rise to atomic-layer-thick MX layers. It's important to note that in certain etching procedures, both X and A atoms are extracted from the initial MAX phase. Concurrently with the removal of A atoms from MXene's active surfaces, surface functional groups T_x are attached, leaving behind hydrophilic electronegative surfaces. Compared to TMDs like MoS_2, the contact forces between MXene layers are two to six times stronger [36].

M atoms are arranged in a close-packed manner in the crystal structure of MXenes, whereas X atoms occupy some of the octahedral interstitial spaces. M atoms can consist of a combination of two elements or a single element. Regarding the latter, the two types of M components can be arranged in the lattice in an in-plane or out-of-plane manner, or they can form solid solutions. M atoms are hexagonally close-packed and stacked in M_2CT_x (n = 1), with an ABABAB ordering. One specific instance of this is when M atoms can have partial vacancies, resulting in $n + 1 = 1.33$. M atoms exhibit a face-centred cubic arrangement with an ABCABC order in $M_3C_2T_x$ and $M_4C_3T_x$ (for $n = 2$ and $n = 3$, respectively) [37]. The unique characteristics of the transition metal (M) are another aspect. The metal M's electronegativity is important in addition to surface groups. The bandgap enhances from Ti to Hf due to the reduction of electronegativity, as can be shown by comparing the band structures of M_2CO_2 MXenes. Here, M can be Zr, Hf, which are in the same column of the periodic table [38].

MXenes, a class of emerging 2D materials, have unique optical properties and a large degree of tunability; as a result, they can be used as materials for a wide range of optical applications. While a transmission valley was discovered around 750–800 nm and was thought to be attributed to the surface plasmon resonance (around 780 nm) and the inherent out-plane inter-band transitions (around 800 nm), MXene films are claimed to exhibit a broadband optical transmittance of 90%. In particular, it has been demonstrated through experimentation that $Ti_3C_2T_x$ has a better optical transmittance than both graphene and reduced graphene oxide – roughly 98% in the visible range [39]. The fascinating characteristic that makes MXenes good electrode candidates in optoelectronics is their

114 ■ 2D Metals

high transparency. The UV, visible, near-infrared, and ultraviolet spectra of $Ti_3C_2T_x$ films produced by coating with spray at varying thicknesses suggest that the material exhibits a broad absorption in the 700–800 nm range and a prominent absorption peak in the UV. The inter-band transition is responsible for the abrupt absorption peak observed in the ultraviolet region [40].

8.2.5 2D Metal-Organic Frameworks

Another family of gaze materials is referred to as MOFs, and they have a crystalline porous structure made up of metallic nodes and organic ligands. It has several promising characteristics. Changing the metal ions and ligands more specifically could enable MOFs to perform distinct functions. More crucially, 2D MOF nanosheets have a large number of easily accessible active sites on their surface, just like other 2D materials. This property may be relevant for applications including catalysis, electrochemistry, and sensing [41]. The two-dimensional topological frameworks are layered and primarily display single, triple, quadruple, and six-connected topological networks as well as mixed multiconnected or multi-node topologies. To create 2D MOF nanosheets, two approaches, such as the top-down and bottom-up approaches can be applied. In the former, bulk MOFs are delaminated, whereas in the latter, 2D MOF nanosheets are directly synthesized. Since sonication or shaking will be enough to break down the weak interlayer connection in MOF, the top-down approach is simpler. On the other hand, the bottom-up approach is recommended for producing highly yielded, well-dispersed MOF nanosheets. While reports of the creation of 2D MOF nanosheets or nanofilms on particular substrates exist. It is uncommon to directly synthesize 2D MOF nanosheets [42].

Due to their large surface area and abundant active sites, which is caused by their high aspect ratio, 2D MOFs are attractive for a variety of uses, most notably sensing. Moreover, its 2D morphology also makes the active areas on its surface easily accessible. Excellent mechanical strength and flexibility are two further appealing qualities of 2D MOFs, which are useful in industries like "gas separation" as well as "sensor production." This is made possible due to the nanosheet's shape and the coordination linkages within the structure. Young's modulus of 5 GPa was found by C. Hermosa et al. while they examined the mechanical characteristics of a two-dimensional Cu MOF (flakes and the layers ranging from 1 to 50) [43]. Moreover, it has been found that 2D MOFs have a significantly higher chemical stability than 3D MOFs. In sensor applications that might necessitate operation in aqueous environments with elevated pH levels, the chemical durability of these MOFs is of paramount importance. Z. Zhang et al. have currently developed extremely thin nanosheets (~3.97 nm in thickness) composed of an "indium-porphyrin" based MOF known as InTCPP that exhibited remarkable durability in aqueous solutions with pH ranges of 2–11. In comparison to 3D InTCPP, the nanosheets demonstrated increased photocatalytic activity toward hydrogen generation [44]. The freshly generated MOF nanosheets' surface energy is lowered by the addition of surfactants, including "cetyltrimethylammonium bromide" (CTAB) or "polyvinylpyrrolidone" (PVP), which inhibits the nanosheets from aggregating. The use of surface plasmon resonance (SPR)-based sensors show great promise for high-sensitivity analyte detection across a broad spectrum. SPR sensors have distinct benefits over electrochemical, colorimetric, or luminescent sensors, such as ease of use, label-free functioning, and real-time assay capabilities.

8.3 APPLICATIONS

8.3.1 Solar Cells

The most effective and widely used solar cells (SCs) to date are made of inorganic materials like silicon. On the other hand, the expense and difficulty of producing pure crystalline inorganic materials led to a significant investigation into substitute photovoltaic modules. On the other hand, organic and perovskite photovoltaics (PV) are attractive emerging technologies due to their affordability and the wide range of material choices available. Furthermore, the creation of semi-transparent flexible devices and compatibility with printing procedures are made possible by these developing SCs. However, there are other barriers that the market for organic solar cells (OSCs) and Perovskites solar cells (PeSCs) is actually facing. Compared to their inorganic equivalents, OSCs have poorer PCE and shorter lifespans. In perovskites, on the other hand, indicate stability difficulties while having PCEs that are comparable to crystalline silicon ($\approx$25%). In order to create organic and perovskite PV devices that are both extremely stable and efficient, researchers are working very hard [45].

With the advent of non-fullerene acceptors, positive outcomes for OSCs with PCE exceeding 16% have been obtained lately, increasing the appeal of commercializing OSCs. Through the modification of their chemical composition and basic research on the impact of the interfaces between charge transport layers and perovskite, more stable perovskite-based devices have been produced in parallel.

The active layer in both types of SCs are squeezed in between the two electrodes. Here, Figure 8.4(a) depicts the usual setup. The transparent conducting electrode (TCE), which permits light transmission, is placed on a substrate (flexible polymer or quartz) upon which the stack is placed. The active layer is usually evaporated on top of the metal electrode, which is usually aluminium, gold, or silver. Furthermore, in order to balance and optimize the charge transmission, interface layers are frequently required. The alignment of energy levels with the active layer determines which device components are selected [46]. In the absence of light and under illumination, a voltage is systematically varied from negative to positive values to evaluate the parameters employed for evaluating the performance of a supercapacitor (SC). The PCE can be derived from the current density–voltage (J–V) profile, which is the outcome of the measurement as depicted in Figure 8.4 (b).

8.3.2 LEDs

LEDs are semiconductor devices that, when a voltage is applied, emit light. Organic LED (OLED) devices boast numerous exceptional features, including their remarkable brightness, full-colour capabilities, lightweight design, and energy-efficient operation. The emitting material of an OLED is inserted between two electrodes in a vertical stack. To improve injecting charge into the organic emitting layer, injection layers are frequently used in between the electrodes and the emissive layer, which has been depicted in Figure

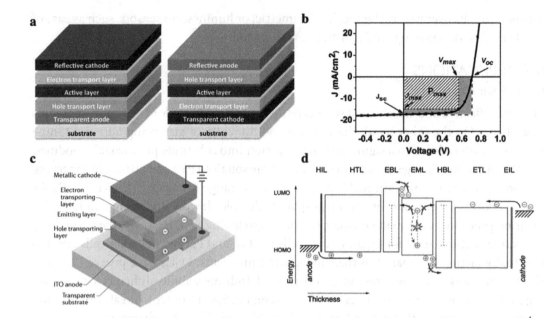

FIGURE 8.4 (a) Schematic depiction of conventional (left) and inverted (right) organic solar cells (OSCs), (b) J–V characteristics of a typical solar cell. (c) Illustration of typical OLEDs. (d) The transfer of charge towards an OLED, accompanied by supplementary transport, injection, and blocking layers. Adapted with permission from reference [46]. Copyright (2020), Wiley.

8.4(c). Precisely because of their exceptional optoelectronic characteristics, which include high absorption coefficients, high colour purity, high photoluminescence quantum yield (PLQY), along with substantial hole and electron mobilities, perovskites are currently becoming a desirable alternative to organic LEDs, which have already made their way into the display market [47]. In recent years, Perovskite LEDs (PeLEDs) have indeed rapidly reached external quantum efficiency (EQE) levels as substantial as 21.6%. PeLEDs and OLEDs share a similar device construction and fundamental mode of operation.

As Figure 8.4(d) illustrates, the three primary functions of an OLED are charge injection, charge transport, and recombination. The most important metric for assessing an LED's performance is its EQE. The ratio of photons released in the user's direction to the quantity of carriers injected is known as the EQE [48].

8.3.3 Photodetectors

2D oxides hold promise for applications in photodetectors, with several varieties of these materials having already been identified and reported for their distinctive photodetection properties. Depending on the 2D materials, Figure 8.5 (a) depicts the schematic presentation of the photodetectors. Meanwhile, based on 2D MgO, the current curves of the photodetector, that is time resolved at vacuum UV light is illustrated in Figure 8.5 (b). The photodetector can even detect a very faint light signal (0.85 pW) when exposed to 150 nm light. This is because excited carriers in 2D MgO have a high charge collecting efficiency. This work offers a fresh approach to the creation of vacuum ultraviolet photodetectors of

2D Metals in Optoelectronics ■ 117

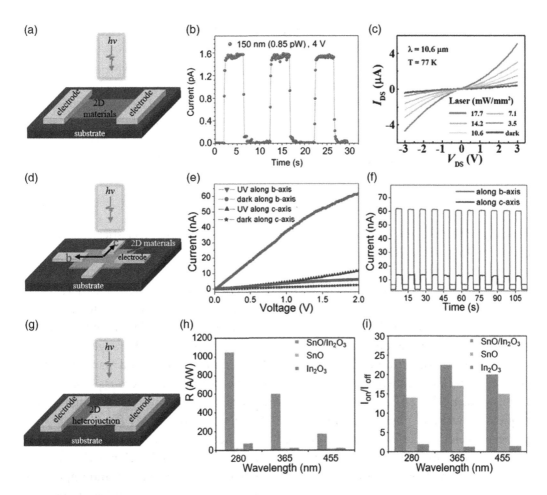

FIGURE 8.5 (a) Schematically presented photodetector. (b) Photodetector having I–T curve. (c) Current–voltage curve based on Fe$_3$O$_4$. (d) Detection with directional sensitivity photodetector. (e) I–V curves of α-MoO$_3$. (f) I–T curves of the α-MoO$_3$. (g) 2D Heterojunction-based photodetector. (h, i) Responsiveness and current ratio of the photodetectors utilizing SnO/In$_2$O$_3$, SnO, and In$_2$O$_3$. Adapted with permission from reference [23]. Copyright (2022), Wiley.

the future generation. Because of their great flexibility, 2D oxides show promise for use in flexible photodetectors as well. Certain 2D oxides also perform exceptionally well in infrared and visible light detection. Figure 8.5 (c) displays the current–voltage (I–V) curves of the photodetector at 77 K for various light power densities. An ideal output is achieved with a laser wavelength of 10.6 μm. B. P. Yalagala used cellulose paper to construct the V$_2$O$_5$ photodetectors [49]. The responsivity levels did not significantly change after 500 bending cycles. There are certain 2D materials that exhibit in-plane optical and electrical anisotropy due to their poor symmetric crystal structure. As research progresses, several 2D oxides exhibiting in-plane anisotropy like α-MoO$_3$ have been discovered. The anisotropic photodetector schematic diagram is displayed in Figure 8.5 (d).

The crystal orientation of the α-MoO$_3$ sample produced by Zhong et al. was ascertained through transmission electron microscopy (TEM) [50]. Electrode pairs were constructed along the α-MoO$_3$ b- and c-axes, respectively. As depicted in Figure 8.5 (e, f), the "dark current," "photocurrent," and "on/off ratio" along the b-axis were found to be 2.1, 4.3, and 1.8 times greater than those along the c-axis. Based on the findings, there is a lot of promise for using α-MoO$_3$ in anisotropic photoelectric detection. Figure 8.5 (g) displays the schematic diagram of a 2D heterojunction-based photodetector with two electrodes. The photodetectors employing this heterostructure are illustrated in Figure 8.5 (h, i). These devices exhibited substantially enhanced responsiveness and current ratios when subjected to illuminations at 280, 365, and 455 nm compared to photodetectors based solely on SnO or In$_2$O$_3$.

8.4. CONCLUSION AND FUTURE PERSPECTIVE

This chapter offers a comprehensive overview of the structure, characteristics, and diverse applications of various 2D materials, with a specific focus on TMDs, oxides, metal-halide perovskites, MXenes, and Metal-Organic Frameworks (MOFs). These materials have gained significant attention in recent years due to their unique properties and versatile applications across various scientific fields. Their unique structural characteristics and exceptional properties, such as high carrier mobility, tuneable bandgaps, and outstanding optoelectronic behaviour, make them highly appealing for a wide range of applications. Based on optoelectronics, 2D materials have a promising future and many appealing applications because of their current great performance features. Numerous possible uses, including solar cells, LEDs, photodetection, and more, have been proven. Owing to their unique chemical and physical characteristics, 2D metals have already found usage in a number of fields, including memory devices, biomedical engineering, and catalysis. Yet there are still issues to be resolved, such as gaining a deeper comprehension of the stability of 2D metals and alloys as their chemical compositions get more complex and improving control over the thickness, morphology, and shape of these materials so that their uses can grow in the future. In the future, research and applications involving 2D metals with customizable size, shape, and chemical composition will prevail over researchers who can produce these metals in large quantities using novel processing techniques and having numerous applications.

REFERENCES

1. T. Wang, M. Park, Q. Yu, J. Zhang, Y. Yang, Stability and synthesis of 2D metals and alloys: a review, *Mater. Today Adv.* 8 (2020) 100092.
2. X. Liu, X. Duan, P. Peng, W. Zheng, Hydrothermal synthesis of copper selenides with controllable phases and morphologies from an ionic liquid precursor, *Nanoscale* 3 (2011) 5090–5095.
3. A. Taffelli, G. Ligorio, L. Pancheri, A. Quaranta, R. Ceccato, A. Chiappini, M. V. Nardi, E. J. W. L. Kratochvil, S. Dire, Large area MoS$_2$ films fabricated via sol-gel used for photodetectors, *Opt. Mater.* 135 (2023) 113257.
4. W. Huang, L. Gan, H. Yang, N. Zhou, R. Wang, W. Wu, H. Li, Y. Ma, H. Zeng, T. Zhai, Controlled synthesis of ultrathin 2D β-In$_2$S$_3$ with broadband photoresponse by chemical vapor deposition, *Adv. Funct. Mater.* 27 (2017) 1702448.
5. A. V. Averchenko, I. A. Salimon, E. V. Zharkova, S. Lipovskikh, P. Somov, O. A. Abbas, P. G. Lagoudakis, S. Mailis, Laser-enabled localized synthesis of Mo$_{1-x}$W$_x$S$_2$ alloys with tunable composition, *Mater. Today Adv.* 17 (2023) 100351.

6. S. Gao, Y. Lin, X. Jiao, Y. Sun, Q. Luo, W. Zhang, D. Li, J. Yang, Y. Xie, Partially oxidized atomic cobalt layers for carbon dioxide electroreduction to liquid fuel, *Nature* 529 (2016) 68–71.
7. N. Zhang, M. Q. Yang, S. Liu, Y. Sun, Y. J. Xu, Waltzing with the versatile platform of graphene to synthesize composite photocatalysts, *Chem. Rev.* 115 (2015) 10307–10377.
8. L. Wang, Y. Zhu, J. Q. Wang, F. Liu, J. Huang, X. Meng, J. M. Basset, Y. Han, F. S. Xiao, Two-dimensional gold nanostructures with high activity for selective oxidation of carbon–hydrogen bonds, *Nat. Commun.* 6 (2015) 6957.
9. R. Xu, T. Xie, Y. Zhao, Y. Li, Single-crystal metal nanoplatelets: cobalt, nickel, copper, and silver, *Cryst. Growth Des.* 7 (2007) 1904–1911.
10. R. Cheng, S. Jiang, Y. Chen, Y. Liu, N. Weiss, H. C. Cheng, H. Wu, Y. Huang, X. Duan, Few-layer molybdenum disulfide transistors and circuits for high-speed flexible electronics, *Nat. Commun.* 5 (2014) 5143.
11. P. Kumbhakar, C. C. Gowda, P. L. Mahapatra, M. Mukherjee, K. D. Malviya, M. Chaker, A. Chandra, B. Lahiri, P. M. Ajayan, D. Jariwala, A. Singh, C. S. Tiwary, Emerging 2D metal oxides and their applications, *Mater. Today* 45 (2021) 142–168.
12. A. Kojima, K. Teshima, Y. Shirai, T. Miyasaka, Organometal halide perovskites as visible-light sensitizers for photovoltaic cells, *J. Am. Chem. Soc.* 131 (2009) 6050–6051.
13. M. Koc, W. Soltanpoor, G. Bektas, H. J. Bolink, S. Yerci, Guideline for optical optimization of planar perovskite solar cells, *Adv. Opt. Mater.* 7 (2019) 1900944.
14. K. Galkowski, A. Mitioglu, A. Miyata, P. Plochocka, O. Portugall, G. E. Eperon, J. T. W. Wang, T. Stergiopoulos, S. D. Stranks, H. J. Snaith, R. J. Nicholas, Determination of the exciton binding energy and effective masses for methylammonium and formamidinium lead tri-halide perovskite semiconductors, *Energy Environ. Sci.* 9 (2016) 962–970.
15. H. Xu, A. Ren, J. Wu, Z. Wang, Recent advances in 2D MXenes for photodetection, *Adv. Funct. Mater.* 30 (2020) 2000907.
16. R. Li, L. Zhang, L. Shi, P. Wang, MXene Ti_3C_2: An effective 2D light-to-heat conversion material, *ACS Nano* 11 (2017) 3752–3759.
17. J. Liu, Y. Chen, X. Feng, R. Dong, Conductive 2D conjugated metal–organic framework thin films: synthesis and functions for (Opto-)electronics, *Small Struct.* 3 (2022) 2100210.
18. T. Tan, X. Jiang, C. Wang, B. Yao, H. Zhang, 2D material optoelectronics for information functional device applications: status and challenges, *Adv. Sci.* 7 (2020) 2000058.
19. C. Gong, Y. Zhang, W. Chen, J. Chu, T. Lei, J. Pu, L. Dai, C. Wu, Y. Cheng, T. Zhai, L. Li, J. Xiong, Electronic and optoelectronic applications based on 2D novel anisotropic transition metal dichalcogenides, *Adv. Sci.* 4 (2017) 1700231.
20. C. Ataca, H. Sahin, S. Ciraci, Stable, single-layer MX_2 transition-metal oxides and dichalcogenides in a honeycomb-like structure, *J. Phys. Chem. C* 116 (2012) 8983–8999.
21. K. F. Mak, J. Shan, Photonics and optoelectronics of 2D semiconductor transition metal dichalcogenides, *Nat. Photon.* 10 (2016) 216–226.
22. M. I. Serna, S. H. Yoo, S. Moreno, Y. Xi, J. P. Oviedo, H. Choi, H. N. Alshareef, M. J. Kim, M. M. Jolandan, M. A. Q. Lopez, Large-area deposition of MoS_2 by pulsed laser deposition with in situ thickness control, *ACS Nano* 10 (2016) 6054–6061.
23. X. Hu, K. Liu, Y. Cai, S. Q. Zang, T. Zhai. 2D oxides for electronics and optoelectronics, *Small Sci.* 2 (2022) 2200008.
24. M. M. Y. A. Alsaif, K. Latham, M. R. Field, D. D. Yao, N. V. Medehkar, G. A. Beane, R. B. Kaner, S. P. Russo, J. Z. Ou, K. K. Zadeh, Tunable plasmon resonances in two-dimensional molybdenum oxide nanoflakes, *Adv. Mater.* 26 (2014) 3931–3937.
25. D. Andres-Penares, M. Brotons-Gisbert, C. Bonato, J. F. Sánchez-Royo, B. D. Gerardot, Optical and dielectric properties of MoO_3 nanosheets for van der Waals heterostructures, *Appl. Phys. Lett.* 119 (2021) 223104.

26. H. S. Choi, J. W. Shin, E. K. Hong, I. Hwang, W. J. Cho, Hybrid-type complementary inverters using semiconducting single walled carbon nanotube networks and In-Ga-Zn-O nanofibers, *Appl. Phys. Lett.* 113 (2018) 243103.
27. J. Zhu, X. Liu, M. L. Geier, J. J. McMorrow, D. Jariwala, M. E. Beck, W. Huang, T. J. Marks, M. C. Hersam, Layer-by-layer assembled 2D montmorillonite dielectrics for solution-processed electronics, *Adv. Mater.* 28 (2015) 63–68.
28. C. Huo, B. Cai, Z. Yuan, B. Ma, H. Zeng, Two-dimensional metal halide perovskites: theory, synthesis, and optoelectronics, *Small Methods* 1 (2017) 201600018.
29. V. G. Pedro, E. J. Juarez-Perez, W. S. Arsyad, E. M. Barea, F. F. Santiago, I. M. Sero, J. Bisquert, General working principles of $CH_3NH_3PbX_3$ perovskite solar cells, *Nano Lett.* 14 (2014) 888–893.
30. G. Li, F. W. R. Rivarola, N. J. L. K. Davis, S. Bai, T. C. Jellicoe, F. de la Peña, S. Hou, C. Ducati, F. Gao, R. H. Friend, N. C. Greenham, Z. K. Tan, Highly efficient perovskite nanocrystal light-emitting diodes enabled by a universal crosslinking method, *Adv. Mater.* 28 (2016) 3528–3534.
31. J. Liu, Y. Xue, Z. Wang, Z. Q. Xu, C. Zheng, B. Weber, J. Song, Y. Wang, Y. Lu, Y. Zhang, Q. Bao, Two-dimensional $CH_3NH_3PbI_3$ perovskite: synthesis and optoelectronic application, *ACS Nano* 10 (2016) 3536–3542.
32. J. Wang, N. Wang, Y. Jin, J. Si, Z. K. Tan, H. Du, L. Cheng, X. Dai, S. Bai, H. He, Interfacial control toward efficient and low-voltage perovskite light-emitting diodes, *Adv. Mater.* 27 (2015) 2311–2316.
33. Y. Fang, H. Wei, Q. Dong, J. Huang, Quantification of re-absorption and re-emission processes to determine photon recycling efficiency in perovskite single crystals, *Nat. Commun.* 8 (2017) 14417.
34. J. Peng, Y. Chen, K. Zheng, T. Pullerits, Z. Liang, Insights into charge carrier dynamics in organo-metal halide perovskites: from neat films to solar cells, *Chem. Soc. Rev.* 46 (2017) 5714–5729.
35. M. Naguib, M. Kurtoglu, V. Presser, J. Lu, J. Niu, M. Heon, L. Hultman, Y. Gogotsi, M. W. Barsoum, Two-dimensional nanocrystals produced by exfoliation of Ti_3AlC_2, *Adv. Mater.* 23 (2011) 4248–4253.
36. T. Zhang, C. E. Shuck, K. Shevchuk, M. Anayee, Y. Gogotsi, Synthesis of three families of titanium carbonitride MXenes, *J. Am. Chem. Soc.* 145 (2023) 22374–22383.
37. H. Kim, Z. Wang, H. N. Alshareef, MXetronics: electronic and photonic applications of MXenes, *Nano Energy* 60 (2019) 179.
38. K. Hantanasirisakul, Y. Gogotsi, Electronic and optical properties of 2D transition metal carbides and nitrides (MXenes), *Adv.Mater.* 30 (2018) 1804779.
39. K. Hantanasirisakul, M. Q. Zhao, P. Urbankowski, J. Halim, B. Anasori, S. Kota, C. E. Ren, M. W. Barsoum, Y. Gogotsi, Fabrication of $Ti_3C_2T_x$ MXene transparent thin films with tunable optoelectronic properties, *Adv. Electron. Mater.* 2 (2016) 1600050.
40. M. Ebrahimi, C. T. Mei, Optoelectronic properties of $Ti_3C_2T_x$ MXene transparent conductive electrodes: microwave synthesis of parent MAX phase, *Ceram. Int.* 46 (2020) 28114–28119.
41. P. Deria, J. E. Mondloch, O. Karagiaridi, W. Bury, J. T. Hupp, O. K. Farha, Beyond post-synthesis modification: evolution of metal–organic frameworks via building block replacement, *Chem. Soc. Rev.* 43 (2014) 5896.
42. A. Gallego, C. Hermosa, O. Castillo, I. Berlanga, C. J. Gómez-García, E. Mateo-Martí, J. I. Martínez, F. Flores, C. Gómez-Navarro, J. Gómez-Herrero, S. Delgado, F. Zamora, Solvent-induced delamination of a multifunctional two-dimensional coordination polymer, *Adv. Mater.* 25 (2013) 2141–2146.
43. C. Hermosa, B. R. Horrocks, J. I. Martínez, F. Liscio, J. Gómez-Herrero, F. Zamora, Mechanical and optical properties of ultralarge flakes of a metal-organic framework with molecular thickness, *Chem. Sci.* 6 (2015) 2553–2558.

44. Z. Zhang, Y. Wang, B. Niu, B. Liu, J. Li, W. Duan, Ultra-stable two-dimensional metal-organic frameworks for photocatalytic H_2 production, *Nanoscale* 14 (2022) 7146–7150.

45. R. Lin, K. Xiao, Z. Qin, Q. Han, C. Zhang, M. Wei, M. I. Saidaminov, Y. Gao, J. Xu, M. Xiao, A. Li, J. Zhu, E. H. Sargent, H. Tan, Monolithic all-perovskite tandem solar cells with 24.8% efficiency exploiting comproportionation to suppress Sn(II) oxidation in precursor ink, *Nat. Energy* 4 (2019) 864.

46. A. G. Ricciardulli, P. W. M. Blom, Solution-processable 2D materials applied in light-emitting diodes and solar cells, *Adv. Mater. Technol.* 5 (2020) 1900972.

47. W. Xu, Q. Hu, S. Bai, C. Bao, Y. Miao, Z. Yuan, T. Borzda, A. J. Barker, E. Tyukalova, Z. Hu, M. Kawecki, H. Wang, Z. Yan, X. Liu, X. Shi, K. Uvdal, M. Fahlman, W. Zhang, M. Duchamp, J.-M. Liu, A. Petrozza, J. Wang, L.-M. Liu, W. Huang, F. Gao, Rational molecular passivation for high-performance perovskite light-emitting diodes, *Nat. Photon.* 13 (2019) 418.

48. T. Leijtens, S. D. Stranks, G. E. Eperon, R. Lindblad, E. M. J. Johansson, I. J. McPherson, H. Rensmo, J. M. Ball, M. M. Lee, H. J. Snaith, Electronic properties of meso-superstructured and planar organometal halide perovskite films: charge trapping, photodoping, and carrier mobility, *ACS Nano* 8 (2014) 7147–7155.

49. B. P. Yalagala, P. Sahatiya, C. S. R. Kolli, S. Khandelwal, V. Mattela, S. Badhulika, V_2O_5 nanosheets for flexible memristors and broadband photodetectors, *ACS Appl. Nano Mater.* 2 (2019) 937–947.

50. M. Zhong, K. Zhou, Z. Wei, Y. Li, T. Li, H. Dong, L. Jiang, J. Li, W. Hu, Highly anisotropic solar-blind UV photodetector based on large-size two-dimensional α-MoO_3 atomic crystals, *2D Mater.* 5 (2018) 35033.

CHAPTER 9

2D Metals as Photocatalysts

Subhadeep Biswas and Anjali Pal

9.1 INTRODUCTION

Photocatalysis is a novel concept that utilizes the potential of the catalytic property of different materials under exposure of light or photon to show different important chemical phenomena. It can solve energy and environmental problems associated with fossil fuels to a considerable extent. Interestingly, this technique serves as a useful tool for various environmental reactions. For example, from an environmental perspective water splitting is one of the most important chemical reactions of recent time. In brief, it includes two chemical reactions represented as: $4H^+ + 4e^- \rightarrow 2H_2$ (reduction reaction) and $2H_2O \rightarrow O_2 + 4H^+ + 4e^-$ (oxidation reaction). It involves the formation of hydrogen which is designated as clean fuel as combustion can take place without the occurrence of CO_2. Moreover, hydrogen can react with methane to form natural gas which also has high demand as fuel for vehicles. A promising photocatalyst with high efficiency is required for this purpose. Kokabi and Touski described that for water splitting reaction, an ideal photocatalyst should fulfill some basic criteria such as structural stability, sufficient bandgap, and proper band-edge position [1]. It should have distinguished features such as efficient utilization of solar light, suitable band structure for separation of charge carrier, etc. Different interesting photocatalysts have been developed for water-splitting reaction worldwide so far. Zhao et al. developed MoS_2/CdS nanocomposite material as a photocatalyst for hydrogen production through water-splitting reaction [2]. The CdS covering prevented the catalyst from photo corrosion. Under visible and simulated solar light irradiation, the maximum evolution rate of H_2 occurred (63.71 $mmolg^{-1}h^{-1}$ and 71.24 $mmolg^{-1}h^{-1}$). Ultrathin porous g-C_3N_4 nanosheet catalyst was explored by Chen et al. [3] for water-splitting reaction. Pure H_2 and O_2 can be obtained from the catalytic action under visible light exposure. The as-prepared g-C_3N_4 catalyst was found to be stable even after 100 h of water-splitting reaction.

Another important application of photocatalysis is the green synthesis of ammonia through N_2 fixation process. Ammonia (NH_3) is one of the important constituents of

122   DOI: 10.1201/9781032645001-9

fertilizers. Besides, it is also considered an efficient renewable fuel due to its high energy density. The traditional method of ammonia production depends on Haber-Bosch process which was invented in the early 1900s. The conventional process involves the reaction of nitrogen and hydrogen under high temperature and pressure with high consumption of fossil fuels. As a result, a considerable amount of CO_2 is released into the atmosphere every year which may lead to the intensification of the greenhouse effect. Photocatalytic N_2 reduction is an alternative green technique for ammonia synthesis. Ammonia can be produced naturally by means of multiple proton and electron migration pathways. The process can be efficiently achieved under the influence of a photocatalyst. Ran et al. [4] reported Penta-B_2C, a semiconductor-based composite and an active photocatalyst for ammonia synthesis. Excellent light absorption capacity along with rich B atoms in the activation site of the catalyst facilitated the reaction.

CO_2 reduction is another hot topic of research at present. As a result of high consumption of fossil fuels globally, the level of CO_2 in the atmosphere has risen considerably. CO_2 reduction is undoubtedly a sustainable technique to curb the rise in pollution. Moreover, it also helps in carbon capture and utilization. Several novel photocatalysts have been reported in the literature for this purpose. He et al. [5] reported the successful application of ZnO/g-C_3N_4 composite catalyst for photoreduction of CO_2. From the experimental results, it was clear that the ZnO loading on the g-C_3N_4 surface did not change the light absorption activity. However, the combination of two (ZnO and g-C_3N_4) resulted in the formation of a heterojunction which prevented the electron-hole recombination. As a result, catalytic efficiency was enhanced. In comparison to g-C_3N_4, the composite catalyst showed 4.9 times higher efficiency.

Various new-age novel materials emerged as promising photocatalyst materials. Among them, graphene and its various analogs have occupied a significant position. Xie et al. [6] categorized graphene analogs into three classes viz., (1) metal-free analogs, (2) transitional metal compound carbides, and (3) metallenes. Metallenes are further subdivided into monometallenes, bimetallenes, and trimetallenes. Metal bonds are characterized by their robustness and isotropic nature. This leads to the formation of a closely packed crystal structure having ultrathin morphology of atom thickness which is also thermodynamically unfavorable. However, despite challenges in synthesis, metallenes are able to procure a significant place among different newest catalyst materials. Like other important materials, metallenes or two-dimensional (2-D) metal analogs also deserve special mention as a photocatalyst material. Their distinguished physicochemical and electronic properties with a very thin surface (<5 nm) have evolved them as one of the wonder catalyst materials. Different research groups of recent times reported about the excellent catalytic property of metallenes. The term "metallene" was first suggested by Duan et al. [7]. The authors synthesized poly(vinylpyrrolidone)-supported monolayered rhodium nanosheets through a facile solvothermal technique. Thus, the newly prepared 2D structure was called metallene. These ultrathin nanostructures opened a new dimension in the field of catalysis. Different research and review articles of the latest times described the efficiency of metallenes in catalytic activities. Anne and Choi in their recent review article, highlighted the recent progress that took place in the field of application of metallenes as active catalyst for hydrogen evolution reaction [8]. Cao et al. [9] in their review article mentioned that the presence of under-coordinated metal atoms in metallenes has made them one of the most promising catalyst materials. Additionally, their ultrathin nature as well as high surface-to-volume ratio helps them with high exposure to coordinatively unsaturated atoms which in turn results in high catalytic activity. The present chapter describes the recent advances that has taken place in the field of application of metallenes as active photocatalyst. It describes the basic properties of metallene, some synthesis procedures that have been reported, and applicability in the field of catalysis. Recent photocatalytic applications are highlighted in detail. Interesting characterization techniques often explored are also discussed. Lastly, challenging issues and bright perspectives are presented before ending the chapter.

9.2 CATALYTIC PROPERTIES OF METALLENE

Metallenes belong to one of the newest members of the 2D materials family. Their tunable electronic structure, modifiable surface state along with high carrier mobility have made them a promising candidate for the catalytic application of the new generation. Noble metals such as Pd, Ir, and Rh constituting metallenes are often used in electrocatalysis for energy conversion purposes.

2D metallene structures possess some extraordinary features derived from structural anisotropy, rich surface chemistry, and efficient mass diffusion capacity [10]. Large surface area, free dangling bonds and plentiful unsaturated surface atoms have made metallenes as one of the most promising catalyst materials. Reported studies unveiled that bandgap structure engineering can be applied to the metallene composite catalysts for controlling the structure of the catalyst. Peng et al. [11] showed that in GaX/As heterogenous catalyst, various interfacial properties can be altered by means of variation of interlayer distance or by application of an external electric field. The authors mentioned that the interlayer distance can be altered by giving pressure through scanning tunneling microscopy tip or vacuum thermal annealing at different time intervals.

Different 2D metals possess unique physical properties as reported by researchers worldwide. In recent times, antimonene has been isolated by means of mechanical and liquid phase exfoliation. Some of the interesting features of antimonene include possession of high carrier mobility, thermal conductivity, and strain-induced direct–indirect bandgap transitions. Moreover, Barrio et al. [12] highlighted that besides single-layered antimonene, there exists multilayered antimonene which represents nontrivial topological character, whose full exploration as a catalyst material is yet to be done. Singh and Ahuja [13] also described the excellent catalytic property of β-antimonene monolayer catalyst. Its high specific surface activity, interesting electronic characteristics, and moderate bandgap have made it a promising catalyst material for various purposes. Moreover, the authors demonstrated that Bi doping of the single layer β-Sb can develop a highly efficient photocatalyst for oxygen reduction reaction (ORR). Further, detailed experimental investigations revealed that doping by single atom replacement has led to the formation of a photocatalyst which is superior in activity in comparison to the reference catalysts such as IrO_2 (110) and Pt (111). Kokabi and Touski [1] investigated the electronic and photocatalytic properties of antimonene. One of the attractive features of the antimonene nanosheet is that it can exhibit catalytic performance throughout a wide pH range. The authors studied the electronic and catalytic properties of different-sized H passivated antimonene nanosheets. All the nanosheets considered for investigation were stable in nature. However, with the increase in size, stability was enhanced. One of the important criteria for water-splitting photocatalyst is that adiabatic electron affinity and adiabatic ionization energy of the photocatalyst are required to be less than the reduction energy and higher than the oxidation energy of water splitting. It has been found experimentally that the as-synthesized metallene-based catalyst satisfied the above criteria in the wide pH range of 0–14.

Like antimonene, arsenene is also explored by researchers as an active photocatalyst. Due to its suitable bandgap of 1.6–2.4 eV, it is a useful semiconductor material in the field of nano-photonics [14]. It has an adjustable electronic structure as well as high carrier mobilities. Moreover, it can be used as a suitable material for composite formation with g-C_3N_4 in developing heterostructure type-II semiconductor-based photocatalyst. But Chai et al. [15] mentioned that high recombination of photogenic electron and hole pair often hinders its catalytic activity. Hence, composite formation is often recommended. Therefore, the authors prepared a composite photocatalyst made of SnC and arsenene for water splitting. It showed improved catalytic efficiency, reduced electron-hole recombination and broad absorption spectrum. Due to the composite formation, a type-II heterostructure is formed, in which photogenerated electrons are transferred from SnC monolayer to arsenene sheet. As a result, an electric field is built up in the interface region which is helpful for preventing electron-hole recombination. Zhu et al. [16] prepared arsenene/HfS_2 Z-scheme photocatalyst for water-splitting reaction. The as-prepared catalyst has intrinsic type-II staggered band alignment with a bandgap of 0.79 eV. The band structure is shown in Figure 9.1.

Lu et al. [17] in their recent review article described interesting properties of metallene-based materials. Among the various distinguishable properties, the authors mentioned that it favors easy functionalization with ligands. Especially, they mentioned a study where it was observed that 2D gold nanostructures showed higher affinity towards C–H oxidation in comparison to the gold nanoparticles [18]. Due to these advantages, metallenes are considered as excellent catalyst materials for organocatalysis, electrocatalysis, and photocatalysis.

Wang et al. [19] mentioned justifiably, that vacancy defects and crystal phase are two essential criteria affecting the surface electronic properties of the catalyst material. However, it is a daunting task for engineers and scientists to manufacture novel catalysts possessing both qualities. However, the authors were successful in fabricating such a composite by incorporating metallene. Apart from these, one of the most interesting properties of the 2D materials is to offer a platform where van der Waals heterostructures can be created where two chemically different types of materials are stacked together by means of van der Waals force. These stacked materials are different from 3D semiconductor materials as each layer can function as the bulk material and interface, and therefore charge displacement within each layer is reduced. As a result, these types of structures find wide applications in photocatalysis, optoelectronic, and other important devices.

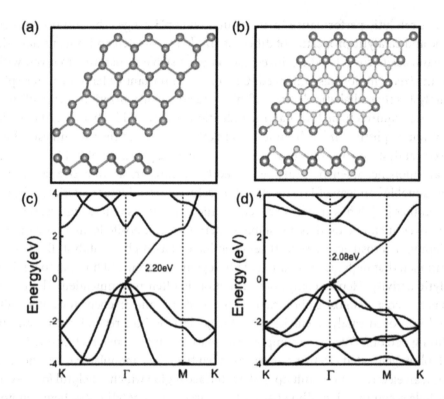

FIGURE 9.1 Top view and side view of (a) arsenene and (b) HfS$_2$ monolayer. Calculated band structures of arsenene (c) and HfS$_2$ monolayer (d), obtained from HSE06 functional. The green, silver, and yellow balls represent As, Hf, S atoms, respectively. Adapted with permission [16], Copyright (2022), Elsevier.

9.3 EMERGENCE OF DIFFERENT METALLENE-BASED CATALYSTS AND THEIR SYNTHESIS PROCEDURE

Different researchers utilized metallene-based catalysts to perform several chemical reactions. Mostly, the reported studies are based on electrocatalytic reactions [20–29]. Zhang et al. [30] developed PdCu metallene electrocatalyst for oxygen reduction and formic acid oxidation purposes. The 2D Cu-doped Pd metallene catalyst was produced by one-pot solvothermal method. The as-synthesized catalyst exhibited superior half-wave potential and mass activity in alkaline solution in comparison to the commercially available Pt/C catalyst. Wang et al. [31] reported the successful application of PdH$_x$ metallenes possessing vacancies for electrocatalytic reduction of N$_2$. Excellent catalytic efficiency was achieved due to the synergism between the vacancy engineering and lattice hydrogen atoms. The metallene catalyst showed 14.97 times higher efficiency in comparison to that of the Pd nanosheet catalyst without vacancy. It was synthesized via a facile wet-chemical method using Pd(acac)$_2$ and W(CO)$_6$ as precursors. During the reaction, CO decomposed from W(CO)$_6$ and acted as a structure guiding agent behind the formation of ultrathin metallene while etching with acetic acid resulted in the production of porous and vacant structure in the catalyst matrix.

PdMo alloy-based bimetallene electrocatalyst was utilized by Luo et al. for ORR [32]. The bimetallene catalyst showed good catalytic performance in alkaline medium which is also a challenging issue for the Pt catalyst. Moreover, the PdMo catalyst provides high electrochemically active surface area as well as high atomic utilization. From the density functional theory analysis, it was clear that alloying effect, strain due to the curved geometry, and quantum size effect owing to the thinness of the bimetallene sheet tuned the electronic configuration of the system for optimized oxygen binding. Wang et al. [33] utilized polyaniline functionalized porous Pd metallene composite for oxygen reduction reaction.

Prabhu and Lee [34] in their recent review article highlighted the unique properties of metallenes which have made them a promising catalyst material for electrocatalysis. Lu et al. [17] in their recent review article summarized various techniques for preparing metallene-based composites. The authors classified the synthesis procedure into top-down and bottom-up approaches. Among top-down approaches, some important methods such as mechanical cleavage, ultrasonic exfoliation, electrochemical exfoliation, plasma-assisted processes etc. have been mentioned. On the other hand, in the category of bottom-up techniques, the authors stated about molecular beam epitaxial method, vapor deposition technique, wet-chemical methods, etc.

Mao et al. [35] in their recent work, prepared amorphous Rh metallene sulfide as an electrocatalyst for cathodic reduction of nitrobenzene to aniline. Simultaneously, in the anodic reaction, the sulfide ion got oxidized. In another work, Prabhu et al. [36] explored Rh metallene incorporated oxygen-bridged single atomic tungsten as the active electrocatalyst for hydrogen evolution reaction. The authors deployed a facile solvothermal method having control over the growth kinetics for the production of the metallene-based catalyst. Extreme high mass activities, high turnover frequencies and excellent stability, and negligible deactivation are some of the notable characteristics of the as-prepared catalyst.

Wang et al. [19] explored Pd_4S metallene catalyst for hydrogen evolution reaction as well as for sulfion oxidation reaction. The composite was prepared by means of hydrothermal sulfuration. Apart from possessing excellent stability and catalytic activity, the as-prepared catalyst was able to reach a current density of 100 $mAcm^{-2}$ corresponding to the potential of 0.776 V which facilitated simultaneous hydrogen production as well as sulfur ion degradation.

Xie et al. [37] in their work developed a reconstruction strategy of Pd metallene at atomic scale to generate a series of nonmetallic atoms intercalated Pd metallene which is represented as M-Pdene. Here M stands for H, N, C. On the other hand, the authors also prepared S-doped metallene with an amorphous structure. The preparation strategy of M-Pdene catalysts is shown in Figure 9.2. Firstly, Pdene was prepared by heating $Pd(acac)_2$ and $W(CO)_6$ in acetic acid and N, N-dimethylformamide at 50°C for 1 hour inside a capped vial. The CO decomposed in the reaction served as the capping agent for the anisotropic growth of 2D ultrathin metallene structures. After intercalation of nonmetallic atoms, Pdene was converted to M-Pdene with lattice expansion and smooth surface. However, due to the intercalation of relatively larger S atom, the face-centered cubic structure of Pd was broken and amorphization in its structure was induced. Among all these metallene-based

128 ■ 2D Metals

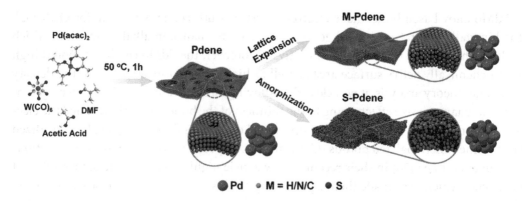

FIGURE 9.2 Schematic illustration for the synthesis of Pdene followed by lattice expansion and amorphization, together with the change of surface porosity and thickness. Adapted with permission [37], Copyright (2022), ACS.

catalysts, N-doped metallene exhibited the highest catalytic efficiency towards methanol oxidation reaction.

Huang et al. [38] prepared Fe-doped ultrathin Pd metallene composite for electrocatalytic reduction reaction of oxygen. In comparison to the Pd metallene, Fe-doped structure resulted in more catalytic efficiency. The authors mentioned that due to the presence of stable Fe dopant, smaller overpotential is delivered during hydrogenation of O* to OH*. Zhao et al. [39] demonstrated ultrathin Pd metallene catalyst as an excellent mimic enzymes. Zhang et al. [40] utilized stanene quantum dots as the active electrocatalyst for CO_2 reduction purpose.

9.4 METALLENE AS ONE OF THE MOST MODERN PHOTOCATALYST MATERIAL

Metallenes have been described by researchers as one of the newest emerging catalyst materials, but the number of reported studies on its photocatalytic activity is very less in comparison to other older materials. However, some reported studies are available in the literature revealing its photocatalytic efficiency. Zhang et al. [41] explored g-C_3N_4/antimonene photocatalyst for CO_2 reduction purpose. The composite was prepared by ultrasonic exfoliation technique. Schematic for the preparation of metallene-based photocatalyst has been shown in Figure 9.3.

In one of the recent studies, Zhang et al. [42] used layered metallene as an efficient electron mediator for the preparation of C_3N_4/bismuthene/BiOCl Z-scheme photocatalyst. The as-synthesized photocatalyst showed excellent performance towards photoreduction of CO_2. Bismuthene acted as the bridge helping in processing superior charge conductibility, exposing abundant metal-semiconductor contact sites and reduced charge diffusion distance. Thus, as a result CO_2 reduction was enhanced greatly.

Zhao et al. [43] explored 2D Sb photocatalyst for N_2 reduction purpose. Due to the prevalence of van der Waals force, 2D Sb nanosheets can be easily produced by exfoliating in a suitable solvent. Various important surface properties such as lateral size, thickness, surface oxygen content can be optimized using cascade centrifugation protocol.

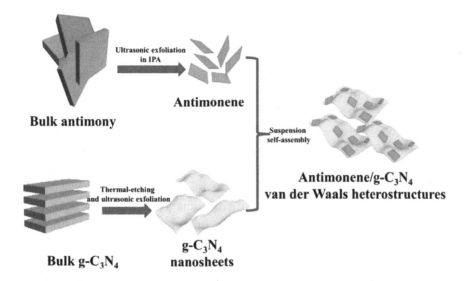

FIGURE 9.3 Schematic diagram of preparation process of Sb/g-C$_3$N$_4$ van der Waals heterostructures. Adapted with permission [41], Copyright (2022), Elsevier.

Two-dimensional nanosheets possessed Sb and oxygen vacancies resulting in excellent NH$_3$ production. There was no requirement for the cocatalysts and the yield obtained was nearly eight times in comparison to the bulk Sb catalyst. All the photoreduction experiments were conducted in a water medium under the exposure of visible light (300 Xe lamp). The experiments were performed at 25°C for 1 hour and the amount of NH$_3$ generated was estimated using the indophenol blue method. It is quite interesting to observe that bulk Sb and Sb$_2$O$_3$ were unable to reduce N$_2$ under irradiation. On the other hand, Sb nanosheets performed well in photoreduction and the efficiency was enhanced with the increase in centrifugation speed. It may be due to the surface oxygen and edge defects. Li et al. [44] applied Fe/Mo bimetallene-covered Bi$_2$Mo$_{0.3}$W$_{0.7}$O$_6$ (BMWO) photocatalyst for the same purpose. The BMWO material was prepared following a facile hydrothermal method. Further, Fe-BMWO, Mo-BMWO, Fe/Mo-BMWO composites were developed using one-pot wet-chemical approach. Experimental findings showed that in comparison to the unmodified BMWO nanocrystals, the catalytic activity of Fe/Mo bimetallene-covered catalyst was about 4.8 times.

Singh and Ahuja [13] explored 2D Bi-doped β-Sb monolayer photocatalyst for the water-splitting purpose. The photocatalyst thus synthesized was found energetically, thermally, and dynamically stable. It possesses a high excitonic binding energy of 0.604 eV which helps to confine both electrons and holes.

Arsenene/HfS$_2$ composite was utilized by Zhu et al. [16] as an efficient metallene-based photocatalyst for water-splitting purpose. From the Gibb's free energy diagram, it was evident that the solar energy can directly take part in water-splitting activity on the catalyst surface. Peng et al. [11] explored chalcogenides/arsenene van der Waals structure heterogeneous photocatalyst for water-splitting reaction. The heterogenous catalyst thus prepared was abbreviated as GaX/As. It was found experimentally that in comparison to the pristine catalyst, the heterostructure was able to undergo indirect–direct bandgap transition by variation of interlayer distances. Besides, the arsenene composite-based catalyst showed high carrier mobility of nearly 2,000 $cm^2V^{-1}s^{-1}$, transport anisotropy and conducive environment for migration and prevention of electron-hole recombination. Moreover, the heterogeneous catalyst revealed extra visible light absorption capacity beyond the absorption capacity of monolayer GaX catalyst. Li et al. [45] prepared a novel heterogeneous catalyst made from monolayered arsenene and Group-III monochalcogenide (GaX, X = S, Se) and applied the same for photocatalytic water-splitting reaction. The photocatalyst thus prepared was found to be thermodynamically stable containing an intrinsic type-II band structure. From the charge density calculation, it was found that As layer was positively charged while the GaX part was bearing negative charge. A potential drop was found across the surface which resulted in the formation of built-in electric field which inhibited the recombination of photogenerated electrons and holes. In one of the studies, Li et al. [46] designed arsenene/$Ca(OH)_2$ van der Waals heterostructures containing different types of vertical-stacking configuration for photocatalytic water-splitting purpose. Among those, it was found that β stacking configuration proved to present the most stable structure and its bandgap and its band-edge position can be tuned by biaxial strain.

In recent times, Qian et al. [47] explored ultrathin Pd metallene/g-C_3N_4 composite as cocatalyst for photocatalytic hydrogen production. The Pd-ene/g-C_3N_4 photocatalyst was prepared by hydrothermal and deposition method. Acetlyacetonate palladium has been used as the Pd source and the resultant composite thus formed represent a 2D/2D structure. The composite catalyst was prepared according to the previously reported study [32]. Briefly, the preparation procedure started with the mixing of 0.1 g $Pd(acac)_2$, 0.04 g of $Mo(CO)_6$, 0.24 g ascorbic acid, and 40 mL oleylamine in a 100 mL sample bottle, stirring at 350 rpm for 10 min followed by sonicating for 1 hour. After transferring the whole solution to a 120 mL reactor, it was heated at 80°C for 12 hours. The obtained product was washed with cyclohexane after centrifugation to eradicate excess oleylamine, and the final product produced after lyophilizing was called Pd-ene. With 2 wt% of Pd on the surface of the catalyst, the composite exhibited highest catalytic activity of hydrogen evolution at the rate of 31,292 μmol/h/g which is more than 98 times in comparison to pure g-C_3N_4 catalyst. The exceptional 2D/2D structure of the composite facilitated the separation and transfer of photogenerated carriers which ultimately enhanced the hydrogen production. Moreover, the authors also concluded that the 2D nature of the Pd-ene catalyst made it more exposed to the reaction.

9.5 DIFFERENT CHARACTERIZATION TECHNIQUES

Often different characterization procedures are performed in order to get more insight regarding the catalytic activity of the metallenes. Optical absorption coefficient is frequently used as an important factor for quantifying the catalytic property of a photocatalyst. From, the optical measurement it was deduced that forming composite between arsenene and HfS_2, visible light utilization capacity was enhanced in comparison to both arsenene and HfS_2 monolayer [16]. Hence, it is successful in constructing the heterocatalyst for better utilization of the visible light. Wu et al. [14] also studied the optical behavior of the arsenene/g-C_3N_4 photocatalyst towards water-splitting reaction. It is evident from Figure 9.4 that there are many absorption peaks in the visible region, indicating high absorption capacity towards visible light. Moreover, it can be seen that in comparison to the monolayered g-C_3N_4, the absorption coefficient of the composite heterostructure was found to be greatly enhancing. The reflection and refraction coefficient of the material is shown in Figure 9.4 (c) and 4 (d). It is seen that the highest reflection coefficient of 5 eV is obtained for the material in the UV region indicating strong absorption towards UV light. However, in case of refraction it is seen that the arsenene/g-C_3N_4 composite has the highest refraction in the infra-red region and it gets lowered in the visible region.

In the present age of computation and digitalization, utilization of density functional theory (DFT) is often done by researchers. Zhao et al. found superior catalytic reducing ability of 2D Sb nanosheets regarding N_2 fixation [43]. In order to gain further insight into the catalytic phenomenon, the authors further performed DFT analysis. Likewise, previous reports, binding free energy of the intermediate state *NNH was compared on various sites of Sb and Sb_2O_3. Results indicate that the defects existing in the basal plane and edges were found favorable for N_2 reduction. Dang et al. [48] also utilized DFT calculation for a better understanding of intrinsic bonding characteristics and catalytic activity of iridium metallene oxide for oxygen evolution.

Morphological characteristics of the catalysts are often confirmed from the microscopic analysis (usually scanning electron microscopy (SEM) and transmission electron microscopy (TEM)). Prabhu et al. [36] reported that from SEM and low-resolution TEM image,

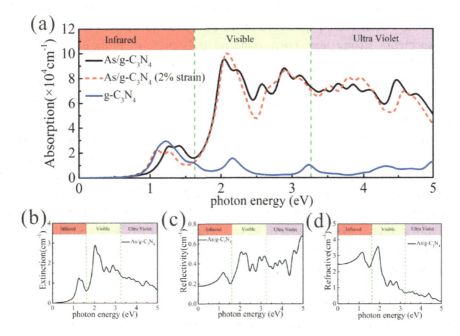

FIGURE 9.4 (a) Absorption coefficient of the arsenene/g-C_3N_4 heterostructure with or without 2% strain and monolayer g-C_3N_4. (b) Extinction coefficient. (c) Reflectivity coefficient. (d) Refractive coefficient of the arsenene/g-C_3N_4 heterostructure. Adapted with permission [14], Copyright (2021), Elsevier.

132 ■ 2D Metals

it is clear that Rh-O-W metallene composite represents a 2D nanostructure. Based on the thickness of the deposition on the copper grid, it was estimated that the average thickness of the entwined sheets was approximately 3.7 nm. However, from atomic force microscopic (AFM) analysis, it was clear that the as-prepared metallene composite catalyst was ultra-thin in nature. Mao et al. [35] concluded from the SEM and TEM images that a-RhS$_{2-x}$ possess a graphene-like morphology and ultrathin thickness. AFM study confirmed that the thickness of the nanosheet was approximately 1.25 nm. Moreover, in the HRTEM images obtained, it was found that there were no visible lattice fringes, which implied the presence of amorphous regions. Further, the selected area electron diffraction (SAED) pattern showed poor halo rings which is also an indication of the amorphous portions in the nanosheet.

Zhang et al. [41] performed in-situ FTIR analysis of Sb/g-C$_3$N$_4$ composite, which is commonly performed to monitor the intermediate products formed during the photoreduction of CO$_2$. As per the DFT analysis, it has been found that the composite has a strong affinity towards the intermediate *COOH. Peaks at 1319 cm^{-1}, 1337 cm^{-1}, 1518 cm^{-1}, 1544 cm^{-1} in the FTIR spectrum correspond to the vibration of *COOH. These peaks were visible for the composite containing 30% by weight of Sb. However, the intensities were enhanced with the increase in duration of the illumination. Hence, the result obtained from the in-situ FTIR analysis showed good corroboration with the DFT study.

Often X-ray photoelectron spectroscopic (XPS) analysis is performed for obtaining information regarding the chemical composition of the photocatalyst. Qian et al. [47] carried out the XPS analysis of Pd-ene, g-C$_3$N$_4$, and Pd-ene/g-C$_3$N$_4$ photocatalyst. Survey spectrum along with deconvoluted spectrum of the photocatalysts is shown in Figure 9.5. In the survey spectrum, strong peaks corresponding to the presence of the elements C, N, and O are observed. However, a faint peak at 340 eV is visible due to the low content of Pd in the matrix. From C1s spectrum of the photocatalyst Pd-ene/g-C$_3$N$_4$ four peaks are visible at 284.6 eV, 286.3 eV, 287.8 eV, and 288.8 eV due to the presence of graphitic C species. On the other hand, from N1s spectrum (Figure 9.5c), four new peaks at 398.3 eV, 399.5 eV, 400.8 eV, and 404.1 eV are observed which correspond to the sp^2 N in C–N=C. Furthermore, the authors also confirmed the chemical states of the elements in the catalyst Pd-ene/g-C$_3$N$_4$. The results indicate the presence of both metallic and oxidized Pd in the photocatalyst matrix. Prominent peaks at 339.8 eV and 334.5 eV for 3d$_{5/2}$ and 3d$_{3/2}$ reveal the existence of Pd^0 species, while the occurrence of feeble peaks at 342 and 336.6 eV denotes the oxidized Pd state. From a comparative analysis of 3d spectrum of Pd metallene and Pd-ene/g-C$_3$N$_4$, the latter shows a red shift in comparison to the former. It indicates that after forming the composite, new surface bonding states are formed.

9.6 EXISTING RESEARCH GAPS, LIMITATIONS

Undoubtedly, metallene is one of the most innovative scientific developments of the new age. It serves as an excellent catalyst material for different purposes and possesses several unique features. Several distinguishing features such as anisotropy, atomic under coordination, intrinsic strain influence, etc. have made them a wonder material of the modern times. There is a huge scope for forming new metallene-based composites for photocatalysis

2D Metals as Photocatalysts ■ 133

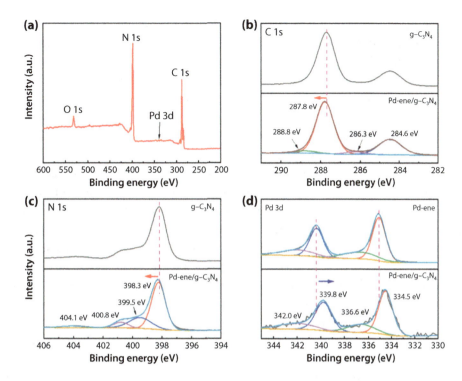

FIGURE 9.5 (a) XPS survey spectrum of the Pd-ene/g-C_3N_4. (b, c) High-resolution C 1s and N 1s XPS spectra of g-C_3N_4 and Pd-ene/g-C_3N_4 samples, severally. (d) High-resolution Pd 3d spectra of Pd-ene and Pd-ene/g-C_3N_4 samples. Adapted with permission [47], Copyright (2023), Elsevier.

by means of varying elemental compositions, surface ligand moieties, etc. They possess the capability of being applicable in various important photocatalytic reactions. Despite having several unique qualities, the term "metallene" represents a field which is still in its infancy. There are various drawbacks behind its feasibility for application at a large scale as identified by the researchers. Some of the important aspects are presented in this section.

Gao et al. [49] highlighted some of the critical challenges that need to be dealt with regarding the large-scale application of metallene catalysts. Till now, metallene-based catalysts have proved to be robust for reactions like photocatalytic water splitting, CO_2 reduction, N_2 fixation, etc. However, there are other important reactions such as methanol, formic acid oxidation, methane conversion etc. for which these catalysts have not been explored to a considerable extent. They also mentioned the fact that the yield, conversion efficiency, and selectivity of these metallene-based catalysts are still a major issue for practical purposes. Further, it is also justified that more in-situ characterization techniques, theoretical calculations, and incorporation of the machine learning process in the catalytic phenomenon will make the system more realistic.

Xie et al. in their latest review article highlighted different existing challenges and future perspectives regarding the application of metallene [6]. Among the several issues mentioned one major importance is that although successful synthesis of metallene having different thickness, composition, lattice constant, and porosity is possible, it is very difficult to have precise control over their formation at an atomic scale. Secondly, in spite of the upcoming advanced characterization techniques, the detailed mechanism of formation of 2D metals needs to be fully cleared to the scientific community using experimental tools and theoretical simulations. Production of metallenes in large quantities for practical purposes is also a challenging issue. It has been observed that in the bottom-up synthesis procedure, by increasing the volume of the reaction mixture, maintenance of other environmental parameters strictly for the anisotropic growth, is often troublesome. Long-term stability of the metallenes is also another alarming issue that needs to be thoroughly addressed prior to the large-scale application.

Prabhu and Lee in their recent article described several promising characteristics of metallenes to serve as new-age catalyst material [34]. However, for the authors, one of the most significant points is that, most of the metallene-based structures are formed of noble metals. But this inclusion has made the composition costly and not economical from a practical point of view. Although less expensive transition metals have been used recently in some studies for supporting the catalyst material, it has also failed to reduce the ultimate cost. Therefore, the authors suggested implementing newer techniques such as topotactic reduction of layered transition metal oxides or hydroxides for synthesizing metallenes made up of earth-abundant metals.

REFERENCES

1. A. Kokabi and S. B. Touski, Electronic and photocatalytic properties of antimonene nanosheets. *Phys. E Low Dimens. Syst. Nanostruct.*, 124 (2020) 114336.
2. H. Zhao, H. Fu, X. Yang, S. Xiong, D. Han and X. An, MoS_2/CdS rod-like nanocomposites as high-performance visible light photocatalyst for water splitting photocatalytic hydrogen production. *Int. J. Hyd. Energy*, 47 (2022) 8247–8260.
3. X. Chen, R. Shi, Q. Chen, Z. Zhang, W. Jiang, Y. Zhu and T. Zhang, Three-dimensional porous g-C_3N_4 for highly efficient photocatalytic overall water splitting. *Nano Energy*, 59 (2019) 644–650.
4. W. Ran, C. He, W. Chen, C. Zhao and J. Huo, Rich B active centers in Penta-B_2C as high-performance catalyst for nitrogen reduction. *Chin. Chem. Lett.*, 32 (2021) 3821–3824.
5. Y. He, Y. Wang, L. Zhang, B. Tang and M. Fan, High efficiency conversion of CO_2 to fuel over ZnO/g-C_3N_4 photocatalyst. *Appl. Catal. B: Environ.*, 168–169 (2015) 1–8.
6. M. Xie, S. Tang, B. Zhang and G. Yu, Metallene-related materials for electrocatalysis and energy conversion. *Mater. Horiz.*, 10 (2023) 407–431.
7. H. Duan, N. Yan, R. Yu, C. Chang, G. Zhou, H. Hu, H. Rong, Z. Niu, J. Mao, H. Asakura, T. Tanaka, P. J. Dyson, J. Li and Y. Li, Ultrathin rhodium nanosheets. *Nat. Commun.*, 5 (2014) 3093.
8. B. R. Anne and S. Choi, Metallene-based catalysts toward hydrogen evolution reaction. *Curr. Opin. Electrochem.*, 39 (2023) 101303.
9. C. Cao, Q. Xu and Q. Zhu, Ultrathin two-dimensional metallenes for heterogeneous catalysis. *Chem. Catal.*, 2 (2022) 693–723.
10. B. Jiang, Y. Guo, F. Sun, S. Wang, Y. Kang, X. Xu, J. Zhao, J. You, M. Eguchi, Y. Yamauchi and H. Li, Nanoarchitectonics of metallane materials for electrocatalysis. *ACS Nano*, 17 (2023) 13017–13043.
11. Q. Peng, Z. Guo, B. Sa, J. Zhou and Z. Sun, New gallium chalcogenides/arsenene van der Waals heterostructures promising for photocatalytic water splitting. *Int. J. Hyd. Energy*, 43 (2018) 15995–16004.

12. J. Barrio, C. Gibaja, J. Tzadikov, M. Shalom and F. Zamora, 2D/2D graphitic carbon nitride/antimonene heterostructure: Structural characterization and application in photocatalysis. *Adv. Sust. Syst.* 3 (2018) 1800138.
13. D. Singh and R. Ahuja, Theoretical prediction of a Bi-doped β-antimonene monolayer as a highly efficient photocatalyst for oxygen reduction and overall water splitting. *ACS Appl. Mater Interfaces*, 13 (2021) 56254–56264.
14. Y. B. Wu, C. He, F. S. Han and W. X. Zhang, Construction of an arsenene/g-C_3N_4 hybrid heterostructure towards enhancing photocatalytic activity of overall water splitting: a first-principles study. *J. Solid State Chem.*, 299 (2021) 122138.
15. J. Chai, Z. Wang and Y. Li, Investigation of the mechanism of overall water splitting in UV-visible and infrared regions with SnC/arsenene vdw heterostructures in different configurations. *Phys. Chem. Chem. Phys.*, 22 (2019) 1045–1052.
16. X. T. Zhu, Y. Xu, Y. Cao, D. F. Zou and W. Sheng, Direct Z-scheme arsenene/HfS_2 van der Waals heterojunction for overall photocatalytic water splitting: first principles study. *Appl. Surf. Sci.*, 574 (2022) 151650.
17. C. Lu, R. Li, Z. Miao, F. Wang and Z. Zha, Emerging metallenes: synthesis strategies, biological effects and biomedical applications. *Chem. Soc. Rev.*, 52 (2023) 2833–2865.
18. L. Wang, Y. Zhu, J. Wang, F. Liu, J. Huang, X. Meng, J. Basset, Y. Han and F. Xiao, Two-dimensional gold nanostructures with high activity for selective oxidation of carbon-hydrogen bonds. *Nat. Commun.*, 6 (2015) 6957.
19. W. Wang, Q. Mao, S. Jiang, K. Deng, H. Yu, Z. Wang, Y. Xu, L. Wang and H. Wang, Heterophase Pd_4S metallene nanoribbons with Pd-rich vacancies for sulfur ion degradation-assisted hydrogen production. *Appl. Catal. B: Environ.*, 340 (2024) 123194.
20. F. Lin, F. Lv, Q. Zhang, H. Luo, K. Wang, J. Zhou, W. Zhang, W. Zhang, D. Wang, L. Gu and S. Guo, Local coordination regulation through tuning atomic scale cavities of Pd metallene toward efficient oxygen reduction electrocatalysis. *Adv. Mat.*, 34 (27) (2022) 2202084
21. K. Chen, Z. Ma, X. Li, J. Kang, D. Ma and K. Chu, Single-atom Bi alloyed Pd metallene for nitrate electroreduction to ammonia. *Adv. Func. Mat.*, 33 (2023) 2209890.
22. H. Yu, T. Zhou, Z. Wang, Y. Xu, X. Li, L. Wang and H. Wang, Defect-rich porous palladium metallene for enhanced alkaline oxygen reduction electrocatalysis. *Ang. Chem.*, 133(21) (2021) 12027–12031.
23. X. Li, P. Shen, Y. Luo, Y. Li, Y. Guo, H. Zhang and K. Chu, PdFe single-atom alloy metallene for N_2 electroreduction. *Ang. Chem.*, 134 (28) (2022) e202205923.
24. K. Chen, F. Wang, Atomically dispersed W_1-O_3 bonded on Pd metallene for cascade NO electroreduction. *ACS Catal.*, 13 (2023) 9550–9557.
25. Q. Mao, K. Deng, H. Yu, Y. Xu, Z. Wang, X. Li, L. Wang and H. Wang, In situ reconstruction of partially hydroxylated porous Rh metallene for ethylene glycol-assisted seawater splitting. *Adv. Func. Mater.*, 32(31) (2022) 2201081.
26. K. Deng, Q. Mao, W. Wang, P. Wang, Z. Wang, Y. Xu, X. Li, H. Wang and L. Wang, Defect-rich low-crystalline Rh metallene for efficient chlorine-free H_2 production by hydrazine-assisted seawater splitting. *Appl. Catal. B: Environ.*, 310 (2022) 121338.
27. J. Wu, J. Fan, X. Zhao, Y. Wang, D. Wang, H. Liu, L. Gu, Q. Zhang, L. Zheng, D. J. Singh, X. Cui and W. Zheng, Atomically dispersed MoO_x on rhodium metallene boosts electrocatalyzed alkaline hydrogen evolution. *Angew. Chem. Int. Ed.*, 61 (2022) e202207512.
28. T. Zeng, X. Meng, S. Sun, M. Ling, C. Zhang, W. Yuan, D. Cao, M. Niu, L. Y. Zhang and C. M. Li, Tensile-strained holey Pd metallene toward efficient and stable electrocatalysis. *Small Methods,* 7(11) (2023) 2300791.
29. L. Y. Zhang, X. Meng, W. Zhang, T. Zeng, W. Yuan and Z. Zhao, Synthesis of palladium-tungsten metallene-constructed sandwich-like nanosheets as bifunctional catalysts for direct formic acid fuel cells. *ACS Appl. Energy Mater.*, 4 (11) (2021) 12336–12344.

30. L. Zhang, Z. Zhao, X. Fu, S. Zhu, Y. Min, Q. Xu and Q. Li, Curved porous PdCu metallene as a high-efficiency bifunctional electrocatalyst for oxygen reduction and formic acid oxidation. *ACS Appl. Mater Interf.*, 15, 4 (2023) 5198–5208.

31. Z. Wang, H. Zhao, J. Liu, D. Zhang, X. Wu, N, Nie, D. Wu, W. Xu, J. Lai and L. Wang, The PdH_x metallene with vacancies for synergistically enhancing N_2 fixation. *Chem. Eng. J.*, 450 (2022) 137951.

32. M. Luo, Z. Zhao, Y. Zhang, Y. Sun, Y. Xing, F. Lv, Y. Yang, X. Zhang, S. Hwang, Y. Qin, J. Ma, F. Lin, D. Su, G. Lu and S. Guo, PdMo bimetallene for oxygen reduction catalysis. *Nature*, 574 (2019) 81–85.

33. H. Wang, W. Wang, H. Yu, Q. Mao, Y. Xu, X. Li, Z. Wang and L. Wang, Interface engineering of polyaniline-functionalized porous Pd metallene for alkaline oxygen reduction reaction. *Appl. Catal. B: Environ.*, 307 (2022) 121172.

34. P. Prabhu and J. Lee, Metallenes as functional materials in electrocatalysis. *Chem. Soc. Rev.*, 50 (2021) 6700–6719.

35. Q. Mao, X. Mu, K. Deng, H. Yu, Z. Wang, Y. Xu, X. Li, L. Wang and H. Wang, Sulfur vacancy-rich amorphous Rh metallene sulfide for electrocatalytic selective synthesis of aniline coupled with efficient sulfion degradation. *ACS Nano*, 17 (2023) 790–800

36. P. Prabhu, V. Do, C. K. Peng, H. Hu, S. Chen, J. Choi, Y. Lin and J. Lee, Oxygen-bridged stabilization of single atomic W on Rh metallenes for robust and efficient pH-universal hydrogen evolution. *ACS Nano*, 17 (2023) 10733–10747.

37. M. Xie, B. Zhang, Z. Jin, P. Li and G. Yu, Atomically reconstructed palladium metallene by intercalation-induced lattice expansion and amorphization of highly efficient electrocatalysis. *ACS Nano*, 16 (2022) 13715–13727.

38. S. Huang, S. Lu, S. Gong, Q. Zhang, F. Duan, H. Zhu, H. Gu, W. Dong and M. Du, Sublayer stable Fe dopant in porous Pd metallene boosts oxygen reduction reaction. *ACS Nano*, 16 (2022) 522–532.

39. X. Zhao, Z. Li, Z. Ding, S. Wang and Y. Lu, Ultrathin porous Pd metallene as highly efficient oxidase for colorimetric analysis. *J. Colloid Interface Sci.*, 626 (2022) 296–304.

40. M. Zhang, S. Zhou, W. Wei, D. Ma, S. Han, X. Li, X. Wu, Q. Xu and Q. Zhu, Few-atom-layer metallene quantum dots toward CO_2 electroreduction at ampere-level current density and Zn-CO_2 battery. *Chem. Catal.*, 2 (2022) 3528–3545.

41. J. Zhang, J. Fu and K. Dai, Graphitic carbon nitride/antimonene van der Waals heterostructure with enhanced photocatalytic CO_2 reduction activity. *J. Mater. Sci. Technol*, 116 (2022) 192–198.

42. D. Zhang, X. Cui, L. Liu, Y. Xu, J. Zhao, J. Han and W. Zheng, 2D bismuthene metal electron mediator engineering super interfacial charge transfer for efficient photocatalytic reduction of carbon dioxide. *ACS Appl. Mater. Interf.*, 13 (2021) 21582–21592.

43. Z. Zhao, C. Choi, S. Hong, H. Shen, C. Yan, J. Masa, Y. Jung, J. Qiu and Z. Sun, Surface-engineered oxidized two-dimensional Sb for efficient visible light-driven N_2 fixation. *Nano Energy*, 78 (2020) 105368.

44. H. Li, H. Deng, S. Gu, C. Li, B. Tao, S. Chen, X. He, G. Wang, W. Zhang and H. Chang, Engineering of bionic Fe/Mo bimetallene for boosting the photocatalytic nitrogen reduction performance. *J. Colloid Interf. Sci.*, 607 (2022) 1625–1632.

45. J. Li, Z. Huang, W. Ke, J. Yu, K. Ren and Z. Dong, High solar-to-hydrogen efficiency in Arsenene/GaX (X = S, Se) van der Waals heterostructure for photocatalytic water splitting. *J. Alloys Compd.*, 866 (2021) 158774.

46. X. Li, B. Wang, X. Cai, W. Yu, L. Zhang, G. Wang and S. Ke, Arsenene/Ca(OH)$_2$ van der Waals heterostructure: strain tunable electronic and photocatalytic properties. *RSC Adv.*, 7 (2017) 44394–44400.

47. A. Qian, X. Han, Q. Liu, L. Ye, X. Pu, Y. Chen, J. Liu, H. Sun, J. Zhao, H. Ling, R. Wang, J. Li and X. Jia, Ultrathin Pd metallenes as novel co-catalysts for efficient photocatalytic hydrogen production. *Appl. Surf. Sci.*, 618 (2023) 156597.
48. Q. Dang, H. Lin, Z. Fan, L. Ma, Q. Shao, Y. Ji, F. Zheng, S. Geng, S. Yang, N. Kong, W. Zhu, Y. Li, F. Liao, X. Huang and M. Shao, Iridium metallene oxide for acidic oxygen evolution catalysis. *Nat. Commun.*, 12 (2021) 6007.
49. W. Gao, Z. Li, Q. Han, Y. Shen, C. Jiang, Y. Zhang, Y. Xiong, J. Ye, Z. Zou and Y. Zhou, State-of-the-art advancements of atomically thin two-dimensional photocatalysts for energy conversion. *Chem. Commun.*, 58 (2022) 9594–9613.

CHAPTER 10

2D Metal-based Electrocatalysts

Properties and Applications

Pandian Lakshmanan, Krishna Kumar M,
Radha D Pyarasani, and John Amalraj

10.1 INTRODUCTION

Two-dimensional (2D) metallic nanomaterials, limited to nano-sized dimensions in one direction and infinitely expansive in the other two, range in thickness from 0.8 to 100 nm. These ultrathin metallic materials boast high-density coordinatively unsaturated metal sites (CUMS) and vast electrochemically active surface areas (EASA). Their surficial atoms, with low coordination numbers, promote efficient hydrogen proton adsorption, crucial for effective hydrogen evolution reaction (HER) due to nearly all metal atoms engaging in catalytic reactions. Such exposed CUMS enhance nanomaterial utilization efficiency while preserving exceptional catalytic properties. Nanoclusters, with upshifted d-band centers, exhibit higher HER activity compared to highly coordinated metal structures, emphasizing the significance of 2D metals rich in low-coordinated atoms as promising HER catalysts.

Atoms at different positions have different catalytic activities typically in the order of corners > edges > flat facets [1]. The high-index planes of nanocrystals exhibit higher activities for breaking chemical bonds than the low-index planes [2]. Enhanced catalytic efficiencies in 2D metals arise from high surface-to-bulk atom ratios and abundant corner/edge atoms, surpassing 0–1D counterparts. Electrocatalysis faces challenges including slow kinetics, high overpotentials, and expensive and unstable electrocatalysts.

Hydrogen demand is predicted to surge 3.3 times by 2050, particularly in power, aviation, and heavy industry sectors. Electrochemical water splitting hinges on HER for clean

138

DOI: 10.1201/9781032645001-10

hydrogen. Seeking advanced catalysts like Platinum Group Metals (PGM)/non-PGM—with large surface areas, high conductivity, low toxicity, and specific catalytic traits (e.g., near-zero DGH*, minimal water decomposition barriers, wide potential windows, rapid charge transfer, and robust stability) is imperative.

Figure 10.1 depicts the typical CO_2 reduction reactor. In general, the electrocatalytic CO_2 conversion takes place in a fuel cell, proceeding via proton-assisted multiple electron transfer processes at the cathode, which yields diverse carbon and multi-carbon molecules such as formic acid (HCOOH), carbon monoxide (CO), methanol (CH_3OH), methane (CH_4), ethanol (C_2H_5OH), ethylene (C_2H_4), propanol (C_3H_7OH), and other products, largely based on the intrinsic properties of electrocatalysts and operation parameters [3, 4]. Nonetheless, the efficiency of CO_2 electroreduction is usually hindered by its considerable activation energy for *COOH formation, competitive hydrogen evolution reaction (HER), moderate selectivity, and poor durability of the electrode materials [5–7]. Therefore, the development of CO_2RR catalysts with high electrocatalytic activity, robustness, and more selectivity is imperative to overcome the existing obstacles.

Several challenges remain in the field of electrocatalysts: **Catalytic Activity** (high activity at low overpotentials (the extra voltage required to drive the reaction), **durability** (highly

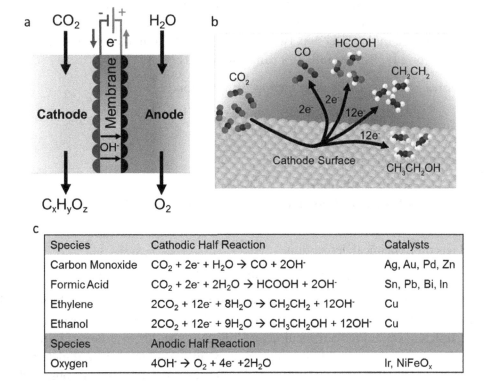

FIGURE 10.1 (a) Typical CO_2 electrolyzer. (b) The CO_2 electroreduction at cathode surface. (c) Half-cell reactions in alkaline aqueous conditions at the cathode (Anodic half-cell reaction is OER) and typical catalysts. Adapted with permission from Ref [4]. Copyright (2022), American Chemical Society.

140 ■ 2D Metals

acidic or alkaline conditions can lead to catalyst degradation), *cost, scalability, selectivity, good mass transport, understanding mechanisms, non-noble metal alternatives*. Here are some key descriptors for electrocatalysts: Overpotential (η), Exchange Current Density (j0), Tafel Slope (β), Faradaic Efficiency, Turnover Frequency (TOF), Active Site Density, Electronic Structure and Band Structure. In this chapter, the properties of 2D metallic materials and their electrocatalytic applications are discussed. The variation of catalytic performance (reaction-wise, i.e., HER, ORR, CO2RR) with structure/morphology is discussed. Lastly, perspectives on the opportunities and challenges in the applications of 2D-metal nanostructures are described.

10.2 HER MECHANISM

Mechanistically, three possible principal steps are involved in the electrochemical HER process for the reduction of protons in acidic media or water molecules in alkaline media to hydrogen molecules (H_2) on the surface of an electrode with a minimum external potential applied [8]. The first step is the Volmer reaction (Eqs 10.1 and 10.2), where a proton reacts with an electron to generate an adsorbed hydrogen atom (H^*) on the electrode material surface (M). The proton sources are the hydronium cation (H_3O^+) and the water molecule in acidic and alkaline electrolytes, respectively. Subsequently, the H_2 formation may occur via the Heyrovsky reaction (Eqs 10.3 and 10.4) or the Tafel reaction (Eq. 10.5) or both. In the Heyrovsky step, another proton diffuses to the H^* and then reacts with a second electron to produce H_2. In the Tafel step, two H^* in the vicinity combine on the surface of the electrode to evolve H_2. The overall HER can be written as:

(1) Volmer reaction: electrochemical hydrogen adsorption:

$$\text{Acidic Medium: } H_3O + M + e^{-1} \leftrightarrow M-H^* + H_2O \tag{10.1}$$
$$\text{Alkaline Medium: } H_2O + M + e^{-1} \leftrightarrow M-H^* + OH- \tag{10.2}$$

(2) Heyrovsky reaction: electrochemical desorption:

$$\text{Acidic Medium: } H^+ + e^{-1} + M-H^* \leftrightarrow H_2 + M \tag{10.3}$$
$$\text{Alkaline Medium: } H_2O + e^{-1} + M-H^* \leftrightarrow H_2 + OH^- + M \tag{10.4}$$
$$\text{(or)}$$

(3) Tafel reaction: chemical desorption

$$\text{Both Acidic and Alkaline Mediums: } 2\ M-H^* \leftrightarrow H_2 + 2M \tag{10.5}$$

The potential difference necessary to increase or decrease the current density by tenfold is designated by the Tafel slope (b). It indicates the mechanism of the HER process.

Three different conditions can occur depending upon the slow rate-determining step:

At 25°C, when the chemical desorption (combination) reaction is the rate-determining step the Volmer or discharge reaction is fast, the (b) of 29 mV dec^{-1} should be observed and given by $b = 2.3RT/2F = 0.029$ V dec^{-1}.

At 25°C, if the electrochemical desorption (Heyrovsky reaction) is the rate-limiting step and the discharge reaction is fast, the (b) should be 39 mV dec^{-1} and given b = 4.6RT/3F = 0.039 V dec^{-1}.

At 25°C, if the discharge reaction is slow, the (b) should be 116 mV dec^{-1} and given by b = 4.6 RT/F = 0.116 V dec^{-1}.

10.2.1 Volcano Plots: Relationship between j_0 and ΔGH^*

The plot of log(j_0) Vs ΔGH^* represents a Volvano curve that is useful to identify the suitable metals for hydrogen evolution. Too weak reactant–catalyst interaction leads to a very less number of intermediates bound to the catalyst surface (very slow reaction). On the other hand, too strong interaction results in dissociation failure (reaction stops due to blockage of the active sites).

The H* adsorption/H$_2$ desorption can be evaluated by measuring the ΔGH^* for the HER reaction processes [9]. Therefore, a suitable intermediate interaction is crucial for hydrogen evolution and such a condition is represented by ΔG_{H^*} = 0, according to the Sabatier principle. In practical devices, the cathode often operates at significantly higher overpotentials, making the Tafel slope of a material a critical factor in determining its practical utility [10]. Unfortunately, the lack of reliable polarization data has hindered a comprehensive analysis of the HER in alkaline environments [11].

10.3 ELECTROCHEMICAL WATER SPLITTING

It is an endothermic reaction that necessitates thermodynamic effort and, potentially, a potential comparable to 1.23 V. The effectiveness of a 2D metal-based material as an electrocatalyst strongly hinges on several critical variables (Figure 10.2) that must be carefully taken into account before designing the catalyst. The introduction of an external voltage

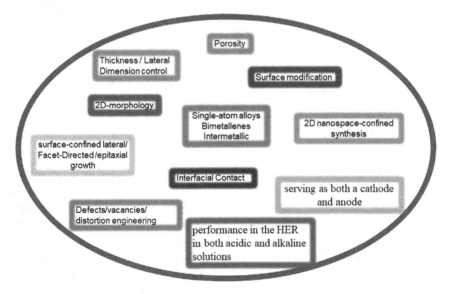

FIGURE 10.2 Various important aspects involved in the design of 2D-metal-based electrocatalytic materials.

in the electrodes results in the OER and HER at the anode and cathode, respectively. Xu et al. [12] designed ultrathin 2D-Pt nanodendrites (Pt NDs) with controlled size using a surfactant-directed solution-phase synthesis. This growth process, termed crystalline facet-directed step-by-step in-the-plane epitaxial growth is akin to the synthesis of organic dendrimers. This generated Pt NDs (ranging from 0 to 25) demonstrated a smaller overpotential of 0.01 V at a current density of 50 mA cm^2 and lower Tafel slope (32.7 mV dec^{-1}) compared to Pt/C.

Hong et al. [13] constructed atomically flat 2D-Pt nanodendrites (2D-PtNDs) using a confined synthesis approach, forming a maximized and tightly bound lateral heterointerface with NiFe-layered double hydroxide (LDH). The well-oriented {110} crystal surface of Pt allowed for strong LDH binding and promoted electronic interactions. The relocated charge interfacial bond in 2D-PtND/LDH accelerated hydrogen generation and achieved a significant Pt mass activity enhancement, approximately 11.2 times higher than that of 20 wt % Pt/C, along with improved long-term stability.

2D Metals can take different shapes (Figure 10.3). Lin et al. [14] fabricated large, high aspect ratio (up to ~107) 2D Pt. It exhibited HER a mass activity of 8.06 mA μg^{-1} at 0.06 V (18 times higher than commercial Pt/C) catalyst. The 3 nm-thick 2D-Pt film exhibited the highest HER area activity, requiring only a minimal overpotential of about 19 mV to

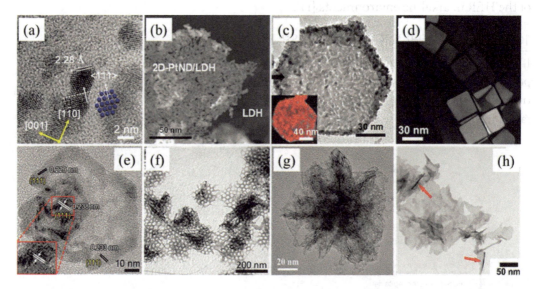

FIGURE 10.3 Various 2D metal-based materials: (a) Pt NDs. Adapted with permission [12], Copyright (2019), American Chemical Society. (b) 2D Pt/LDH, Adapted with permission from Ref [13]. Copyright (2022), American Chemical Society. (c) NiFe-LDH/2D-Pt hybrid, Adapted with permission [21]. Copyright (2020), American Chemical Society. (d) AL-Pt/Pd3Pb, Adapted with permission [22]. Copyright (2019), American Chemical Society. (e) PtAgCo Nanosheets, Adapted with permission [24]. Copyright (2019), American Chemical Society. (f) Mesoporous Metallic Iridium Nanosheets. Adapted with permission [17]. Copyright (2018), American Chemical Society. (g) IrRh Nanosheet assemblies, Adapted with permission [27]. Copyright (2019), American Chemical Society. (h) Ru nanosheets. Adapted with permission [20]. Copyright (2016), American Chemical Society. h):

achieve a current density of 10 mA cm^{-2}. In contrast, Pt films with thicknesses of 10 nm and 30 nm showed larger overpotentials of approximately 44 mV and 69 mV, respectively, indicating a significant decrease in HER catalytic activity as the thickness of the 2D Pt increased. Further reducing the thickness to 1 nm or 2 nm also resulted in decreased HER area activity compared to the 3 nm-thick sample. This is attributed to increased resistivity, which hinders electron transport during the HER process. The Tafel slopes were 46.5 mV/dec for the 1 nm-thick Pt, 40.9 mV dec^{-1} for the 3 nm-thick Pt, and 34.1 mV dec^{-1} for the commercial Pt/C catalyst. These values indicate that the Volmer-Tafel mechanism dominates the HER process.

Xu et al. [15] used amphiphilic surfactants in nanoconfined lamellar mesophases to guide PdNS growth in aqueous solution. The 2D PdNSs exhibited specific surface facets. The study revealed that {100}-exposed PdNSs exhibit superior catalytic activity and stability compared to {110}- and {111}-exposed ones and bulk Pd counterparts. The lowest overpotential of 67 mV was seen for PdNSs{100} at a current density of 10 mV cm^2, which is 91 mV and 160 mV lower than that of PdNSs{110} and PdNSs{111}, respectively. Under a constant overpotential of 42 mV at 50 mA cm^2, PdNSs{100} and commercial PdB exhibited 96.7% and 65.4% retention after 20 h.

Stepwise controllable synthesis of 2D-Pt has been reported (Figure 10.4), which is very useful for understanding growth.

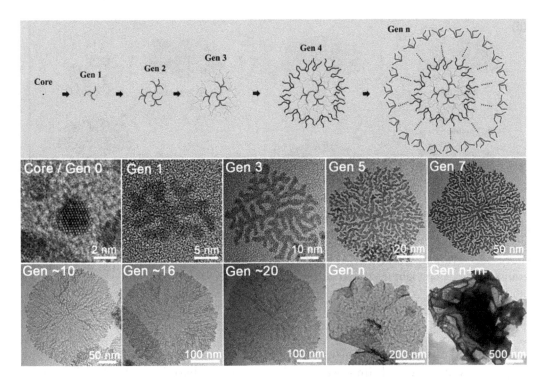

FIGURE 10.4 Schematic representation and corresponding TEM images of 2D inorganic PtNSs at different generations. Adapted with Permission from Ref. [12]. Copyright (2019), American Chemical Society.

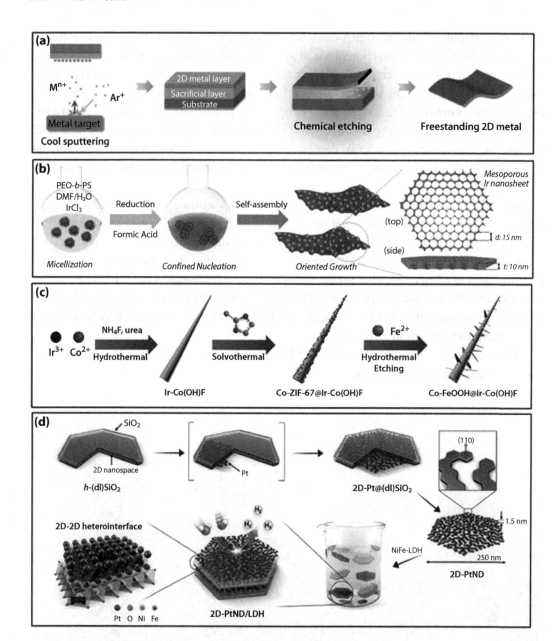

FIGURE 10.5 Various synthetic schemes of 2D-Metal-based materials: (a) Pt NSs. Adapted with Permission [14]. Copyright (2023), American Chemical Society. (b) Mesoporous Ir NSs. Adapted with Permission [17]. Copyright (2018), American Chemical Society. (c) Co-FeOOH@Ir-Co(OH)F. Adapted with Permission [33]. Copyright (2023), American Chemical Society. (d) Pt NDs. Adapted with Permission [13]. Copyright (2022), American Chemical Society.

2D-Ir-based catalysts have shown promising HER performances with low Tafel slopes and good stability [16, 17]. Improved synthesis methods have been developed recently (Figure 10.5).

Atomically thin Rh-NSs [18] required only 0.4 V for hydrogen (H_2) production (in the presence of isopropanol, and acetone was produced at the anode). Porous Rh NSs [19] required a remarkably low overpotential of only 37.8 mV. Furthermore, Rh NSs exhibit significantly smaller Tafel slopes when compared to Rh/C and Pt/C [20].

Generally, Pt-containing catalysts have exhibited high mass activity and low overpotentials [21–24]. Gu et al. [25] selectively modified specific facets and regions with other materials. An epitaxial deposition of a 0.5-nm-thick Pd or Ni layer was done exclusively on the flat {110} surface of 2D Pt within the nanoreactors. In the absence of nanoreactors, a non-epitaxial deposition occurred at the {111/100} edges. Importantly, these distinct locations of Pd/Pt and Ni/Pt heterointerfaces had varying effects on electrocatalytic synergy for the hydrogen evolution reaction (HER). The Pt{110} facet with 2D–2D interfaced e-Pd deposition exhibited enhanced H_2 generation, while edge-located n-Ni facilitated faster water dissociation (7.9 times higher catalytic activity than that of commercial Pt/C), effectively overcoming sluggish alkaline HER kinetics.

Yang et al. [26] reported single atom alloy (SAA) (Co atoms on a Pd metallene (Pdm) support. The Co/Pdm electrocatalyst exhibits exceptional performance in the HER in both acidic and alkaline solutions. It achieves a Tafel slope of 8.2 mV dec^{-1} and a low overpotential of 24.7 mV at 10 mA cm^{-2} in acidic conditions, surpassing the performance of commercial Pt/C and Pd/C catalysts.

Li et al. [27] developed IrRh NSAs. Specifically, IrRh NSAs require an impressively low overpotential of only 35 mV (Ir NSs- 270 mV, Rh NPs- 171 mV, and Pt/C - 46 mV). The IrRh NSAs exhibit a smaller Tafel slope of 60.5 mV/dec^{-1}, indicating a faster OER rate compared to Ir NSs (70.7 mV/dec^{-1}), Rh NPs (108.0 mV/dec^{-1}), and IrO$_2$ (73.9 mV/dec^{-1}).

Recently, the alloy compositions such as RuRh [28], RuCu [29], RhCo [30], Co/FeOx [31], Co-FeOOH@Ir-Co(OH)F [32], NiMo [33], RuNi [34], and PdTe [35] have shown interesting HER performances due to synergistic effects.

The alloying effects on the current density of HER catalysts are demonstrated in Figures 10.6 and in Table 10.1.

The effect of alloying on the HER performance is quite obvious. As can be seen from Figure 10.6, the combination of alloys such as Ru-Ni, Ru-Rh, Pd-Cu-Pt, Co-Pd, Ir-Rh, and Ni-Pd-Pt shows the highest current density and HER performances due to synergetic effects. Compared to single metal counterparts, alloys reach a closer value to 0.0 V (Vs RHE), revealing the importance of various metal combinations in 2D metals.

10.3.2 Challenges and Conclusions

Making effective interfacial contact is important to achieve good synergistic benefits in the 2D materials. Significant advances have been made through 2D nano-confined synthesis and utilization of nano-reactors/layered double hydroxide materials. Development of more simplified and scalable synthesis would be beneficial. Performance enhancement through the creation of porous structure and defects/vacancy engineering opens more possibilities. Non-noble metal-based materials have shown promising results in terms of overpotential and stability. Hence, materials based on Ru, Ir, and Rh need more attention. Effective formation of alloys and intermetallic materials can also be beneficial towards HER/OER

146 ■ 2D Metals

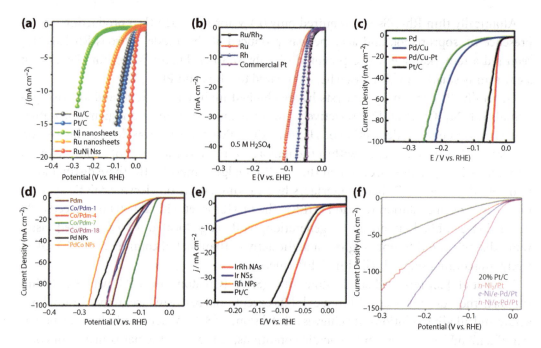

FIGURE 10.6 Effects of alloying on the HER activity indicated by the current density values. (a) Adapted with permission [30]. Copyright (2019), Elsevier. (b) Adapted with permission [28]. Creative Commons 4.0 CC BY Published by Wiley-VCH GmbH. (c) Adapted with permission [23]. Copyright (2017), Wiley-VCH Verlag GmbH & Co. KGaA, Weinheim. (d) Adapted with permission [26]. Copyright (2023), Wiley-VCH GmbH. (e) Adapted with permission [27]. Copyright American Chemical Society (2019). (f) Adapted with permission [25]. Copyright (2023), Wiley-VCH GmbH.

reactions. In the case of the ability to be used as both a cathode and anode, utilization of non-precious metals will be more economical. Enhancement of HER/OER performance through surface modification of 2D materials needs more attention.

10.4 OXYGEN REDUCTION REACTION (ORR)

Electrocatalysis has been an important research domain in recent years. In the ORR, molecular oxygen is electrochemically reduced by four protons and four electrons to form water molecules, which is accompanied by the formation of electrical potential. Sometimes it ends with undesirable partial reduction to H_2O_2 which involves only a two-electron process.

$$O_2 + 4H^+ + 4e^- \rightarrow 2H_2O$$
$$O_2 + 2H^+ + 2e^- \rightarrow H_2O_2$$

Huang et al. have developed Pt porous nanosheets with high surface distortion for enhanced ORR and they claimed that Pt NSs displayed a much-enhanced ORR mass and specific activity of 2.07 A mgPt^{-1} and 3.1 mA cm^{-2} at 0.90 V versus reversible hydrogen electrode, 9.8 and 10.7 times higher than those of commercial Pt/C. To obtain this Pt porous nanosheet, first PtTe$_2$ nanosheets were prepared by a wet chemical process and then the atoms were eliminated from the structure by electrochemical or acid erosion method, this

TABLE 10.1   Summary of Some 2D Metal Nanomaterials as Electrocatalysts in Various Electrochemical Reactions

2D-Material	Electrocatalytic Performance		Thickness (nm)	Shape	Phase/facet	LateralDimensions (nm)	Ref
	η HER [mV@10 mA cm^{-2}]	η OER [mV@10 mA cm^{-2}]					
Pt	< 0.01*&	328$	2.3	ND	Fcc	2–1000	[11]
Pt	19	266*	3	NS	fcc	> 500	[13]
Pd	67	240	2.5	NS	Fcc	> 100	[14]
Ir	20$	260	1.3	NS	FCC	< 100	[15]
Ir	60*		10	Mesoporou NS	fcc	--21.4 ± 7.6	[16]
Rh	42		1.3	NS	hcp/VBP	--7 nm	[17]
Rh	37.8		3.7 ± 1.1	Hierarchical NS	Hcp	–	[18]
Ru	20		1.2	NS			[19]
Pt/NiFe-LDH	31*		~1.5 nm	NS			[20]
Pt/Pd3Pb	13.8*		Submololayer	NS			[21]
CuPt/Pd	22.8		Atomic dispersion	Nanorings			[22]
PtAgCo	27$		< 5 nm	ND	FCC	50–100 nm	[23]
Pd+Ni/Pt	24*		1.5 nm	NS	FCC	240 ±10	[24]
Co-SAC/pD	24.7		1.375 nm	assemblies		width 600 nm	[25]
IrRh	35		– 0.83 nm	bimetallene		– –	[26]
RuRh2	34 24* 12^						[27]
RuCu	19$ 20*	236 234	6 nm	snowflake-like NS	hcp	–	[28]
RuNi	15* 12$	250$ 240*	11.6 ± 3.2 nm	nanostructures	hcp	80.7 ± 14.1 nm	[29]
RhCo	32* 31^	310^	1.3 nm	NS aggregates	fcc		[30]
CoFe	≈ 30*	200*	5.8 nm	NS	fcc		[31]
2D Co-doped FeOOH on 1D Ir-doped Co(OH)F nanorods	38		1.3 nm (CoFeOOH NSs)	NS		–	[32]
Ni$_4$Mo	35* 1.0M KOH		6.1 nm	NS	bct	– 260 ± 10	[33]
Pt/NiFeLDH	70*		1.5	ND		–	[12]
PdTe$_2$	76		< 3				[34]

* – 1.0M KOH; $-0.5M H$_2SO_4$; ^1.0M PBS; @ &50 mA cm^{-2}, NS – Nanosheet; ND – nanodentrite.

148 ■ 2D Metals

process increased the porosity of the obtained Pt nanosheets. The reusability of this catalyst was checked, and it was found that even after 30,000 cycles the structure of the material remained unchanged, and the activity too [36].

The same research group in another investigation has developed Platinum-Lead/ Platinum (PtPb/Pt) core/shell nanoplate catalysts. This particular catalyst was designed to exhibit large biaxial strains which is an important property to boost the ORR catalytic activity. Interestingly it was found that the stable Pt nanoplates with (110) facets have high ORR specific and mass activities that reach 7.8 mA/cm^2 and 4.3 A/mg of platinum at 0.9 volts versus the reversible hydrogen electrode (RHE), respectively. It shows better reusability than the above-mentioned Pt porous nanosheets with negligible activity lost and without composition and structural defect even after 50,000 cycles [37].

With a very similar approach, Wang et al. reported Pd nanosheets with (110) termination. The relationship between the thickness of Pd nanosheets and the ORR activity was further analyzed by calculation of the free-energy barriers of the ORR on (110)-terminated Pd nanosheets with thicknesses of 3 to 8 ML and it was found that each surface has different extent of compressive strain, with the thinner slabs experiencing the greatest compression [38].

Huang et al. have developed hexagonal PtBi nanoplates with better tolerance over organic molecules such as methanol, formic acid, and carbon monoxide. Generally, they are associated with the fuel cell operating conditions. The PtBi nanoplates displayed much potential electrocatalytic activity for ORR compared with PtBi nanoparticles due to their unique 2D nanostructure [39]. Guo et al. have developed a highly curved and sub-nm thick PdMo bimetallene (EASA =138.7 m^2 /one gram of palladium) which achieved a better catalytic activity towards the ORR of 16.37 A/mg of palladium at 0.9 volts versus the reversible hydrogen (mass activity 78 times and 327 times higher than those of commercial Pt/C and Pd/C catalysts, respectively, durability with 30,000 potential cycles) [40].

Wei et al have reported for the very first time the preparation of freestanding and single atom layer (SAL) Pd and Pd-Co films. These SAL films displayed a strong 2D coordinated metallic bond with the dangling bond in the z-direction. It was found that Pd-Co SAL film is six times better than Pt NPs and eight times better than Pd NPs towards the ORR process. SAL structure is unique, and it allows atoms to have zero coordination in 1D but allows them to retain their coordination of metallic crystal structure in other 2Ds [41]. Table 10.2 provides a summary of ORR performance of various catalysts.

10.5 CHALLENGES

Metal-based electrocatalysis for ORR is an important aspect of green energy production, several research groups are actively engaged in this research and have achieved significant results. However, the development of highly active, durable, and ORR catalysts with low cost is still a challenge. Most of the catalysts are sensitive to organic molecules and have poor performance at alkaline conditions and expensive Pt-based catalysts are still a challenge in the development of effective ORR electrocatalysts.

TABLE 10.2   Summary of ORR Performance of Various Catalysts

2D Material	ORR performance	Thickness	Shape	Phase/facet	Lateral dimensions	Ref
Porous Pt Nanosheets	mass activities (2.07 A mg_{Pt}^{-1}) and specific activities (3.1 mA cm^{-2})	0.9 nm	nanosheets	{111}		[35]
PtPb/Pt	mass activities that reach 7.8 mA/cm^2 and 4.3 A/mg of platinum at 0.9 volts versus the reversible hydrogen electrode (RHE)	4.5 ± 0.6 nm	nanoplates	Two types of phases were observed as {010}PtPb//{110}Pt between the PtPb and the edge-Pt layer, and {001}PtPb//{110}Pt between PtPb and top (bottom)-Pt layer	Core/shell nanoplates	[36]
Pd nanosheets		0.2–2.5 nm with different ML		{110}	nanosheets	[37]
PtBi		18 nm	nanoplates	{100}	Hexagonal nanoplates	[38]
PdMo	mass activities 16.37 ± 0.60 A mg_{Pt}^{-1}) specific activities 11.64 ± 0.40 mA cm^{-2})	0.88 nm	nanosheets	cubic phase with a dominant {111}facet	Bimetallene nanosheets	[39]
Pd and Pd-Co alloy		0.45 nm	Single atom layer (SAM)	perfect hexagonal structure with {111} facet	Single atom layer	[40]

10.6 CO$_2$ REDUCTION REACTION (CO$_2$RR)

Utilizing electrical or solar energy for converting CO$_2$ into valuable products offers a promising solution for capturing and utilizing atmospheric CO$_2$. Yang et al. prepared [42] mesoporous bismuth nanosheets through the cathodic transformation of atomic-thick bismuth oxycarbonate nanosheets. These nanosheets enable highly selective CO$_2$ reduction to formate, offering a wide potential window, exceptional Faradaic efficiency (≈100%), and remarkable operational stability. An in situ porous Zn catalyst was prepared by Liu et al. [43]. In a study by Liu et al., an in situ porous Zn catalyst was developed, and it demonstrated the ability to achieve both high faradaic efficiency and a large CO partial current density. The porous structure and favorable adsorption energy of the key intermediate, promoted the electroreduction of CO$_2$ to CO. Typical reactor design is presented in Figure 10.7.

Cao et al. [44] reported the creation of atomically thin bismuthene (Bi-ene) through an in situ electrochemical transformation of ultrathin bismuth-based metal-organic layers. The resulting material displayed ~100% selectivity (stability within a potential window exceeding 0.35 V) for formate production. In a flow-cell reactor, it can achieve current densities exceeding 300.0 mAcm^{-2} without compromising selectivity. Won et al. [45] worked in

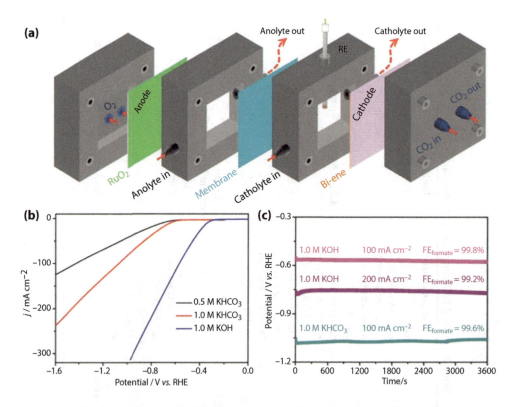

FIGURE 10.7 Electrocatalytic CO$_2$RR performance of Bi-ene in a flow cell: (a) Diagram of the self-designed flow cell. (b) LSV curves in different electrolytes at a scan rate of 10 mζs^{-1}. (c) Chronopotentiometric curves at 100 and 200 mAζm^{-2} in 1.0 mKHCO$_3$ and KOH. Adapted with Permission [44]. Copyright (2020), Wiley-VCH Verlag GmbH & Co. KGaA, Weinheim.

a hierarchical hexagonal Zn catalyst in which, the Zn (101) facet is favorable for CO formation, while the Zn (002) facet favors H_2 evolution during CO_2 electrolysis. Density functional theory (DFT) calculations supported these observations, showing that the (101) facet effectively lowers the reduction potential for CO_2 to CO by stabilizing a COOH intermediate more efficiently than the (002) facet. This suggests that tuning the crystal structure to control the (101)/(002) facet ratio of Zn could serve as a key design principle for achieving the desired product outcomes with Zn catalysts. Liu et al. [46] observed that triangular silver nanoplates (Tri-Ag-NPs) exhibited improved current density, significantly higher Faradaic efficiency (96.8%), enhanced energy efficiency (61.7%), and durability over 7 days. CO formation was observed at a remarkably low overpotential of 96 mV, further highlighting the effectiveness of Tri-Ag-NPs as a catalyst for CO_2RR.

Gao et al. [47] created two types of four-atom-thick layers: pure cobalt metal and cobalt metal with co-existing domains of cobalt oxide. Cobalt primarily produces formate (HCOO–) during CO_2 electroreduction. It was revealed that the surface cobalt atoms in the atomically thin layers have higher intrinsic activity and selectivity for formate production at lower overpotentials compared to surface cobalt atoms on bulk samples. It exhibited stable current densities of about 10 milliamperes per square centimeter over 40 hours and 90% formate selectivity at an overpotential of only 0.24 V. Han et al. [48], created ultrathin bismuth nanosheets which demonstrated high selectivity (approximately 100%) and a large current density over a wide potential range, along with excellent durability lasting over 10 hours. Yang et al. [49] reported the first large-scale synthesis of free-standing Bismuthene, demonstrating its exceptional electrocatalytic efficiency in generating formate (HCOO–) from CO_2 reduction. The catalytic performance of Bismuthene is evidenced by its high Faradaic efficiency (99% at −580 mV vs. the Reversible Hydrogen Electrode), low onset overpotential (less than 90 mV), and remarkable durability (no performance decay even after 75 hours and annealing at 400°C).

Wang et al. [50] reported that 4H Au nanostructures are introduced as advanced electrocatalysts for highly active and selective CO_2 reduction to CO. The researchers compared pure 4H phase Au nanoribbons, hybrid 4H/fcc phase Au nanorods, and fcc phase Au nanorods for CO_2 electroreduction. Their findings revealed that the 4H nanoribbons outperformed the others, achieving over 90% Faradaic efficiency in producing CO. Achieving high selectivity for desirable C_2 products, such as ethylene, is a significant challenge. Zhang et al. [51] reported Cu nanosheets designed with nanoscale defects ranging from 2 nm to 14 nm were employed for the electrochemical production of ethylene from carbon dioxide. Impressively, they achieved a high Faradaic efficiency of 83.2% for ethylene production. The effectiveness of these nanoscale defects was attributed to their ability to enhance the presence of reaction intermediates and hydroxyl ions on the electrocatalyst. This enrichment promoted the C–C coupling reactions necessary for ethylene formation.

10.7 CONCLUSION

The electrochemical reduction of CO_2 to valuable fuels is an appealing strategy for addressing CO_2 capture and energy storage. However, developing electrocatalysts with high activity and selectivity over a broad potential range is a challenging task. The conversion of CO_2

into value-added products using electrical or solar energy is seen as a promising approach for mitigating atmospheric CO_2 levels. Formate is a common product of CO_2 reduction, but achieving high reaction selectivity has been difficult. Electrocatalytic CO_2 reduction to produce fuels is part of a sustainable carbon cycle. Nevertheless, enhancing the catalytic performance of CO_2 electrocatalysts remains a significant challenge, particularly in achieving high product selectivity and large partial current density, which are vital for future electrochemical CO_2 reduction applications.

The electroreduction of CO_2 into useful fuels, particularly when driven by renewable energy, offers an environmentally friendly alternative to fossil feedstocks, addressing the growing issue of CO_2 emissions and their impact on climate. The main challenge is activating CO_2 into intermediate forms, often requiring impractically high overpotentials. Electrocatalytic carbon dioxide reduction to formate is a desirable but complex process, with the outcome being highly sensitive to the surface structures of electrocatalysts. Achieving high selectivity for specific C_2 products, like ethylene, is a significant challenge in this field.

ACKNOWLEDGMENTS

The author, J.A. acknowledges ANID, Fondecyt Regular N° 1230561 and Proyecto 20CEIN2-142146 Consorcio Sur-Subantártico CI2030.

REFERENCES

1. B. An, M. Li, J. Wang, C. Li, Shape/size controlling syntheses, properties and applications of two-dimensional noble metal nanocrystals, *Front. Chem. Sci. Eng.* 10 (2016) 360–382. https://doi.org/10.1007/s11705-016-1576-0.
2. N. Tian, Z.Y. Zhou, S.G. Sun, Platinum metal catalysts of high-index surfaces: from single-crystal planes to electrochemically shape-controlled nanoparticles, *J. Phys. Chem. C.* 112 (2008) 19801–19817. https://doi.org/10.1021/jp804051e.
3. Y. Pei, H. Zhong, F. Jin, A brief review of electrocatalytic reduction of CO_2—Materials, reaction conditions, and devices, *Energy Sci. Eng.* 9 (2021) 1012–1032. https://doi.org/10.1002/ese3.935.
4. S. Overa, B.H. Ko, Y. Zhao, F. Jiao, Electrochemical approaches for CO_2 conversion to chemicals: A journey toward practical applications, *Acc. Chem. Res.* 55 (2022) 638–648. https://doi.org/10.1021/acs.accounts.1c00674.
5. D.D. Zhu, J.L. Liu, S.Z. Qiao, Recent advances in inorganic heterogeneous electrocatalysts for reduction of carbon dioxide, *Adv. Mater.* 28 (2016) 3423–3452. https://doi.org/10.1002/adma.201504766.
6. S. Rasul, D.H. Anjum, A. Jedidi, Y. Minenkov, L. Cavallo, K. Takanabe, A highly selective copper-indium bimetallic electrocatalyst for the electrochemical reduction of aqueous CO_2 to CO, *Angew. Chem. Int. Ed.* 54 (2015) 2146–2150. https://doi.org/10.1002/anie.201410233.
7. J.H. Montoya, C. Shi, K. Chan, J.K. Nørskov, Theoretical insights into a CO dimerization mechanism in CO_2 electroreduction, *J. Phys. Chem. Lett.* 6 (2015) 2032–2037. https://doi.org/10.1021/acs.jpclett.5b00722.
8. C.G. Morales-Guio, L.A. Stern, X. Hu, Nanostructured hydrotreating catalysts for electrochemical hydrogen evolution, *Chem. Soc. Rev.* 43 (2014) 6555–6569. https://doi.org/10.1039/c3cs60468c.

9. T.F. Jaramillo, K.P. Jørgensen, J. Bonde, J.H. Nielsen, S. Horch, I. Chorkendorff, Identification of active edge sites for electrochemical H_2 evolution from MoS_2 nanocatalysts, *Science*. 317 (2007) 100–102. https://doi.org/10.1126/science.1141483.

10. B.E. Conway, L. Bai, M.A. Sattar, Role of the transfer coefficient in electrocatalysis: Applications to the H_2 and O_2 evolution reactions and the characterization of participating adsorbed intermediates, *Int. J. Hydrogen Energy*. 12 (1987) 607–621. https://doi.org/10.1016/0360-3199(87)90002-4.

11. O.A. Petrii, G.A. Tsirlina, Electrocatalytic activity prediction for hydrogen electrode reaction: intuition, art, science, *Electrochim. Acta*. 39 (1994) 1739–1747. https://doi.org/10.1016/0013-4686(94)85159-X.

12. D. Xu, H. Lv, H. Jin, Y. Liu, Y. Ma, M. Han, J. Bao, B. Liu, Crystalline facet-directed generation engineering of ultrathin platinum nanodendrites, *J. Phys. Chem. Lett.* 10 (2019) 663–671. https://doi.org/10.1021/acs.jpclett.8b03861.

13. Y.R. Hong, S. Dutta, S.W. Jang, O.F. Ngome Okello, H. Im, S.Y. Choi, J.W. Han, I.S. Lee, Crystal facet-manipulated 2D Pt nanodendrites to achieve an intimate heterointerface for hydrogen evolution reactions, *J. Am. Chem. Soc.* 144 (2022) 9033–9043. https://doi.org/10.1021/jacs.2c01589.

14. B. Lin, Y. Zhang, H. Zhang, H. Wu, J. Shao, J. Shao, K. Liu, S. Chen, C. Zhou, X. Cheng, J. Lu, K. Hu, Y. Huang, W. Zhao, P. Liu, J. Zhu, H.J. Qiu, Z. Chen, Centimeter-scale two-dimensional metallenes for high-efficiency electrocatalysis and sensing, *ACS Mater. Lett.* 5 (2023) 397–405. https://doi.org/10.1021/acsmaterialslett.2c01066.

15. D. Xu, X. Liu, H. Lv, Y. Liu, S. Zhao, M. Han, J. Bao, J. He, B. Liu, Ultrathin palladium nanosheets with selectively controlled surface facets, *Chem. Sci.* 9 (2018) 4451–4455. https://doi.org/10.1039/c8sc00605a.

16. Z. Cheng, B. Huang, Y. Pi, L. Li, Q. Shao, X. Huang, Partially hydroxylated ultrathin iridium nanosheets as efficient electrocatalysts for water splitting, *Natl. Sci. Rev.* 7 (2020) 1340–1348. https://doi.org/10.1093/nsr/nwaa058.

17. B. Jiang, Y. Guo, J. Kim, A.E. Whitten, K. Wood, K. Kani, A.E. Rowan, J. Henzie, Y. Yamauchi, Mesoporous metallic iridium nanosheets, *J. Am. Chem. Soc.* 140 (2018) 12434–12441. https://doi.org/10.1021/jacs.8b05206.

18. Y. Zhao, S. Xing, X. Meng, J. Zeng, S. Yin, X. Li, Y. Chen, Ultrathin Rh nanosheets as a highly efficient bifunctional electrocatalyst for isopropanol-assisted overall water splitting, *Nanoscale*. 11 (2019) 9319–9326. https://doi.org/10.1039/c9nr02153a.

19. Z. Zhang, G. Liu, X. Cui, Y. Gong, D. Yi, Q. Zhang, C. Zhu, F. Saleem, B. Chen, Z. Lai, Q. Yun, H. Cheng, Z. Huang, Y. Peng, Z. Fan, B. Li, W. Dai, W. Chen, Y. Du, L. Ma, C.J. Sun, I. Hwang, S. Chen, L. Song, F. Ding, L. Gu, Y. Zhu, H. Zhang, Evoking ordered vacancies in metallic nanostructures toward a vacated Barlow packing for high-performance hydrogen evolution, *Sci. Adv.* 7 (2021) 1–10. https://doi.org/10.1126/sciadv.abd6647.

20. X. Kong, K. Xu, C. Zhang, J. Dai, S. Norooz Oliaee, L. Li, X. Zeng, C. Wu, Z. Peng, Free-standing two-dimensional ru nanosheets with high activity toward water splitting, *ACS Catal.* 6 (2016) 1487–1492. https://doi.org/10.1021/acscatal.5b02730.

21. S.W. Jang, S. Dutta, A. Kumar, Y.R. Hong, H. Kang, S. Lee, S. Ryu, W. Choi, I.S. Lee, Holey Pt nanosheets on NiFe-hydroxide laminates: synergistically enhanced electrocatalytic 2D interface toward hydrogen evolution reaction, *ACS Nano.* 14 (2020) 10578–10588. https://doi.org/10.1021/acsnano.0c04628.

22. Y. Yao, X.K. Gu, D. He, Z. Li, W. Liu, Q. Xu, T. Yao, Y. Lin, H.J. Wang, C. Zhao, X. Wang, P. Yin, H. Li, X. Hong, S. Wei, W.X. Li, Y. Li, Y. Wu, Engineering the electronic structure of submonolayer Pt on intermetallic Pd_3Pb via charge transfer boosts the hydrogen evolution reaction, *J. Am. Chem. Soc.* 141 (2019) 19964–19968. https://doi.org/10.1021/jacs.9b09391.

154 ■ 2D Metals

23. T. Chao, X. Luo, W. Chen, B. Jiang, J. Ge, Y. Lin, G. Wu, X. Wang, Y. Hu, Z. Zhuang, Y. Wu, X. Hong, Y. Li, Atomically dispersed copper–platinum dual sites alloyed with palladium nanorings catalyze the hydrogen evolution reaction, *Angew. Chem. Int. Ed.* 56 (2017) 16047–16051. https://doi.org/10.1002/anie.201709803.

24. A. Mahmood, H. Lin, N. Xie, X. Wang, Surface confinement etching and polarization matter: a new approach to prepare ultrathin PtAgCo nanosheets for hydrogen-evolution reactions, *Chem. Mater.* 29 (2017) 6329–6335. https://doi.org/10.1021/acs.chemmater.7b01598.

25. B.S. Gu, S. Dutta, Y.R. Hong, O.F. Ngome Okello, H. Im, S. Ahn, S.Y. Choi, J. Woo Han, S. Ryu, I.S. Lee, Harmonious heterointerfaces formed on 2D-Pt nanodendrites by facet-respective stepwise metal deposition for enhanced hydrogen evolution reaction, *Angew. Chem. Int. Ed.* 62 (2023) e202307816. https://doi.org/10.1002/anie.202307816.

26. S. Yang, Z. Si, G. Li, P. Zhan, C. Liu, L. Lu, B. Han, H. Xie, P. Qin, Single cobalt atoms immobilized on palladium-based nanosheets as 2D single-atom alloy for efficient hydrogen evolution reaction, *Small.* 19 (2023) 2207651. https://doi.org/10.1002/smll.202207651.

27. C. Li, Y. Xu, S. Liu, S. Yin, H. Yu, Z. Wang, X. Li, L. Wang, H. Wang, Facile construction of IrRh nanosheet assemblies as efficient and robust bifunctional electrocatalysts for overall water splitting, *ACS Sustain. Chem. Eng.* 7 (2019) 15747–15754. https://doi.org/10.1021/acssuschemeng.9b03967.

28. X. Mu, J. Gu, F. Feng, Z. Xiao, C. Chen, S. Liu, S. Mu, RuRh bimetallene nanoring as high-efficiency pH-universal catalyst for hydrogen evolution reaction, *Adv. Sci.* 8 (2021) 1–8. https://doi.org/10.1002/advs.202002341.

29. Q. Yao, B. Huang, N. Zhang, M. Sun, Q. Shao, X. Huang, Channel-rich RuCu nanosheets for pH-universal overall water splitting electrocatalysis, *Angew. Chem. Int. Ed.* 58 (2019) 13983–13988. https://doi.org/10.1002/anie.201908092.

30. Y. Zhao, J. Bai, X.R. Wu, P. Chen, P.J. Jin, H.C. Yao, Y. Chen, Atomically ultrathin RhCo alloy nanosheet aggregates for efficient water electrolysis in broad pH range, *J. Mater. Chem. A.* 7 (2019) 16437–16446. https://doi.org/10.1039/c9ta05334d.

31. P. Li, Q. Xie, L. Zheng, G. Feng, Y. Li, Z. Cai, Y. Bi, Y. Li, Y. Kuang, X. Sun, X. Duan, Topotactic reduction of layered double hydroxides for atomically thick two-dimensional non-noble-metal alloy, *Nano Res.* 10 (2017) 2988–2997. https://doi.org/10.1007/s12274-017-1509-3.

32. Z. Xu, Y. Jiang, J.L. Chen, R.Y.Y. Lin, Heterostructured ultrathin two-dimensional co-FeOOH Nanosheets@1D Ir-Co(OH)F nanorods for efficient electrocatalytic water splitting, *ACS Appl. Mater. Interfaces.* 15 (2023) 16702–16713. https://doi.org/10.1021/acsami.2c22632.

33. Q. Zhang, P. Li, D. Zhou, Z. Chang, Y. Kuang, X. Sun, Superaerophobic ultrathin Ni–Mo alloy nanosheet array from in situ topotactic reduction for hydrogen evolution reaction, *Small.* 13 (2017) 1–7. https://doi.org/10.1002/smll.201701648.

34. G. Liu, W. Zhou, B. Chen, Q. Zhang, X. Cui, B. Li, Z. Lai, Y. Chen, Z. Zhang, L. Gu, H. Zhang, Synthesis of RuNi alloy nanostructures composed of multilayered nanosheets for highly efficient electrocatalytic hydrogen evolution, *Nano Energy.* 66 (2019) 104173. https://doi.org/10.1016/j.nanoen.2019.104173.

35. Y. Zuo, N. Antonatos, L. Děkanovský, J. Luxa, J.D. Elliott, D. Gianolio, J. Šturala, F. Guzzetta, S. Mourdikoudis, J. Regner, R. Málek, Z. Sofer, Defect engineering in two-dimensional layered $PdTe_2$ for enhanced hydrogen evolution reaction, *ACS Catal.* 13 (2023) 2601–2609. https://doi.org/10.1021/acscatal.2c04968.

36. Y. Feng, B. Huang, C. Yang, Q. Shao, X. Huang, Platinum porous nanosheets with high surface distortion and Pt utilization for enhanced oxygen reduction catalysis, *Adv. Funct. Mater.* 29 (2019) 1–10. https://doi.org/10.1002/adfm.201904429.

37. L. Bu, N. Zhang, S. Guo, X. Zhang, J. Li, J. Yao, T. Wu, G. Lu, J.Y. Ma, D. Su, X. Huang, Biaxially strained PtPb/Pt core/shell nanoplate boosts oxygen reduction catalysis, *Science.* 354 (2016) 1410–1414. https://doi.org/10.1126/science.aah6133.

38. L. Wang, Z. Zeng, W. Gao, T. Maxson, D. Raciti, M. Giroux, X. Pan, C. Wang, J. Greeley, Tunable intrinsic strain in two-dimensional transition metal electrocatalysts, *Science.* 363 (2019) 870–874. https://doi.org/10.1126/scienceaat8051.
39. Y. Feng, Q. Shao, F. Lv, L. Bu, J. Guo, S. Guo, X. Huang, Intermetallic PtBi nanoplates boost oxygen reduction catalysis with superior tolerance over chemical fuels, *Adv. Sci.* 7 (2020) 1800178. https://doi.org/10.1002/advs.201800178.
40. M. Luo, Z. Zhao, Y. Zhang, Y. Sun, Y. Xing, F. Lv, Y. Yang, X. Zhang, S. Hwang, Y. Qin, J.Y. Ma, F. Lin, D. Su, G. Lu, S. Guo, PdMo bimetallene for oxygen reduction catalysis, *Nature.* 574 (2019) 81–85. https://doi.org/10.1038/s41586-019-1603-7.
41. J. Jiang, W. Ding, W. Li, Z. Wei, Freestanding single-atom-layer pd-based catalysts: oriented splitting of energy bands for unique stability and activity, *Chemistry.* 6 (2020) 431–447. https://doi.org/10.1016/j.chempr.2019.11.003.
42. H. Yang, N. Han, J. Deng, J. Wu, Y. Wang, Y. Hu, P. Ding, Y. Li, Y. Li, J. Lu, Selective CO_2 reduction on 2D mesoporous Bi nanosheets, *Adv. Energy Mater.* 8 (2018) 1–6. https://doi.org/10.1002/aenm.201801536.
43. K. Liu, J. Wang, M. Shi, J. Yan, Q. Jiang, Simultaneous achieving of high faradaic efficiency and CO partial current density for CO_2 reduction via robust, noble-metal-free Zn nanosheets with favorable adsorption energy, *Adv. Energy Mater.* 9 (2019) 1–6. https://doi.org/10.1002/aenm.201900276.
44. C. Cao, D.D. Ma, J.F. Gu, X. Xie, G. Zeng, X. Li, S.G. Han, Q.L. Zhu, X.T. Wu, Q. Xu, Metal–organic layers leading to atomically thin bismuthene for efficient carbon dioxide electroreduction to liquid fuel, *Angew. Chem. Int. Ed.* 59 (2020) 15014–15020. https://doi.org/10.1002/anie.202005577.
45. D.H. Won, H. Shin, J. Koh, J. Chung, H.S. Lee, H. Kim, S.I. Woo, Highly efficient, selective, and stable CO_2 electroreduction on a hexagonal Zn catalyst, *Angew. Chem. Int. Ed.* 128 (2016) 9443–9446. https://doi.org/10.1002/ange.201602888.
46. S. Liu, H. Tao, L. Zeng, Q. Liu, Z. Xu, Q. Liu, J.L. Luo, Shape-dependent electrocatalytic reduction of CO_2 to CO on triangular silver nanoplates, *J. Am. Chem. Soc.* 139 (2017) 2160–2163. https://doi.org/10.1021/jacs.6b12103.
47. S. Gao, Y. Lin, X. Jiao, Y. Sun, Q. Luo, W. Zhang, D. Li, J. Yang, Y. Xie, Partially oxidized atomic cobalt layers for carbon dioxide electroreduction to liquid fuel, *Nature.* 529 (2016) 68–71. https://doi.org/10.1038/nature16455.
48. N. Han, Y. Wang, H. Yang, J. Deng, J. Wu, Y. Li, Y. Li, Ultrathin bismuth nanosheets from in situ topotactic transformation for selective electrocatalytic CO_2 reduction to formate, *Nat. Commun.* 9 (2018) 1–8. https://doi.org/10.1038/s41467-018-03712-z.
49. F. Yang, A.O. Elnabawy, R. Schimmenti, P. Song, J. Wang, Z. Peng, S. Yao, R. Deng, S. Song, Y. Lin, M. Mavrikakis, W. Xu, Bismuthene for highly efficient carbon dioxide electroreduction reaction, *Nat. Commun.* 11 (2020) 1088. https://doi.org/10.1038/s41467-020-14914-9.
50. Y. Wang, C. Li, Z. Fan, Y. Chen, X. Li, L. Cao, C. Wang, L. Wang, D. Su, H. Zhang, T. Mueller, C. Wang, Undercoordinated active sites on 4H gold nanostructures for CO_2 reduction, *Nano Lett.* 20 (2020) 8074–8080. https://doi.org/10.1021/acs.nanolett.0c03073.
51. B. Zhang, J. Zhang, M. Hua, Q. Wan, Z. Su, X. Tan, L. Liu, F. Zhang, G. Chen, D. Tan, X. Cheng, B. Han, L. Zheng, G. Mo, Highly electrocatalytic ethylene production from CO_2 on nanodefective Cu nanosheets, *J. Am. Chem. Soc.* 142 (2020) 13606–13613. https://doi.org/10.1021/jacs.0c06420.

CHAPTER 11

2D Metals as Photocatalysts

Asma Rafiq, Misbah Naz, Shehnila Altaf, and Saira Riaz

11.1 INTRODUCTION

The tremendous advancements in agriculture, industry, and the social economy have given rise to significant hurdles in the form of environmental degradation and energy crisis. These issues pose significant obstacles to the achievement of sustainable development in human civilization [1–3]. Fossil fuels represent a limited and nonrenewable source of energy. One further limitation is the process of extracting energy from these fuels, which has the inherent risk of causing environmental contamination and intensifying the challenges posed by climate change, particularly considering the expanding worldwide populace. The current scenario imposes substantial demands to explore renewable energy sources as feasible substitutes for the limited fossil fuel resources [4].

Artificial photosynthesis has the potential to serve as a sustainable method for producing hydrogen, which may be used as an alternative to clean fuel. This approach might contribute somewhat to addressing the escalating global energy needs, considering the abundant availability of water and sunlight [2, 3]. The photocatalytic method for water splitting has been proposed as a promising and reliable process for the production of clean hydrogen fuel [1]. The achievement of a stable and efficient photocatalyst is the goal in this context. A photocatalyst refers to a substance that can absorb photons emitted by sunshine and subsequently generate electron-hole pairs, so initiating redox processes [4].

Since the first experimental proof of H_2O splitting on a TiO_2 electrode in 1972, other photocatalyst materials have been subsequently synthesized [5]. The photocatalysts synthesized throughout the last 43 years may be classified into three distinct categories. The first category comprises metallic materials that alone exhibit absorption in the ultraviolet (UV) range. The subsequent category encompasses metallic materials that demonstrate absorption in the visible light spectrum. Lastly, the third category pertains to 2D metallic and nonmetallic materials that possess a capacity to absorb extended visible light. One notable drawback of the first generation was their limited capacity to absorb photons since

156

DOI: 10.1201/9781032645001-11

they were only capable of absorbing UV light. Consequently, this led to suboptimal conversion efficiencies in the process of converting solar energy into hydrogen, often referred to as sun-to-hydrogen (STH) conversion efficiencies. The second generation of photocatalysts demonstrated enhanced STH efficiency due to their increased ability to absorb photons within the visible light spectrum [4].

The first and subsequent generations of photocatalysts mostly comprised of Ti, Cd, W, and transition metal dichalcogenides (TMD), in the form of oxides, sulfides, and nitrides. One significant limitation pertains to the human toxicity and corrosiveness of metal-based photocatalysts. These can work in the presence of metals co-catalysts [Pt, Au, Ag etc.]. The study has mostly focused on photocatalysts that are driven by extended visible light and comprised of either a few layers or a single layer of 2D materials. 2D materials exhibit typical structural and electro-optical characteristics [4]. It is important to highlight that the term "2D" encompasses not only monolayer or atomically thin materials but also extends to include freestanding ultrathin metallic nanosheets. The properties of these nanosheets primarily rely on their surface characteristics [6].

This chapter gives a comprehensive overview of the advancements made in the field of 2D metals-based nano-photocatalytic materials. The discussion primarily focuses on the structural features, photocatalytic mechanism, and design strategies used to increase the light-driven ability of these materials. In this chapter, we provide a comprehensive summary of the advancements, difficulties, and potential opportunities pertaining to metal-based photocatalysts in the context of environmental applications.

11.2 SUPERIORITY OF 2D MATERIALS AS PHOTOCATALYSTS

Materials that have a thickness of only one or a few atoms, exceeding 5 nm, possess notable characteristics such as a high ratio of surface atoms, a significant surface area, and inherent quantum confinement of electrons. These properties give rise to exceptional optical, mechanical, and electronic behavior, making them highly promising for various research applications including transistors, catalysis, optoelectronics, energy conversion, and energy storage devices [7–12]. The aforementioned materials exhibit distinctive physiochemical characteristics, including electronic anisotropy, planar conductivity, elevated surface activity, and adjustable energy structure [13].

The reduction of bulk material to atomic dimensions leads to observable changes in atomic structures. Consequently, 2D materials demonstrate not just their inherent bulk properties but also novel characteristics. Semiconductor photocatalytic materials have attained noteworthy consideration because of their potential in addressing environmental pollution and energy storage challenges [14–19]. These materials facilitate the separation of H_2O into its components i.e. hydrogen and oxygen, resulting in the removal of contaminants and a decrease in CO_2 levels via the use of solar light as an external stimulus [20–23]. When subjected to irradiation, the photocatalysts undergo a process whereby they absorb light and generate electron-hole pairs in the conduction band (CB) and valence band (VB). The electron-hole pairs that are created during photoexcitation undergo diffusion towards the surface of the material, followed by migration to the active sites prior to surface reactions. The process of charge recombination occurs, and the separation effectiveness is influenced by factors such as texture, crystal structure, particle size, and crystallinity. Ultimately, the target molecules adhere to the surface of the material and undergo processes of charge injection and desorption, resulting in the formation of final products [24–28]. At present, there is a wide range of semiconductors available for photocatalysis, characterized by tunable electrical and crystal structures.

Despite significant advancements in optimizing the process of photocatalysis, several photocatalysts still exhibit very low photocatalytic activity. This performance is contingent upon the rational design of these materials. The use of ultrathin materials has introduced a novel dimension to this field due to their favorable band structure. The use of ultrathin 2D configurations enables enhanced absorption of UV-visible radiation and provides a significantly increased specific surface area. Nevertheless, the ability of bulk materials to absorb photons is severely restricted as a result of the reflection and transmission occurring at grain boundaries [29]. Moreover, when the thickness of the material drops to the atomic scale, the migration distance of charge carriers is significantly reduced, resulting in rapid movement towards the surface region in 2D materials. The reduction of recombination potential is observed, resulting in an enhancement in photocatalytic activity. Finally, the presence of distinctive 2D structures that possess a significant ratio of surface area to the number of atoms provides numerous active sites that enhance the rate of reaction.

11.3 ENGINEERING PROTOCOLS FOR IMPROVEMENT OF PHOTOCATALYTIC PERFORMANCE

Various defects may be engineered to modify the inherent features of the material. In addition to the use of doping, the manipulation of defects has consistently had a substantial influence on 2D materials, as shown by its effects on the field of photocatalysis. The fundamental characteristics of 2D materials may be significantly influenced by even the smaller defects, owing to their atomic-scale dimensions. Hence, the introduction of surface imperfections such as ionic vacancies, pits, related vacancies, and distortions may be readily used to enhance the electrical properties of 2D materials.

11.3.1 Defect Engineering

The electronic and chemical properties of 2D metal-based photocatalysts may be influenced by surface imperfections. The imperfections inside the crystal may be broadly classified into two categories: deep defects and surface defects. In the crystal, the recombination of photogenerated electrons and holes inside the deep defect significantly diminishes the photocatalyst activity. Therefore, it is crucial to minimize the creation of deep defects during the synthesis. Surface defects in the crystal lattice have the potential to catch charges, although temporarily. However, these charges are promptly released, and this transient capture process may promote the separation of electrons and holes, so augmenting the performance of the photocatalyst. The presence of surface defects in 2D metals-based photocatalysts plays a significant role in shaping the energy band structure and excitation process. These defects lead to a partial overlap between the shallow energy level and the state density of the VB near the upper region of the forbidden zone. Consequently, this overlap results in the expansion of the VB width and a reduction in the width of the forbidden band. The production of oxygen defects in Bi-based photocatalysts has been extensively explored due to their prevalence and comparatively low formation energy. The presence of oxygen vacancies (OVs) may give rise to localized states that broaden the range of light response and facilitate the efficient uptake of charge carriers. This phenomenon has a significant impact on the dynamics, energetics, and subsequent catalytic processes [30].

Recently, Li et al., explored that defects on the surface of BiOBr may significantly accelerate the process of N_2 adsorption and activation, leading to the successful achievement of effective photocatalytic nitrogen fixation (Figure 11.1 (a)) [31]. However, the current understanding of the impact of OVs on the thermodynamics of photocatalytic N_2 fixation remains uncertain. Li et al. then employed BiOCl-OV as a model catalyst to explore the kinetics and thermodynamics of nitrogen N_2. According to the findings shown in Figure 11.1 (b), it can be seen that H_2O_2 exhibits spontaneous dissociation on BiOCl-001-OV [32]. This dissociation results in the fixation of one –OH group at the defect site, while the other –OH group remains away from the surface. The BiOCl-010-OV surface exhibited a comparable dissociation of H_2O_2, wherein the dissociated –OH group preferentially adsorbed onto the Bi atom in close proximity to OV, resulting in the formation of OH$^\bullet$ [32]. The formation of a novel plasma photocatalyst was seen by the deposition of Au metal on the surface of O_2-deficient BiOCl, as shown in Figure 11.1 (c). The findings indicate that oxygen defects possess the ability to trap high-energy electrons, even when the energy level is inadequate for direct penetration of the Schottky barrier [33]. In contrast to extensively investigated OVs, there exists a scarcity of literature on photocatalysts with metal vacancies and their impact on photocatalytic capabilities. Di et al. investigated single-unit-cell ultrathin Bi_2WO_6 nanosheets with Bi vacancies using template-directed technique. The presence of Bi vacancies leads to an increment in the concentration of charge carriers and enhances electronic conductivity. This is achieved by the creation of a new defect level, which improves the density of states at the VB maximum. Furthermore, the vacancy structure of Bi might potentially augment the process of adsorption and activation of H_2O molecules, hence promoting the occurrence of water oxidation processes (Figure 11.1 (d)). Hence, the Bi_2WO_6 material exhibiting a higher concentration of Bi vacancies showed a significant improvement in its photocatalytic oxygen evolution performance compared to Bi_2WO_6 [34]. In their study, Gao et al. presented findings on the presence of a defect level in the form of vanadium vacancy inside single-unit-cell o-$BiVO_4$ layers, as seen in Figure 11.1 (e). Furthermore, they observed that the concentration of holes in o-$BiVO_4$ is elevated near the Fermi level. Consequently, it is possible to boost the absorption of light and improve electrical conductivity [35]. Despite the increasing intensity of research on faulty nanomaterials in the field of photocatalysis, the underlying mechanism of defects in this process remains elusive. Defects may serve as centers for photogenerated carrier recombination, hence exerting a detrimental influence. However, it is important to note that defects might potentially enhance the process of adsorption and activation of substrates, contribute to the occurrence of photocatalytic reactions, and have a beneficial impact also [36].

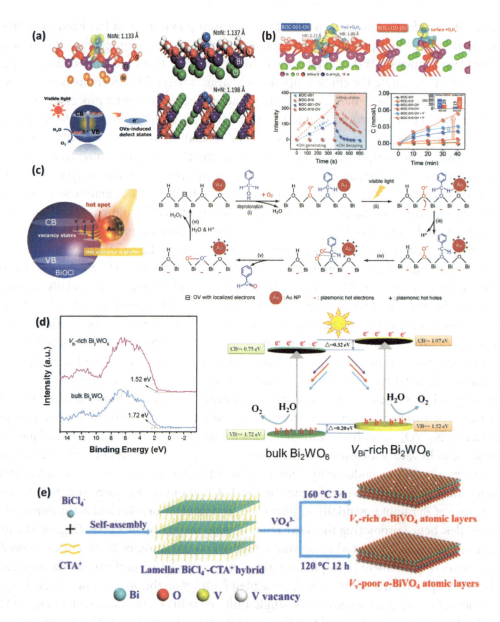

FIGURE 11.1 (a) Defects have a crucial role in enhancing the N$_2$ adsorption and activation to attain photocatalytic N$_2$ fixation. Adapted with permission [31]. Copyright 2015, American Chemical Society. (b) The process of H$_2$O$_2$ dissociation on various crystal surfaces of defective BiOCl. Adapted with permission [32]. Copyright 2017, American Chemical Society. (c) The morphological analysis of the Au-BiOCl-001-OV catalyst and its influence on thermal electron transport, specifically the impact of defects. Adapted with permission [33]. Copyright 2017, American Chemical Society. (d) The X-ray photoelectron spectroscopy valence spectra of the acquired samples were analyzed, together with the schematic representation of the band structure of both bulk Bi$_2$WO$_6$ and V$_{Bi}$-rich Bi$_2$WO$_6$. Adapted with permission [34]. Copyright 2018, Elsevier. (e) The proposed methodology for synthesizing atomic layers of V$_v$-rich and V$_v$-poor o-BiVO$_4$ is outlined in this scheme. Adapted with permission [35]. Copyright 2017, American Chemical Society.

11.3.2 Hybridization Engineering

2D materials possess a significant surface area, leading to an enhanced significance of the surface state compared to the bulk state. The photons generate charge carriers that are then distributed at the surface to participate in redox processes. Therefore, the enhancement of exciton absorption by surface hybridization is desirable. This section includes a discussion on surface hybridization, focusing on several 2D hybridization approaches. The techniques are revealed via the use of many case studies, such as the hybridization of quantum dot sheets with 2D materials, the combination of single atoms with 2D materials, the integration of molecular species with 2D materials, and the hybridization of layered 2D materials.

11.3.2.1 Single Atoms/2D Materials Hybridization

To enhance photocatalytic efficiency, it is plausible to consider the reduction of nanomaterials to individual atoms. Zhang et al. conducted revolutionary research in the field of photocatalysis, particularly in relation to monoatomic-based catalysis. The primary emphasis of the photocatalyst that relies on monoatomic species is the dispersion of isolated monoatoms on the surface of the supporting material. However, the ability to create a higher photocatalytic reaction by the hybridization of single atom/2D materials is limited. Zhang et al. utilized several methods for the creation of individual Rh atoms dispersed over uniform 2D-TiO_2 nanosheets. The findings from High-Angle Annular Dark-Field Scanning Transmission Electron Microscopy and extended X-ray absorption fine structure (EXAFS) indicated that the Rh atoms in the samples under investigation had a chemical environment like Rh_2O_3, which suggests that Rh atoms formed bonds with oxygen atoms and underwent oxidation. In this study, artificially synthesized co-catalysts were used as the active sites for the photocatalytic production of hydrogen gas, as anticipated based on density functional theory simulations. These findings revealed that there has been a substantial tenfold increment in the rate of H_2 evolution in comparison to TiO_2 nanosheets.

Similarly, Wu et al. in their study, used platinum (Pt) atoms as co-catalysts with the aim of improving the hydrogen production performance of C_3N_4 nanosheets when exposed to irradiation. The synthesis of monoatomic Pt/C_3N_4 was carried out via liquid phase reaction using H_2PtCl_6 and C_3N_4, followed by low-temperature annealing. The investigation of the local atomic configuration of Pt/C_3N_4 has been conducted by extended EXAFS spectroscopy and the results supported the notion that the efficiency of photocatalytic hydrogen evolution was greatly improved by the creation of atomic structure consisting of a Pt atom supported on C_3N_4 substrate. The Pt/C_3N_4 catalyst, exhibited a significantly enhanced production rate of about fifty times higher compared to the production rate of pure C_3N_4. Simultaneously, the produced structures exhibited notable stability over several cycles of photocatalytic H_2 production. Therefore, it is possible to fabricate a thin two-dimensional structure with a high density of surface defects secured by monoatomic species, which may effectively improve the photocatalytic performance [37].

162 ■ 2D Metals

11.3.2.2 Quantum Dots/2D Materials Hybridization

The nanoparticles exhibit unpaired valence electrons at their coordinated unsaturated surface atoms. To further decrease the size of nanoparticles, it is necessary to get a bigger percentage of atoms on the surface relative to the total number of atoms. Additionally, the binding energy of these surface atoms should be increased. Therefore, using the size of nanoparticles as a means of monitoring quantum dots (QDs) and adjusting 2D materials, it becomes feasible to generate an interfacial strong connection between these entities. Furthermore, the diminutive dimensions of QDs might lead to a pronounced dispersion effect when interacting with 2D materials, hence augmenting their photocatalytic characteristics.

To enhance the utilization efficiency of Ag, silver quantum dots (Ag QDs) with a size less than 5 nm were synthesized. The photocatalytic efficiency for the degradation of TCH, CIP, and RhB was significantly improved with the incorporation of BiOBr. The research findings indicate that Ag-QDs exhibit simultaneous adsorption, charge separation, and reaction center functionalities, leading to enhanced photocatalytic performance. One example involves the fabrication of nitrogen-doped carbon quantum dots (N-CQDs) with a diameter of 3 nm using a hydrothermal method. These N-CQDs are then deposited onto the atomically thin surface of BiOI nanostructures [374]. The atomic-level configuration and the arrangement of conjugated N-CQDs, significantly enhance the separation of these charges, hence extending the lifespan of excitons. Consequently, the N-CQDs/BiOI photocatalyst exhibited a significant increase in both the concentration of active species and its photocatalytic efficiency. Moreover, there have been investigations into the utilization of several systems that include the hybridization of QD/2D materials to enhance photocatalytic activity [375].

11.3.2.3 Molecular/2D Materials Hybridization

In relation to the use of independent atoms to manipulate electronic structure, it is worth noting that single molecular structures may also play a role in modulating electronic characteristics. Specifically, they can operate as co-catalysts to enhance the efficiency of photocatalytic processes. The incorporation of molecular TiO_2 into a C_3N_4 catalyst has been achieved using polycondensation process involving TiO_2 ion precursors and dicyandiamide inside C_3N_4 sub-nano holes [38]. Lu et al. investigated the electronic structure of 2D subnanopore nanosheets using molecular titanium-oxide incorporation. Figure 11.2 (a) signifies the steps involved in preparing TiO-CN through bottom-up polycondensation of the targeted precursors. The morphology of TiO_2-C_3N_4 nanostructures with an estimated thickness of 3 nm was examined and confirmed using transmission electron microscopy (TEM) analysis as shown in Figure 11.2 (b). The reliable dispersion of TiO_2 on the C_3N_4 framework in an isolated format was determined by high-angle annular dark-field scanning transmission electron microscopy (HAADF-STEM) as seen in Figure 11.2 (c). The results of this study suggest that there has been comparable progress in the development of molecular TiO_2 inside the C_3N_4 system as indicated in atomic force microscopy (AFM) image (Figure 11.2 (d)). Figure 11.2 (e) shows that the overall intensity of the XRD results

2D Metals as Photocatalysts ■ 163

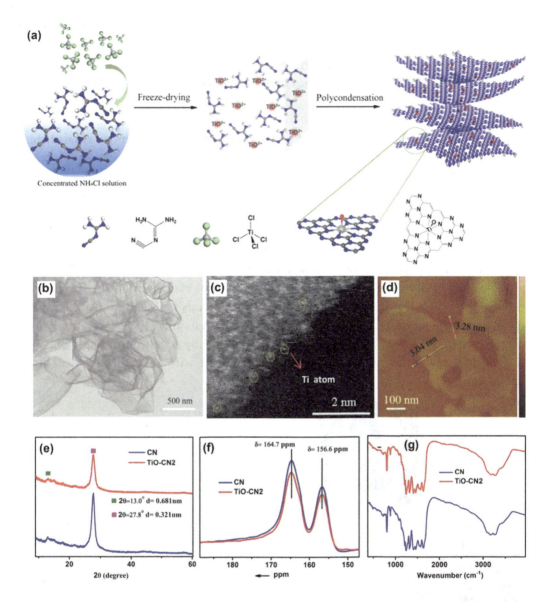

FIGURE 11.2 (a) The diagram shows the steps involved in preparing TiO-CN through bottom-up polycondensation of the targeted precursors. (b) TEM. (c) HAADF-STEM. (d) AFM micrograph of TiO-CN2 nanosheet. (e) XRD analysis of CN and TiO-CN2 nanosheets. (f) solid-state ^{13}C NMR spectra of 2D CN and TiO-CN2. (g) IR spectra of CN and TiOCN2. Adapted with permission [39]. Copyright 2016, Royal Society of Chemistry.

decreases progressively by increasing TiO integration. The XRD analysis demonstrates that TiO inclusion in 2D CN does not destroy the basic building structure of the C–N framework. This analysis was also verified via solid-state ^{13}C NMR spectra as depicted in Figure 11.2 (f). The FTIR spectra of the structure of 2D TiO-CN is depicted in Figure 11.2 (g) [39]. The observed phenomenon arises owing to the additional presence of Ti-O electrons inside the lattice, leading to an enhanced delocalization of π-electrons in the conjugated structure.

In addition, it is worth noting that hybrid catalysts possess an electronic structure that may effectively aid in the separation and isolation of charge carriers. Hence, evolved molecular co-catalyst techniques may serve as a viable approach to effectively segregate photoexcited carriers, hence augmenting the efficiency of photocatalysis [37].

11.3.2.4 2D/2D Stacking Hybridization

Considerable efforts have been devoted to the preparation of 2D Van der Waals heterostructures, either by vertical stacking or in-plane assembly of unique 2D elemental components [40]. The method of synthesizing 2D–2D stacks to enhance photocatalytic efficiency is a commonly used procedure in academic research and practical applications. The reduction of lattice mismatch has been seen as a notable outcome resulting from the analogous layered structures shown by 2D materials. Li et al. conducted a study on single-layer $Bi_{12}O_{17}Cl_2$ that had surface OVs using a Li-intercalation-based exfoliation process [41]. Following that, a deposition process was used to accumulate single layers of MoS_2 onto a single layer of $Bi_{12}O_{17}Cl_2$, facilitated by the presence of vanadium on the surface. This deposition process resulted in the formation of Bi-S bonds between the two nanostructures, as seen in Figure 11.3. Based on the mapping analysis and TEM observations, it was determined that the small MoS_2 nanostructures were firmly attached to a larger $Bi_{12}O_{17}Cl_2$ sheet, leading to the formation of a hybridized structure with 2D characteristics (Figure

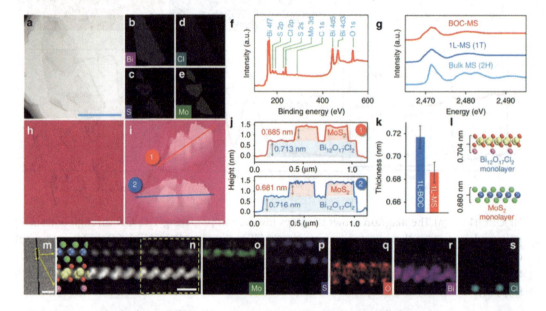

FIGURE 11.3 (a) A top-view TEM micrographs. (b–e) Micrographs depicting elemental mapping. (f) X-ray photoelectron spectroscopy spectra. (g) The XANES spectra at the S K-edge of BOC-MS, 1L-MS, and bulk MS. (h, i) Atomic force microscopy micrographs, A side-view TEM micrograph (m) HAADF-STEM image at atomic resolution (n) and (o–s) the corresponding EELS elemental maps of BOC-MS. (j) Height profiles along the lines in i. (k) Comparison of the average thicknesses of 1L-BOC and 1L-MS in BOC-MS. (l) The theoretical thicknesses of MoS2 and $Bi_{12}O_{17}C_{12}$ monolayers. Adapted with permission [41]. Copyright 2016, Nature Publishing Group.

11.3 (a)). Elemental mapping micrographs (Figures 11.3 (b–e)) coupling with XPS (Figure 11.3 (f)) and X-ray diffraction analysis showed the small and large nanosheets as MoS_2 and $Bi_{12}O_{17}Cl_2$, respectively. X-ray absorption near-edge structure spectra (XANES) (Figure 11.3 (g)) showed that the MoS_2 monolayers in the bilayer revealed a distorted 1T metallic phase. Moreover, nanostructures of varying sizes, namely large-sized nanostructures with an approximate diameter of 0.717 nm and small-sized nanostructures with a diameter of 0.686 nm were observed by AFM micrographs (Figures 11.1 (h,i)) and their corresponding height profiles (Figures 11.3 (j,k)). These nanostructures exhibited a correlation between MoS_2 and $Bi_{12}O_{17}Cl_2$ single layers as shown in Figure 11.3 (l). It was also observed that all MoS_2 sheets were attached on the same surface in $Bi_{12}O_{17}Cl_2$, which was further proved by their TEM micrograph as depicted in Figure 11.3 (m). The aberration-corrected high-angular annular dark-field scanning TEM (HAADF-STEM) illustration (Figure 11.3 (n)) and energy loss spectroscopy (EELS) elemental maps (Figure 11.3 (o–s)) of their cross-sectional atomic structures revealed direct, atomic-resolution indications that this oriented assembly ensued in 2D Janus bilayer junctions of (Cl_2)–$(Bi_{12}O_{17})$–(MoS_2) [41]. Hence, the incorporation of hybrid materials has been shown to enhance the light-harvesting capabilities of photocatalysts. Nevertheless, it is essential to modify the 2D components in order to enhance the interfacial tension between the layers, so creating a 2D/2D hybridized photocatalyst that exhibits better activity [20].

11.4 APPLICATIONS OF 2D METALS-BASED PHOTOCATALYTIC MATERIALS

Currently, the use of 2D metals as photocatalysts has gained significant traction in the domains of purifying environmental contamination. The primary focal points within the realm of environmental studies include the process of photocatalytic degradation of contaminants, which encompasses a wide range of applications. The energy application under consideration primarily revolves around the process of photocatalytic hydrogen production, as well as the reduction of CO_2 to small molecule organic compounds, including methane and methanol. Additionally, the application also involves N_2 fixation [42–44]. In this section, we provide a concise overview of the current state of application for photocatalysts based on 2D metals, as seen in recent years [36].

11.4.1 Environmental Photocatalysis

11.4.1.1 Photocatalytic Purification of Atmospheric Contaminants

The advancement of industrialization has led to a significant concern over the pollution of the environment, posing a substantial risk to public health. Photocatalysis technique is widely acknowledged as a cost-effective and efficient approach to mitigating environmental pollution. The degradation of organic pollutants into H_2O, CO_2, and other inorganic compounds may be achieved by oxidizing reactive oxygen species (ROS) that are produced during the photocatalytic process. These ROS include $O_2^{\bullet-}$, hydroxyl radical (OH$^\bullet$), hydrogen peroxide (H_2O_2), and singlet oxygen ($^1O_2$). Hence, it is important to consider the intrinsic redox capability of photocatalysts, together with the facilitation of photoelectron generation and migration, to enhance the efficiency of ROS production during the design and synthesis of photocatalysts for degradation purposes. The ultrathin 2D nanoparticles possess distinct advantages, including their remarkable ability to adsorb impurities and their amazing capability to harness light. As a result, these nanomaterials have significant potential for various applications in the degradation of pollutants [45].

11.4.1.2 Photocatalytic Decolorization of Liquid Phase Contaminants

The photocatalytic approach eliminates secondary pollutants and is ecologically preferable to conventional methods of treating water contamination. Photocatalysis has the capability to decrease the concentration of dangerous heavy metal ions in water reservoirs by converting them into less expensive ions, hence diminishing their harmful effects. The process of photocatalytic removal of environmental contaminants primarily entails the oxidation of organic contaminants via reactive oxygen species. Guan et al. conducted a study whereby they synthesized a nanostructure of ultrathin BiOCl with a 2D structure and included defect states. This design led to the achievement of high efficiency in the photocatalytic reduction of RhB. Additionally, the study revealed the link between the structure of the nanostructure and its performance, as seen in Figure 11.4 (a) [46]. Furthermore, Liu et al. used a selective deposition technique to coat Ag_2O nanoparticles onto the (040)

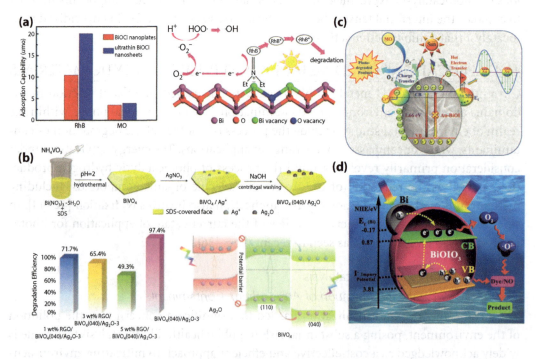

FIGURE 11.4 (a) The adsorption of RhB and MO on the BiOCl photocatalyst, along with a representation of the photosensitization method. Adapted with permission [46]. Copyright 2013, American Chemical Society. (b) The synthesis of $BiVO_4$ (040)/Ag_2O and its use in the photodegradation of MB solution. Adapted with permission [47] C.opyright 2019, Elsevier. (c) The photocatalytic detoxification of methyl orange and Cr(VI) on Au-BiOI, focusing on the underlying mechanism. Adapted with permission [48]. Copyright 2020, American Chemical Society. (d) The potential photocatalytic mechanism and charge separation process in $BiOIO_3$. Adapted with permission [49]. Copyright 2015, American Chemical Society.

facet of $BiVO_4$. The $BiVO_4(040)/Ag_2O$ heterostructures exhibit remarkable light absorption capability and efficient charge separation, resulting in improved photocatalytic MB reduction potential and O_2 evolution, as demonstrated in Figure 11.4 (b) [47]. In a separate study, Chatterjee et al. prepared 3D flower-like BiOI microspheres decorated with plasmonic Au nanoparticles. This unique structure displayed exceptional photocatalytic detoxification properties for both organic (MB) and inorganic (hexavalent chromium) water pollutants, as illustrated in Figure 11.4 (c) [48]. In another study, Yu et al. successfully accomplished the simultaneous incorporation of I-doping and plasmonic Bi-metal deposition into the $BiOIO_3$ material. Bi/I-co-decorated $BiOIO_3$ had a significant photo-oxidizing capability in the degradation of several pharmaceutical and industrial contaminants (Figure 11.4 (d)) [49].

11.4.1.3 Photocatalytic Disinfection

Currently, there has been an emergence of antibiotic resistance among bacteria, mostly attributed to the excessive and inappropriate use of antibiotics. The ongoing process of genetic mutation and adaptation in bacteria that renders them resistant to antibiotics, often known as antibiotic-resistant bacteria (ARB), has the potential to give rise to highly resilient strains, sometimes referred to as superbugs. These superbugs may subsequently disseminate antibiotic resistance genes (ARGs) into the surrounding environment, therefore posing significant risks to both human health and the environment. The phenomenon of photocatalytic disinfection has garnered significant interest due to its ability to disrupt bacterial structure and subsequent death. Additionally, this process can further destroy antibiotic resistance genes (ARGs), hence amplifying its significance in the scientific community. Hence, the $Ag/TiO_2/$graphene oxide 2D ternary composite photocatalysts (referred to as STG) were synthesized using the solvothermal approach as seen in Figure 11.5 (a) and then used for photocatalytic sterilization. STG exhibits the most effective photocatalytic sterilization action relative to pure TiO_2 and $TiO_2/$graphene oxide (TG). Also, STG is capable of completely inactivating ARB within 30 minutes as depicted in Figure 11.5 (b). The observed phenomenon may be related to the efficient and quick separation and migration of photo-induced charges that occur during the photocatalytic process of STG, as shown in Figure. 11.5 (c). This process leads to the generation of a higher quantity of hydroxyl radicals, which possess a potent oxidation capability, as illustrated in Figure 11.5 (d). The hydroxyl radicals possess the ability to undergo oxidation and subsequently disrupt the integrity of the cell membrane of ARB. This disruption leads to the release of cellular contents, ultimately resulting in the deactivation of the bacteria. It is noteworthy to mention that the separation efficiency of photo-induced electron-hole pairs in STG is comparatively lower relative to $Ag/TiO_2/$-reduced graphene oxide (STrG). The observed phenomenon may be attributed to the higher presence of oxygen-containing groups inside graphene oxide, which facilitate interactions with antimicrobial-resistant bacteria and subsequently lead to bacterial deposition. Moreover, it has been shown that STG photocatalysts possess the ability to deactivate intracellular ARGs. These ARGs are responsible for transmitting antibiotic resistance to bacteria via horizontal and vertical gene transfer mechanisms, hence leading to environmental toxicity as seen in Figure 11.5 (e). The

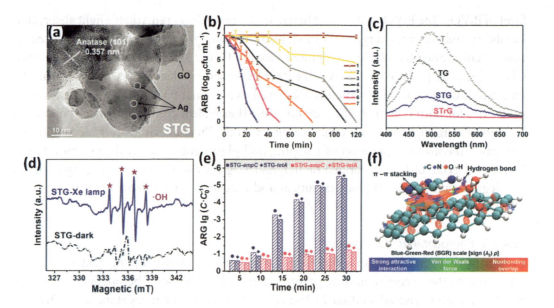

FIGURE 11.5 (a) HRTEM micrographs of STG. (b) The sterilization efficiency of the synthesized materials (100 mg·L^{-1}) towards antibiotic-resistant bacteria (10^7 CFU·mL^{-1}) was evaluated under various intensities of light irradiation [1-Xe lamp control, 2-sunlight control, 3-T+Xe lamp, 4-TG+Xe lamp, 5-STG+Xe lamp, 6-STrG+Xe lamp, and 7-STG+sunlight]. (c) Steady-state PL spectra. (d) ESR signals of DMPO-•OH in an aquatic dispersion. (e) Intracellular ARGs elimination by STG and STrG. (f) The optimized cytosine-GO structure. Adapted with permission [50]. Copyright 2020, Science China Press.

findings indicate that graphene oxide effectively captures ARGs via π–π interactions and then attaches them to the catalyst's surface by hydrogen bonding (Figure 11.5 (f)) [50]. Following this, the adsorbed antibiotic resistance genes (ARGs) undergo degradation by the action of hydroxyl radicals produced during TiO$_2$ photocatalysis [45].

11.5 CONCLUSIONS AND FUTURE PERSPECTIVE

Recently, there have been significant advancements in the development of a diverse set of 2D metals as photocatalysts that exhibit high efficiency in degrading organic compounds when exposed to solar irradiation. These photocatalysts have shown promise in effectively removing ecologically significant substances from aquatic environments. The use of 2D metals as photocatalysts and/or co-photocatalysts has shown promising results in the field of environmental remediation. These materials possess a significant surface area and a high concentration of active sites, leading to enhanced performance in this application. The primary objective of this chapter is to discuss the notable advancements made in the utilization of 2D metals for the purpose of photocatalytic solar conversion. The pursuit of novel materials is a continuous and unceasing endeavor. In this chapter, we have conducted a comprehensive evaluation of the efficacy of 2D metals as photocatalysts. Our objective has been to appraise their aptness and durability as possible catalysts for water splitting, with a focus on their viability for future applications.

In view of the widespread availability of sunlight on a worldwide scale, the use of 2D metals as a photocatalyst for water splitting holds promise for the sustainable production of hydrogen as a clean alternative fuel. This approach might potentially serve as a partial solution to the depletion of global fossil fuel energy sources. Although photocatalysis has shown significant promise in many applications, there remains a need for more optimization and enhancement of the catalysts. This encompasses enhancing the efficiency, selectivity, stability, and cost-effectiveness of the catalysts. There exists a need for a more basic investigation to comprehend the underlying processes governing photocatalysis. This encompasses the examination of the surface chemistry, kinetics of reactions, and mechanisms of charge transfer that are implicated in these reactions. The advancement of photocatalysis has resulted in novel prospects in the fields of environmental remediation and chemical synthesis. Nevertheless, it is essential to foster more interdisciplinary cooperation among researchers hailing from many domains to fully harness the potential of these catalysts.

REFERENCES

1. Q. Zhang, E.G. Xu, J. Li, Q. Chen, L. Ma, E.Y. Zeng, H. Shi, A review of microplastics in table salt, drinking water, and air: direct human exposure, *Environmental Science & Technology*, 54 (2020) 3740–3751.
2. M. Li, Y. Liu, L. Dong, C. Shen, F. Li, M. Huang, C. Ma, B. Yang, X. An, W. Sand, Recent advances on photocatalytic fuel cell for environmental applications—the marriage of photocatalysis and fuel cells, *Science of the Total Environment*, 668 (2019) 966–978.
3. S.C. Peter, Reduction of CO_2 to chemicals and fuels: a solution to global warming and energy crisis, *ACS Energy Letters*, 3 (2018) 1557–1561.
4. M.Z. Rahman, C.W. Kwong, K. Davey, S.Z. Qiao, 2D phosphorene as a water splitting photocatalyst: fundamentals to applications, *Energy & Environmental Science*, 9 (2016) 709–728.
5. K. Vikrant, K.-H. Kim, A. Deep, Photocatalytic mineralization of hydrogen sulfide as a dual-phase technique for hydrogen production and environmental remediation, *Applied Catalysis B: Environmental*, 259 (2019) 118025.
6. T. Wang, M. Park, Q. Yu, J. Zhang, Y. Yang, Stability and synthesis of 2D metals and alloys: a review, *Materials Today Advances,* 8 (2020) 100092.
7. Y. Cao, S. Guo, C. Yu, J. Zhang, X. Pan, G. Li, Ionic liquid-assisted one-step preparation of ultrafine amorphous metallic hydroxide nanoparticles for the highly efficient oxygen evolution reaction, *Journal of Materials Chemistry A*, 8 (2020) 15767–15773.
8. W. Li, X. Qian, J. Li, Phase transitions in 2D materials, *Nature Reviews Materials*, 6 (2021) 829–846.
9. J. Zhang, Y. Xie, Q. Jiang, S. Guo, J. Huang, L. Xu, Y. Wang, G. Li, Facile synthesis of cobalt cluster-CoN x composites: synergistic effect boosts electrochemical oxygen reduction, *Journal of Materials Chemistry A*, 10 (2022) 16920–16927.
10. Y. Cao, Y. Su, L. Xu, X. Yang, Z. Han, R. Cao, G. Li, Oxygen vacancy-rich amorphous FeNi hydroxide nanoclusters as an efficient electrocatalyst for water oxidation, *Journal of Energy Chemistry*, 71 (2022) 167–173.
11. S. Ali, A. Raza, A.M. Afzal, M.W. Iqbal, M. Hussain, M. Imran, M.A. Assiri, Recent advances in 2D-MXene based nanocomposites for optoelectronics, *Advanced Materials Interfaces*, 9 (2022) 2200556.
12. A. Raza, U. Qumar, A.A. Rafi, M. Ikram, MXene-based nanocomposites for solar energy harvesting, *Sustainable Materials and Technologies*, 33 (2022) e00462.

13. J.Z. Hassan, A. Raza, U. Qumar, G. Li, Recent advances in engineering strategies of Bi-based photocatalysts for environmental remediation, *Sustainable Materials and Technologies*, 33 (2022) e00478.
14. Q. Shi, A. Raza, L. Xu, G. Li, Bismuth oxyhalide quantum dots modified sodium titanate necklaces with exceptional population of oxygen vacancies and photocatalytic activity, *Journal of Colloid and Interface Science*, 625 (2022) 750–760.
15. X. Zhong, Y. Liu, S. Wang, Y. Zhu, B. Hu, In-situ growth of COF on BiOBr 2D material with excellent visible-light-responsive activity for U (VI) photocatalytic reduction, *Separation and Purification Technology*, 279 (2021) 119627.
16. Q. Shi, X. Zhang, X. Liu, L. Xu, B. Liu, J. Zhang, H. Xu, Z. Han, G. Li, In-situ exfoliation and assembly of 2D/2D g-C3N4/TiO2 (B) hierarchical microflower: enhanced photo-oxidation of benzyl alcohol under visible light, *Carbon*, 196 (2022) 401–409.
17. M. Ikram, A. Raza, M. Imran, A. Ul-Hamid, A. Shahbaz, S. Ali, Hydrothermal synthesis of silver decorated reduced graphene oxide (rGO) nanoflakes with effective photocatalytic activity for wastewater treatment, *Nanoscale Research Letters*, 15 (2020) 1–11.
18. M. Ikram, E. Umar, A. Raza, A. Haider, S. Naz, A. Ul-Hamid, J. Haider, I. Shahzadi, J. Hassan, S. Ali, Dye degradation performance, bactericidal behavior and molecular docking analysis of Cu-doped TiO$_2$ nanoparticles, *RSC Advances*, 10 (2020) 24215–24233.
19. A. Raza, M. Ikram, M. Aqeel, M. Imran, A. Ul-Hamid, K.N. Riaz, S. Ali, Enhanced industrial dye degradation using Co doped in chemically exfoliated MoS$_2$ nanosheets, *Applied Nanoscience*, 10 (2020) 1535–1544.
20. A. Raza, A. Rafiq, U. Qumar, J.Z. Hassan, 2D hybrid photocatalysts for solar energy harvesting, *Sustainable Materials and Technologies*, 33 (2022) e00469.
21. U. Qumar, J.Z. Hassan, R.A. Bhatti, A. Raza, G. Nazir, W. Nabgan, M. Ikram, Photocatalysis vs adsorption by metal oxide nanoparticles, *Journal of Materials Science & Technology*, 131 (2022) 122–166.
22. G. Wang, J. Chang, W. Tang, W. Xie, Y.S. Ang, 2D materials and heterostructures for photocatalytic water-splitting: a theoretical perspective, *Journal of Physics D: Applied Physics*, 55 (2022) 293002.
23. A. Raza, J.Z. Hassan, A. Mahmood, W. Nabgan, M. Ikram, Recent advances in membrane-enabled water desalination by 2D frameworks: graphene and beyond, *Desalination*, 531 (2022) 115684.
24. A. Raza, X. Zhang, S. Ali, C. Cao, A.A. Rafi, G. Li, Photoelectrochemical energy conversion over 2D materials, *Photochemistry*, 2 (2022) 272–298.
25. A. Raza, S. Altaf, S. Ali, M. Ikram, G. Li, Recent advances in carbonaceous sustainable nanomaterials for wastewater treatments, *Sustainable Materials and Technologies*, 32 (2022) e00406.
26. A. Raza, M. Ikram, U. Qumar, A. Rafiq, Hybrid 2D nanomaterials for photocatalytic degradation of wastewater pollutants, in: *Innovative Nanocomposites for the Remediation and Decontamination of Wastewater*, IGI Global, 2022, pp. 101–125.
27. A. Raza, Y. Zhang, A. Cassinese, G. Li, Engineered 2D metal oxides for photocatalysis as environmental remediation: a theoretical perspective, *Catalysts*, 12 (2022) 1613.
28. A. Raza, M. Ikram, S. Guo, A. Baiker, G. Li, Green synthesis of dimethyl carbonate from CO$_2$ and methanol: new strategies and industrial perspective, *Advanced Sustainable Systems*, 6 (2022) 2200087.
29. Y. Xie, J. Yang, Y. Chen, X. Liu, H. Zhao, Y. Yao, H. Cao, Promising application of SiC without co-catalyst in photocatalysis and ozone integrated process for aqueous organics degradation, *Catalysis Today*, 315 (2018) 223–229.
30. B. Xu, Y. Gao, Y. Li, S. Liu, D. Lv, S. Zhao, H. Gao, G. Yang, N. Li, L. Ge, Synthesis of Bi$_3$O$_4$Cl nanosheets with oxygen vacancies: the effect of defect states on photocatalytic performance, *Applied Surface Science*, 507 (2020) 144806.

31. H. Li, J. Shang, Z. Ai, L. Zhang, Efficient visible light nitrogen fixation with BiOBr nanosheets of oxygen vacancies on the exposed {001} facets, *Journal of the American Chemical Society*, 137 (2015) 6393–6399.

32. H. Li, J. Shang, Z. Yang, W. Shen, Z. Ai, L. Zhang, Oxygen vacancy associated surface Fenton chemistry: surface structure dependent hydroxyl radicals generation and substrate dependent reactivity, *Environmental Science & Technology*, 51 (2017) 5685–5694.

33. H. Li, F. Qin, Z. Yang, X. Cui, J. Wang, L. Zhang, New reaction pathway induced by plasmon for selective benzyl alcohol oxidation on BiOCl possessing oxygen vacancies, *Journal of the American Chemical Society*, 139 (2017) 3513–3521.

34. J. Di, C. Chen, C. Zhu, M. Ji, J. Xia, C. Yan, W. Hao, S. Li, H. Li, Z. Liu, Bismuth vacancy mediated single unit cell Bi2WO6 nanosheets for boosting photocatalytic oxygen evolution, *Applied Catalysis B: Environmental*, 238 (2018) 119–125.

35. S. Gao, B. Gu, X. Jiao, Y. Sun, X. Zu, F. Yang, W. Zhu, C. Wang, Z. Feng, B. Ye, Highly efficient and exceptionally durable CO_2 photoreduction to methanol over freestanding defective single-unit-cell bismuth vanadate layers, *Journal of the American Chemical Society*, 139 (2017) 3438–3445.

36. P. Chen, H. Liu, W. Cui, S.C. Lee, L.a. Wang, F. Dong, Bi-based photocatalysts for light-driven environmental and energy applications: structural tuning, reaction mechanisms, and challenges, *EcoMat*, 2 (2020) e12047.

37. J. Di, J. Xiong, H. Li, Z. Liu, Ultrathin 2D photocatalysts: electronic-structure tailoring, hybridization, and applications, *Advanced Materials*, 30 (2018) 1704548.

38. G. Liu, C. Zhen, Y. Kang, L. Wang, H.-M. Cheng, Unique physicochemical properties of two-dimensional light absorbers facilitating photocatalysis, *Chemical Society Reviews*, 47 (2018) 6410–6444.

39. X. Lu, K. Xu, S. Tao, Z. Shao, X. Peng, W. Bi, P. Chen, H. Ding, W. Chu, C. Wu, Engineering the electronic structure of two-dimensional subnanopore nanosheets using molecular titanium-oxide incorporation for enhanced photocatalytic activity, *Chemical Science*, 7 (2016) 1462–1467.

40. Y. Zhao, W. Yu, G. Ouyang, Size-tunable band alignment and optoelectronic properties of transition metal dichalcogenide van der Waals heterostructures, *Journal of Physics D: Applied Physics*, 51 (2017) 015111.

41. J. Li, G. Zhan, Y. Yu, L. Zhang, Superior visible light hydrogen evolution of Janus bilayer junctions via atomic-level charge flow steering, *Nature Communications*, 7 (2016) 11480.

42. P. Tang, G. Hu, M. Li, D. Ma, Graphene-based metal-free catalysts for catalytic reactions in the liquid phase, *ACS Catalysis*, 6 (2016) 6948–6958.

43. R. He, D. Xu, B. Cheng, J. Yu, W. Ho, Review on nanoscale Bi-based photocatalysts. *Nanoscale Horizons*, 3 (2018) 464–504.

44. X. Jin, L. Ye, H. Xie, G. Chen, Bismuth-rich bismuth oxyhalides for environmental and energy photocatalysis, *Coordination Chemistry Reviews*, 349 (2017) 84–101.

45. M. Li, H. Zhang, Z. Zhao, P. Wang, Y. Li, S. Zhan, Inorganic ultrathin 2D photocatalysts: modulation strategies and environmental/energy applications, *Accounts of Materials Research*, 4 (2022) 4–15.

46. M. Guan, C. Xiao, J. Zhang, S. Fan, R. An, Q. Cheng, J. Xie, M. Zhou, B. Ye, Y. Xie, Vacancy associates promoting solar-driven photocatalytic activity of ultrathin bismuth oxychloride nanosheets, *Journal of the American Chemical Society*, 135 (2013) 10411–10417.

47. T. Liu, X. Zhang, F. Zhao, Y. Wang, Targeting inside charge carriers transfer of photocatalyst: selective deposition of Ag_2O on $BiVO_4$ with enhanced UV–vis–NIR photocatalytic oxidation activity, *Applied Catalysis B: Environmental*, 251 (2019) 220–228.

48. A. Chatterjee, P. Kar, D. Wulferding, P. Lemmens, S.K. Pal, Flower-like BiOI microspheres decorated with plasmonic gold nanoparticles for dual detoxification of organic and inorganic water pollutants, *ACS Applied Nano Materials*, 3 (2020) 2733–2744.

49. S. Yu, H. Huang, F. Dong, M. Li, N. Tian, T. Zhang, Y. Zhang, Synchronously achieving plasmonic Bi metal deposition and I–doping by utilizing $BiOIO_3$ as the self-sacrificing template for high-performance multifunctional applications, *ACS Applied Materials & Interfaces*, 7 (2015) 27925–27933.
50. Z. Zhou, Z. Shen, Z. Cheng, G. Zhang, M. Li, Y. Li, S. Zhan, J.C. Crittenden, Mechanistic insights for efficient inactivation of antibiotic resistance genes: a synergistic interfacial adsorption and photocatalytic-oxidation process, *Science Bulletin*, 65 (2020) 2107–2119.

CHAPTER **12**

Advancement in 2D Metals as Photocatalysts

Azam Aslani, Hadiseh Masoumi, Ahad Ghaemi, and Ram K. Gupta

12.1 INTRODUCTION

Global concerns regarding energy scarcity, pollution, and environmental deterioration have recently intensified significantly. The rapid population growth and increasing industrialization in various countries have led to the excessive use of fossil fuels and the discharge of industrial wastewater without proper treatment, which has resulted in water and air pollution. The absence of clean drinking water and pure air has emerged as one of the most significant challenges in the world today, posing a severe threat to human health and the environment. The environment is contaminated with a range of pollutants, including conventional and emerging organic pollutants; hence, it is crucial to adopt cost-effective and eco-friendly techniques and conduct thorough research in the field of materials engineering. One innovative and sustainable solution that shows great promise in addressing these challenges is the conversion of solar energy into chemical energy or fuel. Photocatalysis, which involves the use of solar energy as a clean, renewable, and abundant energy source, is a particularly promising area of study. Essentially, photocatalysis is an accelerated photoreaction process in the presence of a semiconductor catalyst. During this process, photons with energy $h v$ equal to or greater than the band gap energy of the catalyst are absorbed, creating a free electron from the valence band to the band gap and a hole in the capacity band. This approach has the potential to quickly transform pollutants into environmentally friendly and "green" energy using cutting-edge technologies [1]. Overall, a photocatalytic process in the attendance of a semiconductor embraces the following main steps: (a) absorption of light energy to generate electron–hole pairs, (b) separation of excited charges and transfer of electrons and holes to the photocatalyst surface, and (c) oxidation–reduction levels (Figure 12.1). Since the discovery of TiO_2's photocatalytic ability

DOI: 10.1201/9781032645001-12

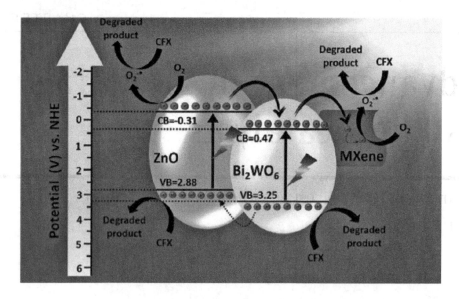

FIGURE 12.1 The photocatalytic mechanism for 2D hybrid nanocomposites under sunlight irradiation. Adapted with permission from [5]. Copyright (2022) Elsevier.

in 1972, research in this field has rapidly expanded. These technologies have proven to be effective in the remediation of environmental pollutants and renewable energy production. Nonetheless, current photocatalytic systems have limitations that hinder their throughput performance during the photocatalytic process. Specifically, the high rate of electron–hole recombination in semiconductor materials with small band gaps poses a significant challenge. This instability causes the excited-state charge carriers to recombine, reducing the efficiency of the photocatalytic process. To achieve the highest separation performance, the recombination of electron–hole pairs should be intercepted. To date, there are diverse scholarship approaches like chemical substitution, introduction of crystal imperfection, design of crystal dimensions, surface morphology and facets, and hybridization with semiconductors that have made the use of 2D materials as photocatalysts increasingly popular. Metals are particularly interesting due to their unique properties, and many types of metal-based compounds fall under the umbrella of 2D materials, such as TMDs, LDHs, TMOs, Mxenes, MOFs, and their hybrid compounds. Surface rectification can also optimize the performance and properties of these materials for specific applications [2–4]. In this chapter, we provide a comprehensive review of 2D metal nanosheets and their potential to advance the field of photocatalyst technology.

12.2 CATEGORIZING OF 2D METALS FOR PHOTOCATALYSTS

12.2.1 2D Layered Double Hydroxide (LDH)

Numerous research studies have been dedicated to exploring 2D LDH nanosheets and their hybrid-type catalysts in the photocatalytic process and environmental protection due to their beneficial performance and the ability to adjust their chemical composition. LDHs comprise a major collection of layered metal hydroxides with the chemical formula $[A^{2+}_{1-x} B^{3+}_{x} (OH)^2]x^+ [C^{n-}_{x-n}]x^-yH_2O$ in which A^{2+} represents divalent metal ions, B^{3+} represents trivalent metal ions, and Cn^- represents soluble anions (organic or inorganic substitution by negative charge (n)), and y is the number of water molecules in the interlayer. Divalent and trivalent metal ions can be represented as (A^{2+}: Mg^{2+}, Fe^{2+}, Cu^{2+}, Ni^{2+}, Zn^{2+}, or Co^{2+}) and (B^{3+}: Al^{3+}, Cr^{3+}, In^{3+}, Fe^{3+}, Mn^{3+}, or Ga^{3+}), respectively. LDHs are promising materials for the extension of visible light-responsive photocatalysts due to their textural, constructional, and acid–base properties, tunable composition, and anion exchange capacity. In addition, the 2D-layered structure of LDH decomposes with heat and causes the formation of non-stoichiometric mixed oxides with unrivaled characteristics suitable for photocatalysis applications [6].

12.2.2 2D Transition Metal Carbides (TMCs)

MXenes are a major class of layered materials with a hexagonal crystal structure consisting of carbides, nitrides, and carbonitrides of transition metals with the general formula $M_{n+1}AX_n$, where M is a primary transition metal (Sc, Y, Ti, Zr, Hf, V, Nb, Ta, Cr, Mo, or W); A represents the elements of groups 12–16 of the periodic table; the letter X represents the two elements carbon or nitrogen, and n is equal to 1, 2, or 3. These materials have high activity for promoting carrier transport, effective charge separation, strong electronic coupling between surfaces, increase of redox reaction rate, high hydrophilicity, and expanded 2D surface and are also exploited as effective hybridization support for disparate photocatalysts. Several MXenes including Ti_3C_2, Ti_2C, Nb_4C_3, Ti_3CN, Ta_4C_3, Nb_2C, V_2C, and Nb_4C_3 were synthesized from MAX phases. The first and best-known MXene synthesized by etching Al from Ti_3AlC_2 is titanium carbide ($Ti_3C_2T_x$) [7].

12.2.3 2D Transition Metal Oxides (TMOs)

In the past four decades, TMOs have been one of the most promising materials for developing 2D nanosheet-based, hybrid-type, high-performance water-splitting photocatalysts. A large number of TMOs have large band-gap energies that provide attractive energy levels for redox reactions, but most of them have poor electron conductivities that reduce photocatalytic efficiencies. Various metal oxides, for example, TiO_2, ZnO, SnO_2, WO_3, Fe_2O_3, In_2O_3, and CeO_2, have been widely investigated as catalysts, with TiO_2 being widely exploited due to its environmental opportuneness, excellent durability, desired electronic structure, and light absorption [8].

12.2.3.1 Perovskites

2D oxide perovskites consist of layers with the generic formula ABO_3 with a cubic structure. Typically, the A and B sites are occupied with two cations of vastly different sizes (dA > dB), while X (typically O) is the anion that attaches to both. Perovskites are defined by a crystal structure resembling calcium titanium oxide ($CaTiO_3$). 2D perovskites are segregated into three phases: Dion–Jacobson phase, Ruddlesden–Popper phase, and Aurivillius phase; these include the general formula ($A' [A_{n1} B_nO_{3n+1}]$), ($A' [A_{n2}B_nO_{3n+1}]$), and ($Bi_2O_2[A_{n-1}B_nO_{3n+1}]$), respectively (Figure 12.2). Ultrathin TMO nanosheets are synthesized by exfoliation of pristine layered TMOs through chemical modification. In the Dion–Jacobson phase, the individuating layer is a layer of alkali metals or lanthanides, which can be easily converted into single-layer nanosheets through ion exchange with bulk organic cations.

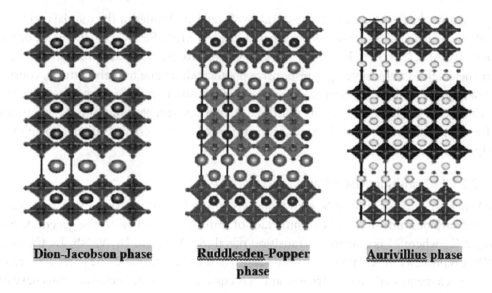

FIGURE 12.2 Crystal structures for variant sorts of layered perovskite phases. Adapted with permission from [10]. Copyright. (2022) Royal Society of Chemistry.

Also, the individuating layer in the Ruddlesden–Popper phase consists of metal ions that are located between the iterating layers of the ABO_3 lattice [9].

12.2.4 2D Transition Metal Dichalcogenides (TMDs)

MX_2 materials have been the subject of extensive research in the field of photocatalysis and their potential applications in technology. These 2D nanoplates consist of intermediate metal ions of groups 4, 5, and 6 (Ti, Zr, and Hf; V, Nb, and Ta; Mo, W) represented by M, and chalcogens (S, Se, and Te) represented by X (Figure. 12.3a). Each monolayer of the transition metal dichalcogenides (TMDs) is composed of three hexagonal atomic layers, creating an "X–M–X" sandwich structure. Weak van der Waals forces hold adjacent monolayers together, with a distance of 6–7 Å apart, thus facilitating their isolation. TMDs exhibit three polyhedral structures, 1T (octahedron), 2H (hexagonal prismatic), and 3R (distorted octahedron), representing different coordination geometries of transition metal atoms)Figure 12.3b.(2D $MX2$ materials have caught the attention of materials scientists owing to their tunable band gap, high carrier mobility, and large surface area, which make them potential catalysts. Depending on the metallic element's d-band filling state, 2D transition metal dichalcogenides (TMDs) can exhibit diverse electronic band structures, such as semiconductors, semimetals, true metals, and even superconductors [11].

12.2.5 2D Metal-Organic Framework (MOF)

Metal-organic frameworks (MOFs) have garnered significant attention in the scientific community owing to their remarkable properties and versatility. These frameworks comprise metal nodes bonded to organic ligands, providing a customizable chemical composition and a suitable structure for various uses. MOFs have a high surface area and porosity, allowing them to accommodate unsaturated metal sites. Additionally, MOFs can act as

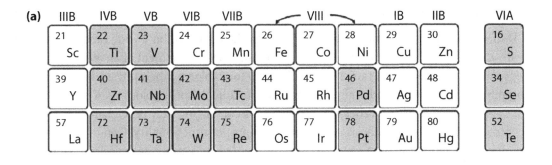

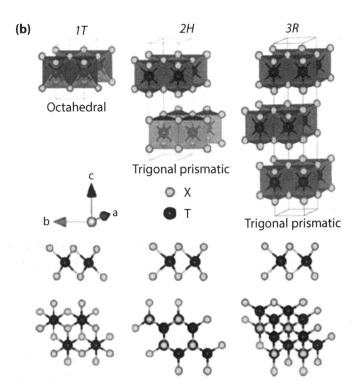

FIGURE 12.3 (a) Layered TMDs in the periodic table highlighted with shadow. (b) Ball-and-stick model of 1T, 2H, and 3R polymorphs of group VITMDs. Blue and yellow balls represent Mo and S atoms, respectively. Adapted with permission from [12].

support structures for metal nanoparticles, thus expanding their versatility. Metal-organic frameworks (MOFs) in the form of 2D nanosheets, with specific shapes, sizes, and compositions, such as 0D MOF nanoparticles, 1D MOF nanowires, and hierarchical mesoporous MOFs, have demonstrated significant potential in the field of photocatalysis. The unique two-dimensional morphology of these MOFs provides several advantages over their three-dimensional counterparts, including a layered structure and coatings that make them more suitable for surface applications. Furthermore, the increased adhesion and contact surface ratio of 2D MOFs with the substrate further enhances their catalytic potential, highlighting their promise for advanced applications in catalysis [13].

12.3 SYNTHESIS APPROACH

2D metals are an intriguing category of 2D materials that exhibit a wide range of samples with at least one dimension between 1 and 100 nm. The development of an efficient synthesis methodology for 2D metals creates a platform for exploring their physical and chemical properties with the desired composition and controllable morphology as well as for various applications. The properties of nanomaterials can be precisely tuned through accurate control of size, shape, synthesis conditions, and proper functionality. Generally, 2D metal synthesis methods for photocatalysis can be divided into two categories: top-down and bottom-up methods [14].

12.3.1 Top-Down Procedure

In the field of nanotechnology, the process of creating ultrathin nanosheets involves physical or chemical methods in a top-down synthesis approach. This method is particularly effective for layered materials with highly anisotropic properties, where the in-plane bonds are strong univalent bonds and the out-of-plane bonds are composed of weak van der Waals interactions or ionic interactions with exchangeable interlayer species. To obtain 2D monolayer and multilayer materials, mechanical force or ultrasonic waves are used to exfoliate van der Waals layered solids. In contrast, chemical reactions by ion exchange or heat are typically used for the top-down chemical approach. The nanosheets are typically obtained by exfoliating the bulk crystal, using methods such as liquid phase exfoliation, mechanical cutting, and etching, either chemically or mechanically. The production of multilayered 2D materials in a solvent-dispersed form, using techniques such as ultrasound, bath sonication, or tip sonication, is almost exclusively based on the LPE technique [15].

12.3.1.1 Mechanical Exfoliation

The mechanical exfoliation method is a widely adopted technique for synthesizing materials that are composed of one or more layers of 2D crystal parts while preserving their structural and crystalline properties. This method involves applying external force through various techniques such as ultrasound, wet ball milling, and freeze melting to process bulk materials into one or more layers of 2D materials with layer-by-layer or interlayer exfoliation. However, during the exfoliation process, the attraction caused by the interaction of van der Waals forces between the layers must be carefully managed to separate one layer from another. In this regard, the lateral and normal forces play vital roles in the separation of the layers, where the van der Waals force is overcome by the normal force during the peeling process, while the lateral force is applied in the lateral direction between the layers. The choice of solvent is also crucial in the efficiency of mechanical exfoliation, with common solvents used in this process including acetone, water, methanol, ethanol, and tetrahydrofuran [16].

12.3.1.2 Liquid Phase Exfoliation

The concept of liquid peeling encompasses various mechanisms depending on the approach used. The most common methods of liquid exfoliation include ion exfoliation,

ion exchange, and ultrasonic exfoliation, which can be classified. Among these, ultrasonic-based exfoliation is the most widely used method, which involves the use of micro-jets and shock waves generated by the collapse of microbubbles due to liquid cavitation to exfoliate layered materials. Through this method, 2D nanosheets can be synthesized by placing bulk materials in suitable solvents and applying appropriate ultrasound treatments. The solvent used plays a critical role in this process, and choosing the right solvent is essential for achieving stable and effective exfoliation. The stability of the mixed solution can be determined by the ionization and polarizability of different components based on the standard solution theory. With the development of new materials and their diverse applications, ultrasonic liquid exfoliation has been used to prepare TMD nanosheets such as MoS_2 and $NbSe_2$ through mechanical exfoliation [17].

12.3.2 Bottom-up Procedure

The bottom-up strategy is opposite and complementary to the top-down approach, which includes the growth of nanosheets directly from atoms, ions, or molecules, which is usually used to synthesize 2D nanosheets with high quality, large lateral dimensions, and high controllability. This process includes methods such as hydrothermal synthesis, hot injection method, and chemical vapor deposition (CVD). One of the important points of this method is to prevent the growth of synthesized nanoplates in the direction perpendicular to the layers [18].

12.3.2.1 Chemical Vapor Deposition (CVD)

Chemical vapor deposition (CVD) is a high-temperature process that involves the controlled deposition of a desired material on substrates. By carefully selecting the reagents, substrates, catalysts, temperature, and gas atmosphere, it is possible to achieve the controlled growth of two-dimensional (2D) nanomaterials with a specific number of layers and degree of crystallinity. Some 2D nanomaterials such as graphene and MOS_2 have been easily prepared using this method. However, the CVD method has some drawbacks, including the requirement for high heat and vacuum conditions [19].

12.3.2.2 Wet Chemical Procedure

Hydrothermal and solvothermal synthesis are widely used "wet" chemical synthesis methods for obtaining various two-dimensional transition metal oxides (2D TMOs). Solvothermal synthesis involves a chemical reaction in a liquid or supercritical fluid at high temperatures, typically carried out in a closed system, such as an autoclave or bomb, capable of withstanding high pressure. However, studying the mechanism of these methods is challenging, as the reactions take place in isolated autoclaves, limiting experimental design for the synthesis of other 2D TMOs. Additionally, synthesis via these methods is sensitive to precursor concentration, choice of solvents, surfactants, and temperature, which hinders precise control over structure, morphology, and reproducibility. Wet chemical techniques allow the removal of the material by introducing various chemicals during development, and ligands can coat the material surface to modify or smooth it [20]. The research by Liang et al. has shown that through the coordination of metal cations with organic anions (such as oleate ions) via electrostatic interactions, atomically thin structures can be produced through a directed self-assembly process for a wide range of 2D ultrathin metal oxides [21].

12.3.2.3 2D-Template Synthetic Procedure

The 2D template synthetic strategy is a highly effective method for producing materials with layered and non-layered nanostructures of varying sizes, morphologies, and charge distributions. The process involves logically selecting 2D templates onto which precursor atoms are drawn, followed by the preparation of the template and the creation of the desired material using the template. If necessary, the template can then be separated from the material. The fabrication process may involve physical methods such as surface coating or chemical methods such as addition, elimination, substitution, or isomerization reactions. After the reaction is complete, the template can be removed using physical methods like dissolution or chemical methods such as calcination. Based on the available templates, template-assisted synthesis (TAS) can be classified further into three methods: hard template method, soft template method, and colloid template method, and Figure 12.4 illustrates the schematic representation of materials constructed using these methods. Another approach to obtain ultrathin 2D photocatalysts in the 2D synthetic strategy is to transform the 2D precursor into a desired phase [22].

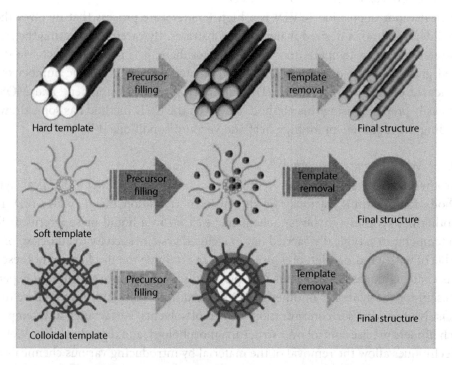

FIGURE 12.4 Schematic representation of the synthesis of materials applying diverse types of templates. Adapted with permission from [23]. Copyright (2020) Royal Society of Chemistry.

12.4 TECHNIQUES TO IMPROVE THE PHOTOCATALYTIC PERFORMANCE OF 2D NANOMETALS

Photocatalysis is a fundamental process where semiconductors generate electron–hole pairs upon light irradiation, which act as the driving force for the redox reaction. In the photocatalyst process, several factors influence the efficiency of photocatalysis, including the appropriate band gap distance, the synthesis of high electron mobility 2D materials, and the architecture of 2D materials to increase surface reaction dynamics. There are numerous approaches to improving the photocatalytic activity of the photocatalyst based on 2D materials. For example, doping elements can modify the band structure of some 2D semiconductors, creating heterojunctions, adjusting dimensions, and modifying the surface to adjust the light absorption properties of the photocatalyst [24].

12.4.1 Elemental Doping

Metal nanocomposites that have 2D structures possess high surface energy. This characteristic makes them easy to adjust their physicochemical and photophysical properties, such as light absorption, charge mobility, charge separation, surface structure, and density through surface engineering. By incorporating appropriate external elements into the host 2D materials, their properties can be altered. Elemental doping, which includes metal doping, non-metal doping, self-doping, and co-doping, can be achieved by pre-treating the substrate catalyst or post-treating the prepared 2D materials [25].

12.4.1.1 Metal Doping

Metal doping is a process of modifying the composition of 2D materials by inserting other metal elements into designated places either before or after their synthesis. Several research groups have employed this technique to augment the photocatalytic performance of 2D semiconductor materials. One way of achieving metal doping is by incorporating other elements into the layered metal oxide matrix during the synthesis of bulk compounds, thereby producing metal-doped metal oxide nanosheets. During the exfoliation process, the composition of the doped metal oxide precursors remains intact. For instance, Wang et al. used cobalt as a metal element to modify BiOCl (Co-BiOCl) nanoplates through the hydrothermal synthesis method [26].

12.4.1.2 Nonmetal Doping

Non-metallic doping can provide an alternative way to reduce the wide band gap of 2D semiconductors towards visible light harvesting, modify the composition of 2D materials, and achieve special properties and advanced applications. In this regard, non-metallic elements N, S, P, and B have been used. In a typical study, Liu et al. in 2016 reported the hierarchical flower-shaped structure of N-doped MoS_2 synthesized using the sol-gel method for the removal of bis Rhodamine B in which thiourea $((NH_2)_2CS)$ is used as the impure nitrogen source. Using non-metallic element nitrogen, the band gap energy of MoS_2 doped with N (2.08 eV) was observed to be slightly lower than that of MoS_2 nanoplates (2.17 eV) [27].

182 ■ 2D Metals

12.4.2 Hybridization

In recent times, there has been a lot of effort put into creating composites using 2D metal compounds, such as metal oxides and chalcogenides, with different compositions. Different methods of synthesis have been developed to hybridize 2D nanomaterials with other low-dimensional functional materials, such as 0D, 1D, or 2D. The functions and properties of composite photocatalysts are highly dependent on the dimensions of their building blocks and the surface contact between them. 2D composites are categorized into 0D/2D, 1D/2D, and 2D/2D systems based on the dimension and size difference of their building blocks. For example, Tan et al. have used Ag–Pd bimetallic nanoparticle alloys on 2D titanium dioxide nanoplates modified with nitrogen to increase the efficiency of CO_2 production for the synthesis of Ag–Pd/TiO_2 hybrid composites with 0D/2D heterogeneity interface. The improved photocatalytic activity of Ag–Pd/TiO_2 rough junctions originated from the abundant Ti^{3+} ions creating oxygen vacancies in the 2D TiO_2 nanosheets due to the nitrogen substitution effect, which contributed to the excellent adsorption of CO_2 reduction [28]. To make hybrid composites with 2D/2D heterogeneous connection, the synthesis of multilayered titania-sandwiched Ti_3C_2 MXenes multilayers with oxygen vacancy-enriched trimetallic CoAlLa-LDH was constructed for enhanced CO_2 photoreduction. The coupling between $Ti_3C_2T_x$ and trimetallic CoAlLa-LDH provides heterogeneous S-scheme pathways for efficient carrier mobility with improved CO_2 reduction performance to produce CO and CH_4 with evolution rates of 46.32 and 31.02 $\mu mol/g^1$ h^1, respectively, compared to low-purity catalysts [29].

12.5 MECHANICS OF PHOTOCATALYSIS

The mechanism of photocatalysis commonly consists of the generation of electron–hole pairs in a photocatalytic semiconductor under appropriate light energy, where they major get transferred to the surroundings and generate the radical sorts to perform the reduction and oxidation reactions into the favorable photocatalytic applications. In this process, the respective band edge position of the photocatalyst with attention to the redox kinds is very significant for achieving the high-grade photocatalytic process. For example, photocatalytic water splitting needs the conduction band to be more positive and the valence band to be more negative. The particular application requires a particular sort of radical kind that essentially generates either electrons or holes. For instance, the dye degradation process occurs via oxidation reactions, where it desires the participation of holes, while the toxic heavy metals can be transformed into nontoxic through reduction reactions that need electrons. Considering the process of degradation of the pollutants, the essential photocatalytic reactions can be explained in the following equations (Eqs (12.1)–(12.9)) [30]:

$$PC + h\,\nu \rightarrow e^- + h^+ \tag{12.1}$$

$$h^+ + H_2O \rightarrow OH^+ + OH^- \tag{12.2}$$

$$h^+ + OH^- \rightarrow OH^\bullet \tag{12.3}$$

$$e^- + O_2 \rightarrow O_2^- \tag{12.4}$$

$$2e^- + O_2 + 2H^+ \rightarrow H_2O_2 \tag{12.5}$$

$$2e^- + H_2O_2 \rightarrow OH^\bullet + OH^- \tag{12.6}$$

$$R + OH \rightarrow \text{Degradation products} \tag{12.7}$$

$$R + h^+ \rightarrow \text{oxidation degraded products} \tag{12.8}$$

$$R + e^- \rightarrow \text{Reduction degraded products} \tag{12.9}$$

12.6 APPLICATION OF PHOTOCATALYST

12.6.1 Dye Degradation

The ongoing expansion of textile industries has led to a significant discharge of organic effluents by various industries, including paper, textile, food, cosmetic, and dye industries, into water bodies. This has resulted in the contamination of water systems by dye pollutants and textile industry waste, which can cause carcinogenic and waterborne diseases. The toxicity, mutagenicity, and carcinogenicity of most dyes further exacerbate this problem. In recent decades, photocatalysis has emerged as a promising technology for degrading dyes, with various photocatalytic nanostructures being employed to degrade different types of dyes, including acid dyes, basic dyes, direct dyes, and reactive dyes [31]. For instance, Bhuvaneswari et al. reported the use of 2D/2D hybrid materials with layered triple hydroxide (LTH) for the photocatalytic degradation of methylene blue dye in aqueous solution under UV–visible light irradiation. The researchers prepared ZnMgAl LTH using a one-step hydrothermal process with the same molar ratio of (Zn), (Mg), and (Al) precursors and a two-dimensional MOF. The 2D/2D hybrid was synthesized using the same method as that used for MOF-5, and its photophysical properties were analyzed through a series of analytical methods. The energy gap of the prepared samples was determined by plotting $(\alpha h\nu)^2$ versus photon energy $(h\nu)$ and was found to be 3.71, 2.93, and 3.15 for MOF-5, LTH, and MOF-5/LTH, respectively. Reducing the band gap from 3.71 to 3.15 may increase methylene blue degradation efficiency. The MOF-5/LTH hybrid sample showed the highest catalytic activity, with degradation efficiencies of 43.3%, 57.7%, and 98.1% for MOF-5, LTH, and MOF-5/LTH hybrid samples, respectively, after 125 min of irradiation. The photocatalytic rate constants for MOF, LTH, and MOF-5/LTH hybrids were reported as 0.0047, 0.0065, and 0.0236, respectively [32].

12.6.2 Pharmaceutical Pollutant Degradation

Pharmaceutical pollutants are another serious concern for the environment. In general, drugs have extreme biological activity, which causes them to affect living organisms. These materials enter the environment through pharmaceutical industries, hospitals, and human and animal excrement [33]. Wang et al. used 0D Bi nanodots/2D Bi_3NbO nanocomposites synthesized by the hydrothermal method for the photocatalytic degradation of ciprofloxacin (CIP) under visible light irradiation. The results showed that the photocatalytic degradation rate of CIP by Bi/Bi_3NbO_7 composites is 4.58 times higher than that by pristine Bi_3NbO_7. The Bi/Bi_3NbO_7 photocatalyst showed high photocatalytic activity after five cycles, which indicates its stability and reusability. Transmission electron microscopy (TEM) analysis showed that Bi nanodots with a diameter of 2–5 nm were uniformly distributed on the surface of Bi_3NbO_7 nanosheets. In addition, experiments and density functional theory (DFT) calculations confirmed a strong covalent interaction between the Bi atom of Bi nanodots and the Bi single bond O layer on the surface of Bi_3NbO_7 nanosheets. This interaction increases the visible light absorption of the composite and leads to improved photocatalytic activity. Despite the significant progress made in the field of density functional theory (DFT), the calculated band gap value of Bi_3NbO_7 (2.56 eV) is lower than the experimental value (2.69 eV) due to the inherent limitations of DFT calculations. However, a comprehensive analysis of the density of states (DOS) of Bi/Bi_3NbO_7 has been performed and compared with the DOS for pure Bi_3NbO_7, which shows that the density of states in the valence band maximum (VBM) of Bi/Bi_3NbO_7 is significant. It increases due to the orbital contribution of the Bi element. Furthermore, the conduction band of Bi/Bi_3NbO_7 shifts to lower energy levels. These findings have great potential to promote the production of photogenerated charge carriers, which in turn contribute to the photocatalytic degradation of CIP [34].

12.6.3 Removal of Heavy Metal

Heavy metal contamination in water, especially drinking water, is a serious global environmental problem due to its toxicity and carcinogenicity. One of the characteristics of heavy metals is a density higher than 5 grams per cubic centimeter and a relative atomic weight greater than 40. Many elements, including first- and second-group metals, transition metals, lanthanides, and actinides in the periodic table, belong to this category. Among heavy metals, chromium (Cr), cadmium (Cd), mercury (Hg), lead (Pb), arsenic (As), vanadium (V), zinc (Zn), copper (Cu), uranium (U), nickel (Ni), and so on cause major problems and serious concerns in the field of health and environment. Many of these elements that are naturally in the earth's crust can enter the atmosphere, water, and soil systems through natural activities such as dust storms and forest fires and industrial, agricultural, and domestic activities. Many of these elements can enter the atmosphere, soil, and water systems through natural, industrial, agricultural, and domestic activities. Industrial activities such as mining, metal plating, painting, microelectronics, paper production, and the use of insecticides and phosphate composts contain various concentrations of heavy metals that can endanger the ecosystem and the health of organisms. These metals can disrupt the proper functioning of organs (for example, the liver, kidney, heart, brain, and lungs) and lead to disorders of the nervous and muscular system. Therefore, the treatment of heavy metal pollution has become a serious global need [35]. Zhang et al. synthesized CdS/$ZnIn_2S_4$ photocatalysts combined with 1D CdS nanorods and 2D $ZnIn_2S_4$ nanosheets for effective Cr(VI) photoreduction. CdS/$ZnIn_2S_4$ composites with a molar ratio of 1:0.33 showed the highest performance with Cr(VI) entirely reduced within 30 min [36].

12.6.4 CO_2 Reduction

Fossil fuel combustion produces CO_2, which, when released into the atmosphere without adequate purification, contributes to the greenhouse effect. To mitigate this problem, converting CO_2 into energy fuels based on selected hydrocarbons and chemicals has emerged as an effective recycling strategy. Multiple electron mechanisms can yield various products, such as CO, HCOOH, CH_3OH, HCHO, and CH_4. However, CO_2's linear structure and high C = O double bond energy (>750 kJ mol^1) render it chemically stable and inert, posing a significant challenge to its conversion. Halmann first performed the reduction process of CO_2 by PEC to CH_3OH, HCOOH, and HCHO using a p-GaP photocathode and carbon anode [37]. The MoS_2/TiO_2 heterogeneous heterojunction composite was used to reduce CO_2 to CH_4 and CO under visible light irradiation, which showed superior performance and stability. First, rose-like MoS_2 was synthesized using the hydrothermal method. Then, 2D/2D MoS_2/TiO_2 nanocomposite was prepared by a one-step hydrothermal experimental method with calcination (300 °C) in an argon atmosphere. The results show that pure MoS_2 exhibits poor photocatalytic activity due to its high electron–hole recombination, and the maximum efficiency of CO and CH_4 in the 10% MoS_2/TiO_2 heterogeneous composite is 268.97 μmol/g cat and 49.93 μmol/g. cat, which are about 5.33 times and 16.26 times that of pure TiO_2 (P25), respectively. It can be concluded that sulfide can increase the photocatalytic activity of TiO_2 for the photocatalytic reduction of CO_2 with high potential [38].

12.6.5 N_2 Fixation

The reduction of N_2 to NH_3, or N_2 fixation, has made it a fundamental and attractive research topic due to the essential role of NH_3 as an energy carrier and its important role in the biological synthesis of agricultural fertilizers. The main method of industrial synthesis of NH_3, which was introduced by Haber-Bosch in the 1900s, has three main limitations: (1) high temperature and pressure (300–500 °C and 150–300 atm) are required to carry out the reaction, (2) requires a lot of energy to carry out the reaction, and (3) causes environmental pollution due to annual CO_2 production of more than 400 million tons. Photocatalysis as an effective solution does not have the limitations of the Haber-Bosch method and therefore has been widely investigated in recent times. The reduction reaction for N_2 fixation is similar to the CO_2 reduction method, but the adsorption of N_2 molecules on the photocatalyst surface is one of the main limitations of the process [39]. Zhao et al. studied two-dimensional very thin layer photocatalysts of AB-LDH hydroxide (A = Mg, Zn, Ni, Cu, and B = Al, Cr), especially CuCr–LDH nanosheets, to investigate their photocatalytic activity for reducing N_2 to NH_3 in water under exposure to visible light (Figure 12.5a–b). The excellent activity of the 2D CuCr–LDH nanometal can be attributed to the highly distorted structure and compressive strain in the LDH nanosheets, which weakens the N_2 triple bond, resulting in a longer N–N distance and conditions for NH_3 generation [40].

12.6.6 H_2 Evolution

Recently, hydrogen has been recognized as a clean, cheap, and renewable energy to replace fossil fuels. Hydrogen production through photocatalytic water splitting is an environmentally friendly strategy for renewable solar energy production with the advantages of improving environmental pollution and reducing traditional fuel consumption [41]. For example, Ou et al. used the 2D/2D $Zn_mIn_2S_{3+m}$/Mxene heterogeneous composite as an effective photocatalyst for the hydrogen evolution reaction (Figure 12.5). Ternary chalcogenides $Zn_mIn_2S_{3+m}$ ($m = 1–3$) have great potential in photocatalytic hydrogen evolution reactions but were modified due to their low efficiency. In this research, MXene ($Ti_3C_2T_x$) nanosheets were used as a cocatalyst substrate to support the in situ growth of very thin $ZnmIn_2S_{3+m}$ nanosheets, accelerating charge separation and mobility, large surface area, and hydrophilicity. In particular, the Schottky junction formed at the $Zn_mIn_2S_{3+m}$/MXene interface induces timely electron transfer to the MXene nanosheets. This mechanism of

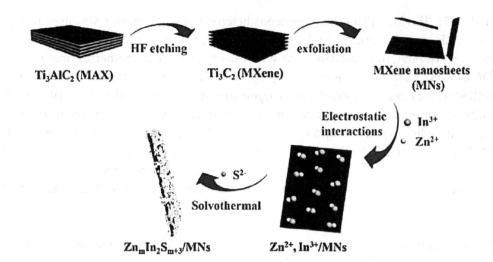

FIGURE 12.5 Synthesis method of ZnmIn$_2$S$_{m+3}$/MNs heterostructure. Reprinted with permission from [42]. Copyright (2021) Elsevier.

H$_2$ photocatalytic evolution was further evidenced by DFT calculations and experimental results. In conclusion, 2D/2D ZnmIn$_2$S$_{3+m}$/MXene shows good stability and excellent H$_2$ yield under visible light irradiation, especially for Zn$_2$In$_2$S$_5$/MXene (2.5 wt% MXene), which is 2.67 times higher than pristine Zn$_2$In$_2$S$_5$ [42].

12.7 CONCLUDING REMARKS AND PERSPECTIVES

Fossil fuels are still the flag bearer of energy solicitation in the world, and it can be forespoken that the percentage of fossil fuels' energy utilization will remain high shortly. Photocatalysis is an inexpensive and environmentally friendly technique that employs solar radiation as an energy source. The efficiency of the photocatalyst strongly depends on the surface characteristics and physical and chemical properties of the photocatalyst. In this chapter, we have introduced 2D nanometals suitable for photocatalytic activity, such as the removal of dye and pharmaceutical industry effluents, CO$_2$ reduction, N$_2$ fixation, reduction of heavy metals, etc. Most of the known photocatalysts have a high band gap, which limits the photocatalytic activity in the UV light range. Therefore, more focus should be applied to developing 2D nanometals with suitable band gaps and no fast recombination of electrons. Various methods such as doping, heterojunction formation, and Z design have been proposed to deal with these problems. However, modifying the surface size of catalysts increases their photocatalytic activity, but the size of nanostructures must be attentively preserved because increasing the size affects the photocatalyst surface and reduces the number of active sites, which leads to a decrease in photocatalytic efficiency. As a result, more research should be done for cost-effective and visible-light-reactionary 2D nanometals.

REFERENCES

1. A. A. Isari, F. Hayati, B. Kakavandi, M. Rostami, M. Motevassel, and E. Dehghanifard, N, Cu co-doped TiO$_2$@ functionalized SWCNT photocatalyst coupled with ultrasound and visible-light: an effective sono-photocatalysis process for pharmaceutical wastewater treatment, *Chemical Engineering Journal* 392 (2020) 123685.

2. S. Barua, D. Sahu, N. Shahnaz, and R. Khan, Chapter 1 - Chemistry of two-dimensional nanomaterials Elsevier, 2020, 1–33.

3. R. Hu, G. Liao, Z. Huang, H. Qiao, H. Liu, Y. Shu, *et al.*, Recent advances of mono elemental 2D materials for photocatalytic applications, *Journal of Hazardous Materials* 405 (2021) 124179.

4. Y. Zhao and H. Duan, *Photocatalysis Using 2D Nanomaterials*, 2022. https://doi.org/10.1039/9781839164620.

5. V. Sharma, A. Kumar, A. Kumar, and V. Krishnan, Enhanced photocatalytic activity of two-dimensional ternary nanocomposites of $ZnO-Bi_2WO_6-Ti_3C_2$ MXene under natural sunlight irradiation, Chemosphere 287 (2022) 132119.

6. B. Song, Z. Zeng, G. Zeng, J. Gong, R. Xiao, S. Ye, *et al.* Powerful combination of $g-C_3N_4$ and LDHs for enhanced photocatalytic performance: A review of strategy, synthesis, and applications, *Advances in Colloid and Interface Science* 272 (2019) 101999.

7. S. Zhang, M. Bilal, M. Adeel, D. Barceló, and H. M. Iqbal, MXene-based designer nanomaterials and their exploitation to mitigate hazardous pollutants from environmental matrices, *Chemosphere* 283 (2021) 131293.

8. K. Kalantar-Zadeh, J. Z. Ou, T. Daeneke, A. Mitchell, T. Sasaki, and M. S. Fuhrer, Two dimensional and layered transition metal oxides, *Applied Materials Today* 5 (2016) 73.

9. W. Guo, Z. Yang, J. Dang, and M. Wang, Progress and perspective in Dion-Jacobson phase 2D layered perovskite optoelectronic applications, *Nano Energy* 86 (2021) 106129.

10. T. Gu, N. Kwon, and S. Hwang, 2D inorganic nanosheet-based hybrid photocatalysts for water splitting, 170, 2022. https://doi.org/10.1039/9781839164620-00170.

11. T. Liang, Y. Cai, H. Chen, and M. Xu, Two-dimensional transition metal dichalcogenides: An overview 1–27, 2019.

12. A. Kuc, Low-dimensional transition-metal dichalcogenides, 2014. https://doi.org/10.1039/9781782620112-00001.

13. S. Li, S. Shan, S. Chen, H. Li, Z. Li, Y. Liang, *et al.* Photocatalytic degradation of hazardous organic pollutants in water by Fe-MOFs and their composites: A review, *Journal of Environmental Chemical Engineering* 9 (2021) 105967.

14. J. Wang, G. Li, and L. Li, Synthesis strategies about 2D materials, *Two-dimensional Materials-Synthesis, Characterization and Potential Applications*, 1–20, 2016. https://doi.org/10.5772/63918.

15. T. Wang, M. Park, Q. Yu, J. Zhang, and Y. Yang, Stability and synthesis of 2D metals and alloys: A review, *Materialstoday Advances* 8 (2020) 100092.

16. S. Das, M. Kim, J.-w. Lee, and W. Choi, Synthesis, properties, and applications of 2-D materials: A comprehensive review, 231–252, *Critical Reviews in Solid State and Materials Sciences* 39 (2014) 231.

17. C. Huo, Z. Yan, X. Song, and H. Zeng, 2D materials via liquid exfoliation: a review on fabrication and applications, 1994–2008, Science Bulletin 60 (2015) 1994.

18. Y. Xue, G. Zhao, R. Yang, F. Chu, J. Chen, L. Wang, *et al.*, 2D metal-organic framework-based materials for electrocatalytic, photocatalytic and thermocatalytic applications, 3911–3936, *Nanoscale* 13 (2021) 3911.

19. J. You, M. D. Hossain, and Z. Luo, Synthesis of 2D transition metal dichalcogenides by chemical vapor deposition with controlled layer number and morphology, *Nano Convergence* 5 (2018) 26.

20. S. Alam, M. A. Chowdhury, A. Shahid, R. Alam, and A. Rahim, Synthesis of emerging two-dimensional (2D) materials–Advances, challenges, and prospects, *FlatChem* 30 (2021) 100305.

21. Y. Li, C. Gao, R. Long, and Y. Xiong, Photocatalyst design based on two-dimensional materials, *Materials Today Chemistry* 11 (2019) 197–216.

22. G. A. Naikoo, F. Arshad, M. Almas, I. U. Hassan, M. Z. Pedram, A. A. Aljabali, *et al.*, 2D materials, synthesis, characterization, and toxicity: A critical review, *Chemico-Biological Interactions* 365 (2022) 110081.

23. R. R. Poolakkandy and M. M. Menamparambath, Soft-template-assisted synthesis: a promising approach for fabricating transition metal oxides, *Nanoscale Advances* 2 (2020) 5015–5045.

24. B. Luo, G. Liu, and L. Wang, Recent advances in 2D materials for photocatalysis, *Nanoscale* 8 (2016) 6904–6920.
25. F. Haque, T. Daeneke, K. Kalantar-Zadeh, and J. Z. Ou, Two-dimensional transition metal oxide and chalcogenide-based photocatalysts, *Nano-Micro Letters* 10 (2018) 1–27.
26. C. Liu, Q. Zhang, W. Hou, and Z. Zou, 2D Titanium/niobium metal oxide-based materials for photocatalytic application, *Solar RRL* 4 (2020) 2000070.
27. P. Liu, Y. Liu, W. Ye, J. Ma, and D. Gao, Flower-like N-doped MoS_2 for photocatalytic degradation of RhB by visible light irradiation, *Nanotechnology* 27 (2016) 225403.
28. D. Tan, J. Zhang, J. Shi, S. Li, B. Zhang, X. Tan, *et al.*, Photocatalytic CO_2 transformation to CH_4 by Ag/Pd bimetals supported on N-doped TiO_2 nanosheet, *ACS Applied Materials & Interfaces* 10 (2018) 24516–24522.
29. E. Grabowska, Selected perovskite oxides: Characterization, preparation, and photocatalytic properties—A review, *Applied Catalysis B: Environmental* 186 (2016) 97–126.
30. Y. Divyasri, Y. Teja, V. N. K. Rao, N. G. Reddy, S. Mohan, M. M. Kumari, *et al.*, "Nanostructures in Photocatalysis: Opportunities and challenges for environmental applications, *Nanostructured Materials for Environmental Applications* (2021) 1–32.
31. S. Vigneshwaran, P. Karthikeyan, C. M. Park, and S. Meenakshi, Boosted insights of novel accordion-like (2D/2D) hybrid photocatalyst for the removal of cationic dyes: Mechanistic and degradation pathways, *Journal of Environmental Management* 273 (2020) 111125.
32. K. Bhuvaneswari, G. Palanisamy, T. Pazhanivel, T. Maiyalagan, P. Shanmugam, and A. N. Grace, In-situ development of metal-organic frameworks assisted ZnMgAl layered triple hydroxide 2D/2D hybrid as an efficient photocatalyst for organic dye degradation, *Chemosphere* 270 (2021) 128616.
33. A. Shahzad, K. Rasool, M. Nawaz, W. Miran, J. Jang, M. Moztahida, *et al.*, Heterostructural $TiO_2/Ti_3C_2T_x$ (MXene) for photocatalytic degradation of antiepileptic drug carbamazepine, *Chemical Engineering Journal* 349 (2018) 748–755.
34. K. Wang, Y. Li, G. Zhang, J. Li, and X. Wu, 0D Bi nanodots/2D Bi_3NbO_7 nanosheets heterojunctions for efficient visible light photocatalytic degradation of antibiotics: Enhanced molecular oxygen activation and mechanism insight, *Applied Catalysis B: Environmental* 240 (2019) 39–49.
35. G. Mahajan and D. Sud, Application of lignocellulosic waste material for heavy metal ions removal from aqueous solution, *Journal of Environmental Chemical Engineering* 1 (2013) 1020–1027.
36. G. Zhang, D. Chen, N. Li, Q. Xu, H. Li, J. He, *et al.*, Preparation of $ZnIn_2S_4$ nanosheet-coated CdS nanorod heterostructures for efficient photocatalytic reduction of Cr (VI), *Applied Catalysis B: Environmental* 232 (2018) 164–174.
37. Y.-X. Pan, Y. You, S. Xin, Y. Li, G. Fu, Z. Cui, *et al.*, Photocatalytic CO_2 reduction by carbon-coated indium-oxide nanobelts, *Journal of the American Chemical Society* 139 (2017) 4123–4129.
38. P.-Y. Jia, R.-t. Guo, W.-g. Pan, C.-y. Huang, J.-y. Tang, X.-y. Liu, *et al.*, The MoS_2/TiO_2 heterojunction composites with enhanced activity for CO_2 photocatalytic reduction under visible light irradiation, *Colloids and Surfaces A: Physicochemical and Engineering Aspects* 570 (2019) 306–316.
39. D. L. T. Nguyen, M. A. Tekalgne, T. H. C. Nguyen, M. T. N. Dinh, S. S. Sana, A. N. Grace, *et al.*, Recent development of high-performance photocatalysts for N_2 fixation: A review, *Journal of Environmental Chemical Engineering* 9 (2021) 104997.
40. Y. Zhao, Y. Zhao, G. I. Waterhouse, L. Zheng, X. Cao, F. Teng, *et al.*, Layered-double-hydroxide nanosheets as efficient visible-light-driven photocatalysts for dinitrogen fixation, *Advanced Materials* 29 (2017) 1703828.
41. R. Shen, L. Zhang, X. Chen, M. Jaroniec, N. Li, and X. Li, Integrating 2D/2D CdS/α-Fe_2O_3 ultrathin bilayer Z-scheme heterojunction with metallic β-NiS nanosheet-based ohmic-junction for efficient photocatalytic H_2 evolution, *Applied Catalysis B: Environmental* 266 (2020) 118619.
42. M. Ou, J. Li, Y. Chen, S. Wan, S. Zhao, J. Wang, *et al.*, Formation of noble-metal-free 2D/2D $ZnmIn_2Sm+ 3$ (m= 1, 2, 3)/MXene Schottky heterojunction as an efficient photocatalyst for hydrogen evolution, *Chemical Engineering Journal* 424 (2021) 130170.

CHAPTER **13**

Two-Dimensional Metallenes for Photocatalysis Applications

Soumita Samajdar and Srabanti Ghosh

13.1 INTRODUCTION

Solar energy harvesting is considered an eco-friendly and renewable approach that would bring sustainable solutions to the energy crisis and pollution of the environment. Photocatalysis using visible light active semiconductors has drawn considerable attention among the researchers of the 21st century. As a substitute for the traditional wide band-gap metal oxides, 2D nanomaterials have garnered widespread interest as photocatalysts owing to their fascinating physicochemical properties such as tuneable layer-dependent bandgaps, flexible regulation of band edge positions, and large surface area with abundant active sites, which make them suitable for photocatalysis applications [1–3].

The enormous success of graphene has encouraged the development of atomically thin 2D nanosheets composed of metal atoms, namely metallenes. This class of newly designed 2D metallenes possesses many unique extraordinary properties such as the presence of coordinatively unsaturated metal atoms that can act as active sites, highly tunable electronic, optical, and structural properties, excellent carrier mobility, and maximum surface area, which makes them highly suitable for catalysis applications. In comparison with metal nanoparticles, the metal atoms in metallenes exhibit superior connectivity, which leads to much better electron transfer during catalysis [4]. Moreover, the ultrathin and flexible nature of metallenes accelerates mass transfer and charge transfer, which also enhances their catalytic performance. Thus, the unique properties of metallenes make them suitable

DOI: 10.1201/9781032645001-13

for various applications such as photocatalytic and photoelectrochemical water splitting, dye degradation, CO_2 photoreduction, and N_2 fixation [5–7].

This chapter begins by providing a brief overview of the various synthesis approaches and characterization techniques of metallenes. Particular attention has been paid to the effect of various strategies such as atomic doping, strain modulation, and defect engineering on the photocatalytic efficiency of metallenes. Special emphasis has been laid on the application of metallenes as cocatalysts and the construction of metallene-based heterostructures. Apart from highlighting the recent progress of metallenes for various photocatalytic applications, this chapter also sheds light on the challenges and prospects of metallenes for further applications in the field of photocatalysis.

13.2 SYNTHESIS METHODS

The catalytic performance of the metallenes, which depends upon the composition, number of layers, and crystal structure, can be varied depending upon the various synthesis methods. The synthesis of metallenes takes place via three major approaches – top-down exfoliation, bottom-up methods, and topotactic metallization.

13.2.1 Top-Down Approach

The top-down method of fabrication of nanomaterials involves reducing the size of bulk materials into nanostructures. Metallenes possess strong covalent interactions within the same plane and weak van der Waals interactions among the various layers which enable facile exfoliation into single or multiple layers. Liquid-phase exfoliation is one of the most effective methods to produce two-dimensional nanosheets. For example, Yang et al. fabricated 2D bismuthene nanosheets via the intercalation of ammonium peroxydisulfate and concentrated sulfuric acid into the bulk Bi granules [8]. The intercalation of the acid molecules dilates the bismuth granules and exfoliates them into layers (Figure 13.1a). In another investigation, Wang and his coworkers successfully fabricated atomically thin 2D antimonene nanosheets through a pre-grinding process followed by consecutive liquid-phase exfoliation assisted with ultrasonication and studied the tunability of bandgap with the layer thickness of metallenes (Figure 13.1b) [9].

13.2.2 Bottom-up Approach

The presence of strong covalent bonds between the metal atoms restricts the exfoliation of metallenes *via* top-down approaches. Therefore, the synthesis of metallenes using wet-chemical bottom-up methods such as ligand-confined growth, space-confined growth, and template-directed growth has attracted wide attention.

13.2.2.1 Ligand-Confined Growth

This has been considered as one of the most powerful strategies for the anisotropic synthesis of metallenes. Generally, CO is used as a gas ligand, which gets strongly adsorbed on the (111) facet of metals and prevents their anisotropic growth along the (111) crystallographic direction, which leads to the formation of 2D nanosheets. Besides directly using CO gas, the decomposition of CO-releasing agents such as formaldehyde, carbonyls, and formic

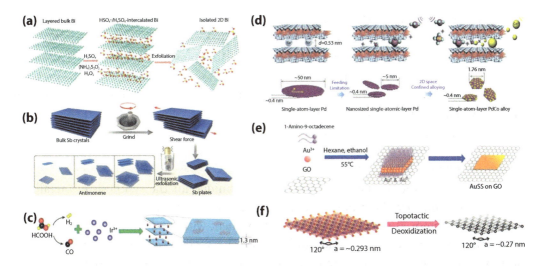

FIGURE 13.1 (a) Synthesis of bismuthene *via* H$_2$SO$_4$-assisted liquid-phase exfoliation and (b) synthesis of antimonene nanosheets via grinding of bulk Sb followed by exfoliation. Adapted with permission from [3], Copyright (2020), American Chemical Society. (c) Synthesis of Ir nanosheets via CO confined growth and (d) preparation of a Pd–Co monolayer by space-confined growth. Adapted with permission from [4], Copyright (2022), Elsevier. (e) Template-directed growth of Au nanosheets on GO and (f) fabrication of Ru nanosheets via topotactic deoxidation. Adapted with permission from [3], Copyright (2020), American Chemical Society.

acid, which can release CO at higher temperatures, can also be used as an alternative. For example, Cheng et al. synthesized iridium nanosheets by using formic acid which decomposes into CO and H$_2$ during the reaction. The CO gets strongly adsorbed on the (111) facet of Ir, and the ligand confinement effect of CO results in the formation of Ir nanosheets of 1.3 nm, as observed in Figure 13.1c [10]. The surface of the nanosheets was hydroxylated, which reduces the surface energy and promotes the preservation of the ultrathin nature of the nanosheets.

13.2.2.2 Space-Confined Growth

In addition to the ligand confinement effect of different ligands, the interlayer spaces between the lamellar structures can also enforce the development of metallenes [11]. The presence of interlayer spaces within the layered double hydroxides (LDHs) can easily be substituted by other anions via ion exchange principle. To utilize this feature of LDHs, Wang et al. prepared 2D [001] plane-oriented Au nanosheets via the intercalation of Au-based anions within the interlayer spaces of Mg- and Al-based LDHs followed by subsequent chemical reduction. This anion exchange not only provided the space-confined effect but also resulted in the stabilization of the as-synthesized Au nanosheets [12]. Besides LDHs, layered montmorillonite can also be used as a space-confined template to synthesize Pd and PdCo NSs with a thickness of 0.5 nm (Figure 13.1d) [13].

13.2.2.3 Template-Directed Growth

2D material-based templates play a pivotal role in directing the growth of metallenes as reported in the literature. For example, Huang et al. employed graphene oxide as a template for the in-situ preparation of Au nanosheets on graphene oxide (GO) in the presence of 1-amino-9-octadecene. After the partial reduction of Au^{3+} to Au^+, it formed a complex with 1-amino-9-octadecene and got absorbed on the surface of GO. Then the as-synthesized complex was reduced to form the Au nanosheets as observed in Figure 13.1e [14].

13.2.3 Topotactic Metallization

Topotactic metallization is another important technique for the synthesis of metallenes in which the non-metallic groups like hydroxide ions and oxide ions present within the layers can be removed under a reductive environment without affecting the original layered structure. For example, Fukuda and his coworkers followed a similar kind of synthesis protocol in order to synthesize Ru nanosheets *via* the topotactic deoxidation of RuO_2 nanosheets under an H_2 atmosphere (Figure 13.1f) [15].

13.3 CHARACTERIZATION METHODS

Due to the highly sensitive ultrathin nature of metallenes, special characterization methods are required to study their morphology, atomic-scale thickness, and surface composition. During characterization using the transmission electron microscopy technique, the ultrathin thermodynamically unstable layers of metallenes are susceptible to structural and morphological transformation on exposure to the electron beam. For example, the changes in the morphology of Rh metallene have been noted during TEM characterization. It was perceived that the porosity of Rh metallene increased when exposed to electron beam irradiation for 28 seconds (Figure 13.2a) [16]. In another investigation, it was found that the Au metallene undergoes a phase transformation from hcp to fcc crystal phase, and the porosity of the Au nanosheets also increases upon exposure to an electron beam for 20 seconds (Figure 13.2b,c) [14]. Hence, morphological and structural changes were observed when exposed to electron beam irradiation for different exposure time durations using TEM.

The atomic-scale thickness of metallenes can be detected with the help of atomic force microscopy (AFM). For example, AFM imaging can be employed to clearly observe the thickness of four atomic-layered Co metallene and PdMo metallenes, which is below 9 Å (Figure 13.2d–g) [14, 17]. Also, the lateral thickness of metallenes can be easily viewed using scanning transmission electron microscopy (STEM). Besides revealing the thickness of metallenes, the number of layers and the arrangement of atoms can also be studied using STEM. Figure 13.2h and 13.2i,j display the lateral STEM and TEM images of Co and Rh metallenes, respectively, which also reveal the crystal structures and arrangement of metal atoms present in metallenes [2, 17].

The coordination number of atoms present on the surface of metallenes can also be revealed by STEM. The perpendicular projection to the basal plane indicates the arrangement of different surface atoms, and the lattice spacings along the various crystallographic

Two-Dimensional Metallenes for Photocatalysis Applications ■ 193

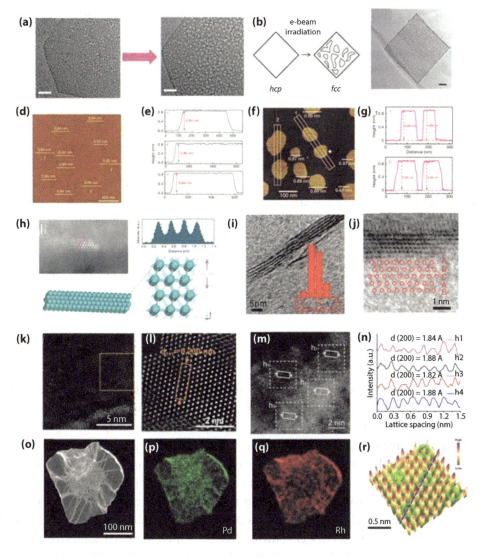

FIGURE 13.2 (a) TEM images of 2D Rh nanosheets. Adapted with permission from [16], Copyright (2014), Springer Nature. (b) Phase transformation of Au nanosheets and (c) TEM image of porous Au metallene. Adapted with permission from [14], Copyright (2011), Springer Nature. (d) AFM image and (e) height profiles of Co metallene. Adapted with permission from [17], Copyright (2016) Springer Nature. (f) AFM image. (g) Height profiles of PdMo bimetallene. Adapted with permission from [7], Copyright (2019), Springer Nature. (h) HAADF-STEM image of Co metallene. Adapted with permission from [17], Copyright (2016), Springer Nature. (i) TEM image of Rh metallene and (j) HRTEM image of Rh monolayer. Adapted with permission from [2], Copyright The Authors, some rights reserved; exclusive licensee [WILEY-VCH]. Distributed under a Creative Commons Attribution License 4.0 (CC BY). (k,l) High-resolution HAADF-STEM image of the RhPd-H. Adapted with permission from [6], Copyright (2020), American Chemical Society. (m) HAADF-STEM image and (n) intensity profile of high-entropy metallene. Adapted with permission from [18], Copyright (2022), American Chemical Society. (o–q) EDX elemental mapping of RhPd-H bimetallene. Adapted with permission from [6], Copyright (2020), American Chemical Society. (r) 3D topographic image of PdFe metallene. Adapted with permission [19], Copyright (2022), Wiley-VCH.

directions can easily be calculated. For example, in the case of RhPd-H bimetallene nanosheets, high-resolution HAADF-STEM images reveal that the average d-spacing between two lattice points in RhPd-H is 0.230 nm (Figure 13.2k,l) [6]. The HAADF-STEM image (Figure 13.2m) displays the arrangement of the surface atoms on the quinary high entropy alloy metallene, and the fast Fourier transformation image reveals that the pentamery metallene consists of (001)-oriented fcc structure. The integrated pixel intensities correspond to the lattice distortions in the pentamery metallenes as the average lattice spacing varies from 18.2 nm to 18.8 nm in the quinary metallenes (Figure 13.2n) [18]. The energy-dispersive X-ray spectroscopy (EDS) analysis helps to detect the distribution of different metals present in polymetallenes. Figure 13.2o–q reveals the EDS elemental mapping of RhPd-H bimetallene, which shows that Rh and Pd have a consistent distribution over the bimetallene surface [6]. Topographic atom imaging analysis can also be used to characterize the metallenes doped with secondary metal atoms. For example, the identification of single Fe atoms in PdFe was carried out by 3D topographic imaging as shown in Figure 13.2r. It was observed that due to the lower atomic number of Fe as compared to Pd, spots having lower intensity appear, which represents the atomic substitution of Pd with Fe [19].

13.4 METALLENES FOR PHOTOCATALYSIS

13.4.1 Surface and Subsurface Composition Control

Introducing a secondary metal atom into the lattice of transition metal atoms leads to changes in the neighboring atoms of the catalytic centers, which can tune their electronic structure, thereby affecting their photocatalytic performance. Doping with other metal atoms can effectively tune the bandgap, which may enhance light harvesting in the visible region and hinder the rate of charge carrier recombination. For example, Singh et al. studied how the effect of doping of β-antimonene with different metal atoms like Bi, As, Te, and Sn influences the electronic properties and modifies the valence band edge and conduction band edge potentials, which regulates their bandgap and affects their photocatalytic performance towards overall water splitting. The orbital-contributed band structures show that the substitution of the Sb atom of monolayer antimonene with Bi and As effectively reduces the bandgap from 1.35 eV to ~1.0 eV, which significantly improves visible light harvesting (Figure 13.3a,b). Sn- and Te-doped antimonene show metallic behavior due to the strong hybridization of Sb orbitals with Sn and Te orbitals at the Fermi level (Figure 13.3c,d). Moreover, the overpotential values for the oxygen evolution reaction using metal-doped antimonene were also studied based on theoretical simulations. It was observed that the Bi-doped antimonene exhibited the lowest overpotential of 0.25 V, as observed in Figure 13.3e, which provides evidence in favor of the better catalytic activity of Bi-doped antimonene. Additionally, it was also observed that the activation energy barrier of the rate-determining step of the hydrogen evolution reaction on the Bi-doped Sb monolayer was 0.06 eV, which is significantly low and promotes facile hydrogen evolution through water splitting [20].

Two-Dimensional Metallenes for Photocatalysis Applications ■ 195

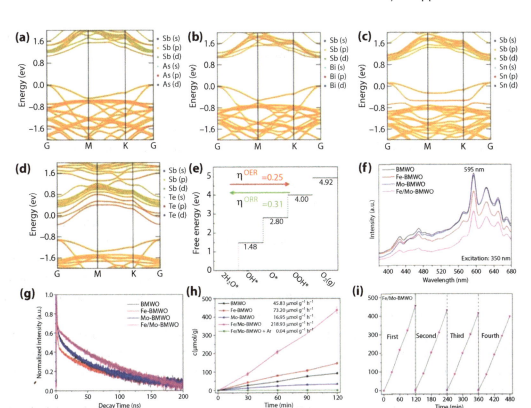

FIGURE 13.3 Orbital-contributed band structures of antimonene nanosheets doped with (a) As, (b) Bi, (c) Sn, (d) Te, and (e) Free energy profile diagram of Bi-doped antimonene nanosheets. Adapted with permission [20]. Copyright The Authors, some rights reserved; exclusive licensee [American Chemical Society]. Distributed under a Creative Commons Attribution License 4.0 (CC BY). (f) Photoluminescence spectra and (g) time–resolved fluorescence decay spectra of bimetallene-coated BMWO. (h) Rate of photocatalytic NH_3 generation using pristine BMWO and BMWO doped with Fe, Mo, and Fe/Mo bimetallene and (i) photostability test for nitrogen reduction using bimetallene coated BMWO over four successive cycles. Adapted with permission [21]. Copyright (2022) Elsevier.

The sublayer doping of secondary metals in monometallenes leads to the formation of bimetallenes, which have also attracted significant attention as photocatalysts. Recently, Li and his coworkers have explored the role of FeMo bimetallene in enhancing the photocatalytic N_2 reduction activity of $Bi_2Mo_{0.3}W_{0.7}O_6$ (BMWO). Figure 13.3f displays photoluminescence (PL) emission spectra, which show that the Fe/Mo bimetallene-coated BMWO exhibited the lowest PL intensity, indicating that it exhibited a lower electron–hole recombination rate. The fluorescence decay spectra show that FeMo-coated BMWO possesses the longest fluorescence lifetime of 10.82 ns, which ensures effective charge separation and migration, thereby leading to an enhancement in the photocatalytic activity (Figure 13.3g). Therefore, bionic Fe/Mo bimetallene shows ~4.8 fold augmentation in the rate of photocatalytic N_2 reduction (~218.93 μmol g^{-1} h^{-1}) compared to unmodified BMWO (Figure

196 ■ 2D Metals

13.3h). Moreover, it also exhibited excellent photocatalytic activity over four successive cycles without any structural and morphological change (Figure 13.3i) [21].

13.4.2 Strain Engineering

Another interesting feature of metallenes is the modulation of the electronic, structural, and optical properties to a considerable extent upon the application of uniaxial and biaxial strain. Lin et al. studied the effects of tensile strain and electrical field on tuning the properties of 2D arsenene-based heterostructures. It was observed that under strain-free conditions the total energy attains minima under equilibrium conditions. The application of tensile strain reduces the interlayer distance, which facilitates stronger interaction between SnS_2 and arsenene (Figure 13.4a) and promotes facile charge transfer. Strain also has a significant influence on the bandgap valence and conduction band edge positions. The variation of bandgap and band edge positions under different strain is displayed in Figure 13.4b. On increasing the compressive strain, both the position of conduction band minima (CBM) and valence band maxima (VBM) rises, but the rate of increase of VBM is much slower than that of CBM, which leads to a reduction in bandgap from 0.870 to 0.083 eV. Moreover, upon increasing the compressive strain, the CBM shows a downward trend while the VBM presents an increasing trend, effectively decreasing the bandgap from 0.870 to 0.347 eV. However, the decrease in bandgap is slightly smaller under tensile strain than under compressive strain.

The influence of strain on the optical properties of arsenene/SnS_2 shows that applying tensile strain shifts the absorption to lower wavelengths (Figure 13.4c). This results in greater visible light absorption, which increases its photocatalytic efficiency. However, the application of compressive strain enhances UV light absorption. The difference in the electronic structures created due to the application of two different kinds of strain results in the alteration of the optical properties. Interestingly, it has also been noted that the charge transfer mechanism undergoes a transition from type-II to type-I due to the change in band alignment upon application of 8% tensile strain [22].

In another investigation, Liu and his coworkers studied the effect of strain engineering on halogen edge-passivated antimonene nanoribbons in both armchair and zigzag configurations denoted by SbNR_X/zSbNR_X (X = F, Cl, Br, and I). It was observed that tensile strains lower than 2% and 4% for SbNR_X in armchair and zigzag configurations, respectively, lead to a slight increase in the bandgap followed by its reduction under higher tensile strains (Figure 13.4d). Under appropriate tensile strain, the band edge potentials for the halogen-passivated SbNR_X nanoribbons in both armchair and zigzag configurations can satisfy the conditions for overall photocatalytic water splitting. However, the band edges of SbNR_H in both armchair and zigzag configurations are not suitable for photocatalytic water splitting even when they are under sufficient tensile strain. Remarkably, a maximum solar-to-hydrogen conversion efficiency of 17.51 % can be achieved by applying 12% tensile strain [23].

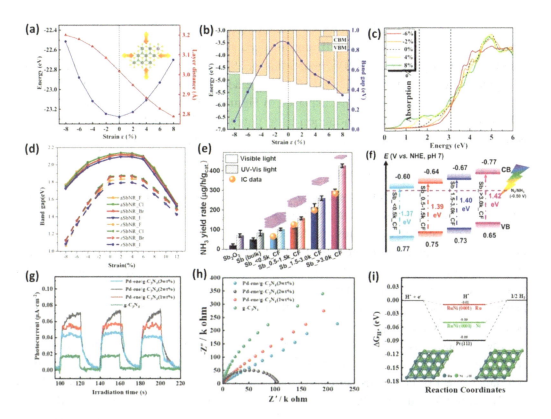

FIGURE 13.4 (a) The total energy and distance between the layers of arsenene/SnS$_2$. (b) Position of band edges and bandgap of arsenene/SnS$_2$ and (c) variation of optical absorption of arsenene/SnS$_2$ with biaxial strain. Adapted with permission from [22], Copyright (2022), Elsevier. (d) Variation of bandgaps of SbNR_X in both armchair and zigzag configurations with strain. Adapted with permission from [23], Copyright (2021), American Chemical Society. (e) Rate of photocatalytic NH$_3$ production under UV or visible light irradiation using exfoliated Sb nanosheets and (f) energy band structures of Sb nanosheets with Sb and O vacancies. Adapted with permission from [24], Copyright (2020), Elsevier. (g) Transient photocurrent and (h) Nyquist plots of g-C$_3$N$_4$ nanosheets with different loading percentages of palladium nanosheets. Adapted with permission from [25], Copyright (2023), Elsevier. (i) Free energy diagram of g-C$_3$N$_4$ nanosheets with RuNi alloy as cocatalyst. Adapted with permission [26], Copyright (2021), Elsevier.

13.4.3 Defect Engineering

Defect engineering of metallenes provides an appealing approach towards enhancing photocatalytic performance by modulating their electronic and chemical properties. The presence of defects creates new active sites for separating photogenerated charge carriers and also improves visible light harvesting. For example, Zhao et al. studied the rate of photocatalytic nitrogen fixation using antimonene nanosheets with Sb and O vacancies. The decrease in the thickness and lateral size of Sb nanosheets leads to an increase in the number of oxygen vacancies. Figure 13.4e shows the variation in the photocatalytic efficiency of the exfoliated Sb nanosheets centrifuged at different speeds. It was observed that photocatalytic activity enhanced with the increasing centrifugation speed. The bulk Sb without

198 ▪ 2D Metals

any vacancy was inactive for photocatalytic nitrogen reduction. Remarkably, the antimonene nanosheets obtained at centrifugation speed greater than 3000 rpm (Sb_>3.0k_CF) resulted in the highest rate of NH_3 generation of 297.5 µg h^{-1} g^{-1} in the visible region due to a greater number of surface oxygen and edge defects of the Sb nanosheets. The band structure of the various materials displayed in Figure 13.4f also reveals that Sb_>3.0k_CF exhibited the most negative conduction band potential, which makes it more suitable for N_2 reduction to ammonia as compared to other materials [24].

13.4.4 Metallenes as Cocatalyst

Cocatalysts lower the activation energy barrier required for photocatalytic reactions, which ameliorates the photocatalytic activity of semiconductors. Metallene-based cocatalysts exhibit greater atomic utilization, higher electrical conductivity, and tunable electronic properties and can provide abundant exposed active sites due to their two-dimensional structure which enables greater visible light absorption and restrains charge carrier recombination. For example, Qian et al. employed ultrathin palladium metallene (Pd-ene) nanosheets as cocatalysts for augmenting the photocatalytic performance of graphitic carbon nitride nanosheets (g-C_3N_4), which showed even superior performance with respect to other noble-metal-based cocatalysts. The combination of 2D g-C_3N_4 with Pd-ene also results in an enhancement of visible light absorption due to the localized surface plasmon resonance effect of Pd. Transient photocurrent spectra shown in Figure 13.4g reveal that g-C_3N_4 with 2% loading of Pd-ene exhibited the highest photocurrent density. The electrochemical impedance spectra also show that Pd-ene/g-C_3N_4 (2 wt%) exhibits the lowest charge transfer resistance, which indicates more facile charge transfer (Figure 13.4h). The combination of the 2D–2D structure of g-C_3N_4 and Pd-ene can effectively promote the separation of photogenerated charge carriers, which enhances the photocatalytic performance. Moreover, the g-C_3N_4 nanosheets with 2 wt% loading of Pd-ene also exhibited the highest rate of photocatalytic H_2 generation (~31.3 µmol g^{-1} h^{-1}) [25].

Apart from monometallenes, bimetallenes may also serve as effective cocatalysts for the augmentation of the photocatalytic efficiency of semiconductors. For example, Han et al. developed 2D RuNi nanosheets as a cocatalyst, which enhanced the photocatalytic performance of g-C_3N_4 nanosheets due to the synergistic effect of Ru and Ni. The interfacial contact at the 2D–2D interface results in much facile charge transfer, thereby suppressing charge carrier recombination. Moreover, bimetallic alloy-integrated g-C_3N_4 showed considerable enhancement in visible light harvesting due to the surface plasmon resonance of metal atoms. The adsorption free energy of H atoms calculated with the help of density functional theory (DFT) simulations (ΔG_H^*) reveals that the ΔG_H^* of RuNi alloys is much more suitable for H_2 adsorption and desorption compared to that of Pt (−0.09 eV), irrespective of whether H_2 is adsorbed on the (0001) surface exposed by Ru or Ni atoms (Figure 13.4i). This suggests that the bimetallic alloys are more capable of acting as active sites for the hydrogen evolution reaction. The bimetallic Ru–Ni alloy-loaded graphitic carbon nitride nanosheets displayed a high rate of photocatalytic hydrogen generation of 35.1 mmol h^{-1} g^{-1} [26].

13.4.5 Construction of Metallene-Based Heterostructures

Although metallenes possess favorable electronic and optical properties, they suffer from the swift recombination of the photogenerated charge carriers, which impedes their photocatalytic performance. The construction of heterojunctions can be considered an effective strategy for promoting charge carrier separation, enhancing visible light absorption, and achieving suitable band alignment necessary for overall water splitting. For example, Li et al. synthesized arsenene-based heterostructures by combining monolayered arsenene and group III monochalcogenides like S and Se. To determine the thermodynamic stability of arsenene-based heterostructures, ab initio molecular dynamics (AIMD) simulations were performed, which revealed that no change in the structure of the Van der Waals heterostructures was observed at 300 K after 5 ps (Figure 13.5a,b). This suggests that As/GaS and As/GaSe structures exhibited excellent thermodynamic stability at room temperature. The fluctuations in temperature and total energy displayed in Figure 13.5c also lead to identical conclusions regarding the thermal stability of the as-synthesized heterostructures. When the arsenene and GaX monolayers come into contact with each other, As monolayer donates electrons to the GaX monolayers which makes As positively charged and GaX monolayers negatively charged. This induces an in-built electric field, which results in a large potential drop of 4.2 and 1.3 eV, across the interface of As/GaS and As/GaSe heterostructures, respectively (Figure 13.5d,e). This large potential drop across the interface can suppress charge carrier recombination. The band alignment of the arsenene and GaX monolayers leads to the development of type-II heterojunction at the interface (Figure 13.5f), which results in a high solar-to-hydrogen conversion efficiency of ~24–26% [27].

The Z-scheme-based heterostructures have gained widespread attention due to their excellent ability to enhance charge separation and strong reducing and oxidizing power of charge species preserved in the conduction band and valence band of two different semiconductors, respectively [28]. For example, Chen and his coworkers combined β-As with SnS_2 in order to form the direct Z-scheme heterojunction, which was capable of overall water splitting. The electronic properties of the heterostructure have been thoroughly explained using DFT simulations. It was observed that the conduction band minima and valence band maxima were composed of orbital contributions from SnS_2 and arsenene nanosheets, respectively (Figure 13.5g). As the conduction and valence bands originated from the orbital contributions of the two different semiconductors better charge separation occurred in the heterostructure. The plane-averaged charge density difference, which has been calculated from a quantitative perspective, also leads to the same conclusion. It was observed that electron-density accumulation and deficiency occur on SnS_2 and β-As, respectively (Figure 13.5h). Based on the band alignment of the two semiconductors, a plausible charge transfer mechanism has been determined, which reveals that the charge transfer occurs through the direct Z-scheme formed at the interface (Figure 13.5i) [29].

13.5 CONCLUSION AND FUTURE PROSPECTS

This chapter begins by highlighting the various synthesis strategies and characterization methods of metallenes. Special emphasis has been placed on the influence of doping, lattice

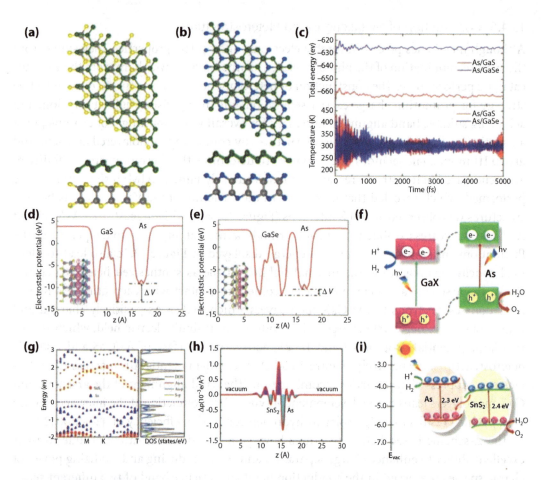

FIGURE 13.5 Ab initio molecular dynamics snapshots of (a) As/GaS, (b) As/GaSe heterostructures, and (c) total energy and temperature fluctuations at 300 K for both heterostructures. The potential drop across the interface of (d) As/GaS and (e) As/GaSe heterostructures. Adapted with permission from [27], Copyright (2021), Elsevier. (f) Charge transfer at the interface of As and GaX *via* type-II mechanism. (g) Projected band structure and density of states of As/SnS$_2$ and (h) planar-averaged charge density difference of heterostructure. Adapted with permission from [29], Copyright (2023), American Chemical Society. (i) Charge transfer *via* direct Z-scheme heterojunction formed between As and SnS$_2$.

strain engineering, defect modulation, and heterostructure construction on enhancing the photocatalytic performance of metallenes. Despite the recent progress of metallenes, their applications as photocatalysts are still in their infancy, and there are a lot of limitations that need to be addressed in the future. Firstly, the large-scale synthesis of metallenes with atomic-scale thickness is really a great challenge. Most of metallenes are generally synthesized via ligand confinement approaches. The ligands have a tendency to partially block the active sites of the metallenes which retards their catalytic performance. Secondly, the unstable nature of metallenes creates a lot of difficulties during their characterization by conventional methods. Therefore, advanced in-situ characterization techniques need to be developed in order to reveal the various physicochemical properties of metallenes. Thirdly, although metallenes have been widely used as electrocatalysts, their applications in the field of photocatalysis are limited. More detailed analysis on the electrochemical properties has to be carried out in order to develop an in-depth understanding of the charge transfer mechanism of metallene-based heterostructures. Moreover, the construction of metallene-based ternary heterostructures and the applications of bimetallenes and trimetallenes as photocatalysts have been rarely reported. Therefore, developing ternary heterojunctions using high entropy metallenes should be highly encouraged in the future, which will pave the way for the design of highly active photocatalysts. Moreover, alternative approaches for designing metallenes with abundant active sites, tunable defects, and thermodynamically stable phases are really essential in order to utilize metallenes for various practical applications.

ACKNOWLEDGMENTS

The authors acknowledge "Science and Engineering Research Board (SERB) POWER Grant" (project no. SPG/2020/000720) for financial support. One of the authors (S.S.) is thankful to CSIR, India, for providing the Junior Research Fellowship award.

REFERENCES

1. M. Xie, S. Tang, B. Zhang, G. Yu, Metallene-related materials for electrocatalysis and energy conversion, *Mater. Horiz.* 10 (2023) 407–432.
2. L. Zhao, C. Xu, H. Su, J. Liang, S. Lin, L. Gu, X. Wang, M. Chen, N. Zheng, Single-crystalline rhodium nanosheets with atomic thickness, *Adv. Sci.* 2 (2015) 1500100–1500105.
3. Y. Liu, K. N. Dinh, Z. Dai, Q. Yan, Metallenes: Recent advances and opportunities in energy storage and conversion applications, *ACS Materials Lett.* 2 (2020) 1148–1172.
4. C. Cao, Q. Xu, Q-L. Zhu, Ultrathin two-dimensional metallenes for heterogeneous catalysis, *Chem. Catal.* 2 (2022) 693–723.
5. D. Zhang, X. Cui, L. Liu, Y. Xu, J. Zhao, J. Han, W. Zheng, 2D bismuthene metal electron mediator engineering super interfacial charge transfer for efficient photocatalytic reduction of carbon dioxide, *ACS Appl. Mater. Interfaces* 13 (2021) 21582–21592.
6. J. Fan, J. Wu, X. Cui, L. Gu, Q. Zhang, F. Meng, B-H. Lei, D. J. Singh, W. Zheng, Hydrogen stabilized RhPdH 2D bimetallene nanosheets for efficient alkaline hydrogen evolution, *J. Am. Chem. Soc.* 142 (2020) 3645–3651.
7. M. Luo, Z. Zhao, Y. Zhang, Y. Sun, Y. Xing, F. Lv, Y. Yang, X. Zhang, S. Hwang, Y. Qin, J-Y. Ma, F. Lin, D. Su, G. Lu, S. Guo, PdMo bimetallene for oxygen reduction catalysis, *Nature* 574 (2019) 1–5.
8. Q-Q. Yang, R-T. Liu, C. Huang, Y-F. Huang, L-F. Gao, B. Sun, Z-P. Huang, L. Zhang, C-X. Hu, Z-Q. Zhang, C-L. Sun, Q. Wang, Y-L. Tang, H-L. Zhang, 2D bismuthene fabricated via acid-intercalated exfoliation showing strong nonlinear near-infrared responses for mode-locking lasers, *Nanoscale* 10 (2018) 21106–21115.
9. X. Wang, J. He, B. Zhou, Y. Zhang, J. Wu, R. Hu, L. Liu, J. Song, J. Qu, Bandgap-tunable preparation of smooth and large two-dimensional antimonene, *Angew. Chem. Int. Ed.* 57 (2018) 8668–8673.
10. Z. Cheng, B. Huang, Y. Pi, L. Li, Q. Shao, X. Huang, Partially hydroxylated ultrathin iridium nanosheets as efficient electrocatalysts for water splitting, *Natl. Sci. Rev.* 7 (2020) 1340–1348.
11. D. Xu, X. Liu, H. Lv, Y. Liu, S. Zhao, M. Han, J. Bao, J. He, B. Liu, Ultrathin palladium nanosheets with selectively controlled surface facets, *Chem. Sci.* 9 (2018) 4451–4455.

12. L. Wang, Y. Zhu, J-Q. Wang, F. Liu, J. Huang, X. Meng, J-M. Basset, Y. Han, F-S. Xiao, Two-dimensional gold nanostructures with high activity for selective oxidation of carbon–hydrogen bonds, *Nat. Commun.* 6 (2015) 6957–6965.

13. J. Jiang, W. Ding, W. Li, Z. Wei, Freestanding single-atom-layer Pd-based catalysts: oriented splitting of energy bands for unique stability and activity, *Chem.* 6 (2021) 431–447.

14. X. Huang, S. Li, Y. Huang, S. Wu, X. Zhou, S. Li, C. L. Gan, F. Boey, C. A. Mirkin, H. Zhang, Synthesis of hexagonal close-packed gold nanostructures, *Nat. Commun.* 2 (2011) 292–298.

15. K. Fukuda, J. Sato, T. Saida, W. Sugimoto, Y. Ebina, T. Shibata, M. Osada, T. Sasaki, Fabrication of ruthenium metal nanosheets via topotactic metallization of exfoliated ruthenate nanosheets, *Inorg. Chem.* 52 (2013) 2280–2282.

16. H. Duan, N. Yan, R. Yu, C-R. Chang, G. Zhou, H-S. Hu, H. Rong, Z. Niu, J. Mao, H. Asakura, T. Tanaka, P. J. Dyson, J. Li, Y. Li, Ultrathin rhodium nanosheets, *Nat. Commun.* 5 (2014) 3093–3101.

17. S. Gao, Y. Lin, X. Jiao, Y. Sun, Q. Luo, W. Zhang, D. Li, J. Yang, Y. Xie, Partially oxidized atomic cobalt layers for carbon dioxide electroreduction to liquid fuel, *Nature* 529 (2016) 68–71.

18. L. Tao, M. Sun, Y. Zhou, M. Luo, F. Lv, M. Li, Q. Zhang, L. Gu, B. Huang, S. Guo, A General synthetic method for high-entropy alloy subnanometer ribbons, *J. Am. Chem. Soc.* 144 (2022) 10582–10590.

19. X. Li, P. Shen, Y. Luo, Y. Li, Y. Guo, H. Zhang, K. Chu, PdFe single-atom alloy metallene for N_2 electroreduction, *Angew. Chem. Int. Ed.* 61 (2022) 202205923–202205932.

20. D. Singh, R. Ahuja, Theoretical prediction of a Bi-Doped β—Antimonene monolayer as a highly efficient photocatalyst for oxygen reduction and overall water splitting, *ACS Appl. Mater. Interfaces* 13 (2021) 56254–56264.

21. H. Li, H. Deng, S. Gu, C. Li, B. Tao, S. Chen, X. He, G. Wang, W. Zhang, H. Chang, Engineering of bionic Fe/Mo bimetallene for boosting the photocatalytic nitrogen reduction performance, *J. Colloid Interface Sci.* 607 (2022) 1625–1632.

22. L. Lin, M. Lou, S. Li, X. Cai, Z. Zhang, H. Tao, Tuning electronic and optical properties of two–dimensional vertical van der Waals arsenene/SnS_2 heterostructure by strain and electric field, *Appl. Surf. Sci.* 572 (2022) 151209–151217.

23. M. Liu, C.-L. Yang, M-S. Wang, X-G. Ma, Halogen edge-passivated antimonene nanoribbons for photocatalytic hydrogen evolution reaction with high solar-to hydrogen conversion, *J. Phys. Chem. C* 125 (2021) 21341–21351.

24. Z. Zhao, C. Choi, S. Hong, H. Shen, C. Yan, J. Masa, Y. Jung, J. Qiu, Z. Sun, Surface-engineered oxidized two-dimensional Sb for efficient visible light-driven N_2 fixation, Nano *Energy* 78 (2020) 105368–105378.

25. A. Qian, X. Han, Q. Liu, L. Ye, X. Pu, Y. Chen, J. Liu, H. Sun, J. Zhao, H. Ling, R. Wang, J. Li, X. Jia, Ultrathin Pd metallenes as novel co-catalysts for efficient photocatalytic hydrogen production, *Appl. Surf. Sci.* 618 (2023) 156597–156604.

26. X. Han, T. Si, Q. Liu, F. Zhu, R. Li, X. Chen, J. Liu, H. Sun, J. Zhao, H. Ling, Q. Zhang, H. Wang, 2D bimetallic RuNi alloy Co-catalysts remarkably enhanced the photocatalytic H_2 evolution performance of g-C_3N_4 nanosheets, *Chem. Eng. J.* 426 (2021) 130824–130833.

27. J. Li, Z. Huang, W. Ke, J. Yu, K. Ren, Z. Dong, High solar-to-hydrogen efficiency in Arsenene/GaX (X = S, Se) van der Waals heterostructure for photocatalytic water splitting, *J. Alloys Compd.* 866 (2021) 158774–158782.

28. S. Ghosh, D. Sarkar, S. Bastia, Y. S. Chaudhary, Band-structure tunability via the modulation of excitons in semiconductor nanostructures: manifestation in photocatalytic fuel generation, *Nanoscale* 15 (2023) 10939–10974.

29. X. Chen, W. Han, Z. Tian, Q. Yue, C. Peng, C. Wang, B. Wang, H. Yin, Q. Gu, Exploration of photocatalytic overall water splitting mechanisms in the Z—scheme SnS_2/β-as heterostructure, *J. Phys. Chem. C* 127 (2023) 6347–6355.

CHAPTER **14**

2D Metals for Fuel Cells

Ashwani Kuma, Jyoti Bala, and Mohd. Shkir

14.1 INTRODUCTION

Fuel cells have emerged as a promising technical frontier in an era characterized by a constant desire for clean and sustainable energy sources. With no influence on the environment, these amazing electrochemical gadgets provide a direct route to transforming chemical energy into electrical power [1]. Fuel cells present a disruptive solution that has the potential to revolutionize energy production and storage, in contrast to current combustion-based power generation, which is frequently rife with inefficiencies and pollution. At their core, fuel cells operate on a simple yet elegant principle. They contribute to the electrochemical reaction of a fuel source and an oxidizing agent, often hydrogen and oxygen. A cell with an electrolyte is where this process takes place, allowing for the passage of ions [2–4].

The variety of applications for fuel cells demonstrates their adaptability. They are useful for powering everything from cars and buses to drones and even spacecraft as well as other types of transportation. For essential infrastructure such as hospitals and data centers, fuel cells provide dependable backup power supplies in stationary environments. They also have the potential to provide clean and effective on-site electricity generation for household and commercial applications. A seamless interaction between renewable energy sources and storage is achieved by research and development projects that integrate fuel cells into grid systems. Additionally, they show promise for providing on-site, clean, and efficient electricity generation for both home and commercial use. To provide a seamless interaction between renewable energy sources and storage, research and development are also being done to integrate fuel cells into grid systems. Two-dimensional (2D) materials have an increasing number of possible energy-related uses in the last few years. Since the discovery of graphene nanosheets, other nanosheet materials have been developed and viable commercial procedures for the manufacturing of nanosheet membranes, such as graphene derivatives, for gas separation and water treatment, have been presented. At least

DOI: 10.1201/9781032645001-14

19 novel 2D materials have been discovered so far, including hexagonal boron nitride [5,6], transition metal dichalcogenides (TMDCs) [7], phosphorene [8], graphitic carbon nitride [9], and 2D transition metal carbides/nitrides or carbonitride (MXenes). Cost reduction, infrastructure development for hydrogen production and distribution, and improving the durability and performance of fuel cell components are among the challenges. Nonetheless, continuous research and increased investment are steadily overcoming these limitations.

14.2 FUEL CELL TYPES AND ADVANCEMENTS

Numerous types of fuel cells have been developed and differ in terms of operating temperature, efficiency, applications, and cost. They are categorized into six major classes based on the fuel and electrolyte used, such as alkaline fuel cell (AFC), phosphoric acid fuel cell (PAFC), solid oxide fuel cell (SOFC), molten carbonate fuel cell (MCFC), proton exchange membrane fuel cell (PEMFC), and direct methanol fuel cell (DMFC).

14.2.1 Alkaline Fuel Cell (AFC)

An AFC is a form of an electrochemical cell that turns chemical energy directly into electrical energy via a chemical interaction between hydrogen and oxygen. AFCs work based on the oxidation of hydrogen (H_2) and reduction of oxygen (O_2) at two distinct electrodes. In AFCs, the electrolyte is an alkaline material, commonly potassium hydroxide (KOH), that permits ions to flow between the anode and the cathode. This separates AFCs from other forms of fuel cells that use various electrolytes such as proton exchange membranes or solid oxides. AFCs typically function at temperatures ranging from 23 °C to 70 °C (73–158°F), which is lower than many other types of fuel cells. As a result, they are suited for a wide range of applications. AFCs offer a high electrical efficiency, particularly at low loads, making them suitable for applications requiring variable power outputs. While AFCs have distinct advantages, the choice of fuel cell technology is dependent on the specific application and its requirements. Certain types of fuel cells may be better suited for certain conditions. Here are some important facts like advantages, disadvantages, applications, and challenges concerning alkaline fuel cells.

Advantages:

- AFCs have been used in space missions and specialty applications for a long time, demonstrating their dependability and capacity.

- They are quick to start and have good load-following characteristics.

Disadvantages:

- They need pure oxygen on the cathode side, which might be difficult to obtain in some applications.

- Intolerance to CO_2 contaminants in the hydrogen supply.

Challenges:

- One of the difficulties with AFCs is that they are sensitive to carbon dioxide (CO_2) and can be poisoned by it. This means that the hydrogen fuel must be exceedingly pure, which can increase the cost of operating.

- AFCs are also competing with other forms of fuel cells, such as proton exchange membrane fuel cells (PEMFCs) and solid oxide fuel cells (SOFCs), which have made substantial advances in recent years.

14.2.2 Phosphoric Acid Fuel Cell (PAFC)

An example of a fuel cell that operates at a somewhat high temperature is a PAFC, which normally operates between 150 °C and 200 °C. Because phosphoric acid has poor ionic conductivity at low temperatures, the Pt electrocatalyst in the anode suffers from severe CO poisoning. Pure phosphoric acid is used as the electrolyte, a viscous, non-conductive liquid that enables ion movement between the cathode and anode. Silicon carbide is the most often employed acid retention matrix, while Pt is the electrocatalyst in both the anode and cathode. The electrical efficiency of PAFCs ranges from 40% to 50%; this is because their operating temperature is higher, allowing for more efficient electrochemical processes. This makes them a popular choice for stationary power generation. They are frequently used in cogeneration systems, which increase system efficiency by capturing waste heat from the fuel cell and using it for heating or other uses. Because of their size and weight, PAFCs have some drawbacks that make them less useful for portable applications. In comparison to some other fuel cell types, they are considerably more expensive to manufacture.

The operation of the PAFCs is depicted in Figure 14.1. PAFC technology has matured and is widely employed in a variety of power plant applications, including the largest fuel cell ever built, an 11 MW PAFC power plant owned by Tokyo Electric Power Co. between 1991 and 1997, and it ran for almost 230,000 hours [10]. PAFC development had halted during the last ten years in favor of polymer electrolyte fuel cells (PEFCs), which were regarded to have higher cost potential. PAFC development, on the other hand, continues. Because water is created as a byproduct, PAFC is considered a clean technology [11].

14.2.3 Solid Oxide Fuel Cell

One kind of electrochemical technology that directly transforms chemical energy into electrical energy at high and intermediate temperatures (500–1000 °C) without a Carnot cycle is the SOFC. Because of its low pollution, low noise, and robust building blocks, SOFCs are regarded as the most efficient and environmentally benign solution for both stationary and distributed power generation. The versatility of fuels that SOFCs may run on, such as hydrogen, natural gas, biogas, and even carbonaceous materials like coal or biomass, is one of their main advantages. They can therefore use the infrastructure that is already in place and are extremely adaptable. Natural gas (about 87%), biogas (40–65%), and coal mine gas (65%) are common sources of methane. Compared to hydrogen, it is

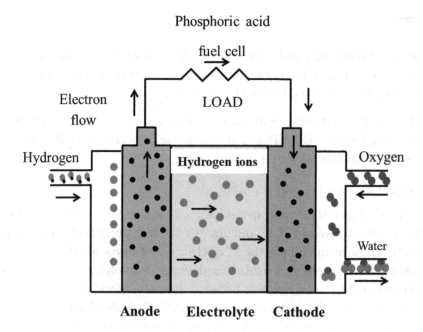

FIGURE 14.1 Schematic diagram of phosphoric acid fuel cell.

the most readily available, high-quality hydrocarbon fuel and is also simpler to store and transport. As a result, methane-fueled SOFCs have received a lot of interest. SOFCs emit only water vapor as a byproduct when using hydrogen as a fuel source, making them a clean and ecologically friendly energy conversion technology [12].

14.2.3.1 Operating Principle

The electrochemical oxidation principle underlies the operation of SOFCs. They are comprised of two porous electrodes, an anode and a cathode, encased in a solid ceramic electrolyte, usually composed of yttria-stabilized zirconia (YSZ). Because the high operating temperature makes it impossible to employ less expensive metals, ceramic materials are used for the electrolyte as well as the cathode and anode. The solid electrolyte and the lack of pumps needed to circulate the heated electrolyte are the main advantages of the SOFC over the MCFC. For improved catalysis and electron conduction, nickel is present in the anode. The fuel cell generation (first, second, and third, with decreasing operating temperature) determines the operating temperature, which ranges from 500 to 1000 °C. The schematic diagram of SOFC is shown in Figure 14.2.

The electrochemical reaction generates a lot of heat that an integrated heat management system can use. The greatest applications for SOFCs are those that employ both the heat and electricity generated, such as stationary power plants and auxiliary power supplies, because it takes a while for them to reach their operational temperature. Usually operating at temperatures between 500 and 1000 °C, SOFCs are hot devices. To enable the passage of oxygen ions through the solid electrolyte, a high temperature is required. Numerous reports on the inkjet printing of SOFC components have been published in the literature.

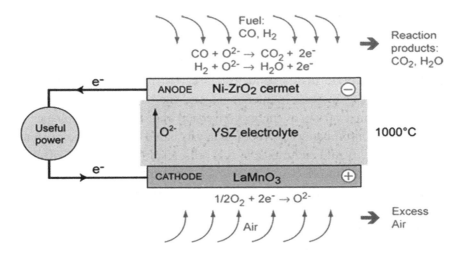

FIGURE 14.2 Schematic diagram of the operational principle of SOFC.

In 2008, the first report on fuel cell inkjet printing was published. The scientists printed a YSZ electrolyte layer and a NiO-YSZ interlayer, both around 6 μm thick, over a commercial NiO-YSZ anode support [13]. Jang and Kelsall claimed in 2022 that 3D NiO-YSZ structures may be printed via inkjet printing to improve SOFC performance [14].

14.2.3.2 Efficiency
SOFCs are renowned for their exceptional ability to transform chemical energy into electrical power, frequently attaining efficiencies greater than 60%. This indicates that 60% of the fuel's stored energy gets transformed into electrical energy that is usable. When compared to coal power plants, this is a significantly better efficiency. To further improve overall efficiency over 80%, they can also be utilized in combined heat and power (CHP) systems to collect and use waste heat.

14.2.3.3 Challenges
Even with all of its benefits, SOFCs have drawbacks. Among them are the elevated working temperatures, which may result in problems with thermal cycling and material deterioration. Furthermore, research is also ongoing in the areas of cost reduction and production scaling up. Lowering the working temperature of SOFCs would significantly increase the possible range of applications for them.

14.3 TWO-DIMENSIONAL MATERIALS FOR FUEL CELLS

2D nanomaterials outperform all others in terms of surface-to-volume ratio, porous structure, greater stability, strong electrical conductivity, good mechanical strength, and flexibility. Because they provide ultrafast and selective molecular transport, 2D materials have gained increased study focus in the field of fuel cells such as graphene oxide (GO), graphene nanosheets, MXenes, and 2D noble metals.

14.3.1 Graphene-Based Anode Catalysts

Since the discovery of graphene by Novoselov et al. [15], it has sparked intense research interest in fields where 2D materials are in high demand. Its exceptional electrochemical qualities allow it to outperform other dimensional carbon nanostructures [16]. Recent years have seen a considerable increase in interest in graphene-based anode catalysts because of their potential uses in a variety of energy conversion and storage technologies, especially fuel cells and lithium-ion batteries. Anode catalysts based on graphene can improve the electrochemical activity and endurance of fuel cells, resulting in a more effective conversion of chemical energy into electrical energy. A large number of studies (both theoretical and experimental) have been conducted on graphene and/or graphene nanocomposites with noble metal nanoparticles [17], nanocrystals [18], metal oxides [19], and so on, as electrocatalysts/supports. Polymer membranes that have been coupled with graphene have excellent ionic conductivity, high tensile strength, and minimal fuel permeability.

The commercial catalyst Pt/C without graphene is not as effective as the Pt-adorned electrocatalyst supported by graphene [20]. Comparing Pd-loaded graphene aerogel to commercial Pd/C, the former shows superior activity [21]. The possibilities and difficulties of fuel cells based on graphene were examined by Iqbal et al. [22]. Because of its low sheet resistance, increased durability, and resistance to corrosion, graphene is a viable option for fuel cell bipolar plates. Additionally, its lower crossover and increased stability contribute to the development of superior polymer electrolyte membranes for PEMFCs. However, graphene-based electrocatalysts have a few technical limitations. Graphene is chemically inert due to the high degree of graphitization of carbon, and it also lacks adequate sites for the deposition of noble metal nanoparticles. Furthermore, the catalytic mechanism of graphene-supported nanocrystals and heteroatom-doped graphene is yet unknown and requires further investigation in order to generate extremely stable and prospective electrocatalysts. To resolve this problem Li and colleagues [23] created a B- and N-doped graphene aerogel-supported Pt catalyst (Pt/BN-GA) in a methanol oxidation reaction (MOR) as depicted in Figure. 14.3.

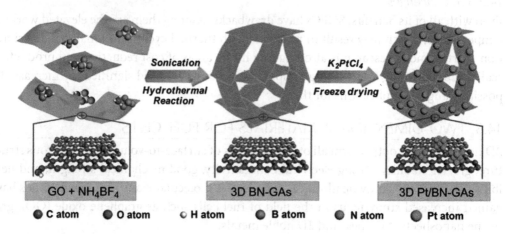

FIGURE 14.3 Schematic of the formation of 3D Pt-doped BN-graphene aerogel. Reproduced with permission from [23].

N-doped graphene support was discovered to promote the dispersion and durability of Pt–Co alloy nanoparticles by improving nanoparticle nucleation and growth kinetics as well as support/catalyst chemical binding [24]. The maximal power density of the PEMFC with Pt–Co/N-doped graphene cathode was four times greater (805 mW cm^{-2} at 60 °C) than that of the commercial Pt/C cathode. Platinum group metal-free (PGM-free) catalysts are more likely to be commercialized on a big scale. According to Liang et al. [25], the Co_3O_4/rGO composite has a similar catalytic activity to Pt in alkaline solutions, with an ORR onset potential of roughly 0.83 V vs reversible hydrogen electrode (RHE), but higher stability (minimal decrease in ORR activity over 25,000 seconds). Its exceptional activity has been attributed to synergetic chemical coupling reactions between metal oxide and graphene. 3D graphene has a huge surface area and porosity, strong electrical conductivity, and linked pore structures, not only offering more anchor sites for immobilizing metal oxide nanoparticles but also boosting reactant mass transport [26, 27]. Zhou and associates developed a unique Pt-Fe/HSG electrode using a modified supercritical fluid approach, and they worked with other organizations to study their 3D honeycomb-structured graphene for ORR [26].

14.3.2 Graphene Oxide (GO)

Another nanomaterial that has been investigated for possible use in fuel cells is GO. GO, which is a derivative of graphene, has been studied due to its high surface area and superior electrical conductivity. GO is a single-layer graphite oxide that is typically created when graphite is chemically oxidized [28]. On the basal plane, oxygen functional groups such as hydroxyl and epoxy groups were detected, whereas carboxy, carbonyl, phenol, lactone, and quinone groups were primarily generated towards the sheet borders. One can modify the oxygen functionality on graphene surfaces to modify the electrochemical behavior of GO. Moreover, GO is a very hydrophilic material that disperses evenly in water, making it a superior electrode material [29]. To make it more compatible with other fuel cell components, it can be functionalized. GO has been investigated for fuel cell applications in the following ways:

- GO can be functionalized to improve its catalytic qualities or used as a support material for different catalysts. The oxygen reduction reaction (ORR) at the fuel cell cathode can be catalyzed more effectively by GO when it is modified with particular functional groups.

- Proton exchange membrane fuel cells (PEMFCs) can use GO as a component of their membranes. PEMs, which are essential parts of PEMFCs, may function better and last longer thanks to GO high surface area and proton conductivity.

- Fuel cell electrodes can have their surfaces modified with GO to increase their electron transport and catalytic activity. The overall goal of this change is to improve the electrochemical reactions that take place at the anode and cathode.

- Membrane electrode assemblies, which comprise a cathode catalyst layer, anode catalyst layer, and membrane, are essential parts of fuel cells. By adding GO to these layers, fuel cell performance can be increased by improving their conductivity and catalytic qualities.

However, GO is only appropriate for low-temperature fuel cell applications since it is susceptible to high temperatures when exposed to oxygen or reducing reagents [30]. It's crucial to remember that even though GO exhibits potential for fuel cell applications, issues including scalability, affordability, and long-term stability still need to be resolved. Scientists are continuously investigating ways to use the special qualities of graphene and its derivatives to develop fuel cell technology as part of this branch of research.

14.3.3 MXenes

The family of 2D materials gained a new member, MXenes, after Naguib et al. MXenes are a new class of 2D nanostructures made of transition metal carbide that have a structure similar to graphene layers. When scientists carefully etched aluminum atoms from stacked ternary carbides or nitrides, or MAX phases, in 2011, they made the first discovery of MXenes. As illustrated in Figure 14.4, there are three potential lattice configurations for MXenes, corresponding to the value of n in MAX: M_2X, M_3X_2, and M_4X_3 (Figure 14.4a). MXene sheets in two dimensions are the result of this etching procedure (Figure 14.4b). A variety of intriguing characteristics are displayed by the resultant MXene materials, including as high electrical conductivity, outstanding mechanical strength, and superior electrochemical capabilities.

MXenes have been identified as a possible support material for applications involving renewable energy sources, such as fuel cell electrocatalysts. MXenes facilitate rapid electron transit and improves the catalytic properties of multi-component systems by

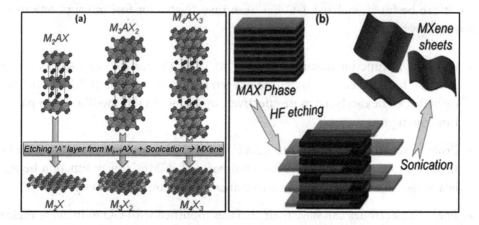

FIGURE 14.4 (a) Atomic structure of different MAX phases and the 2D corresponding MXenes obtained by etching the "A" layer. Reproduced with permission from Ref. [36]. (b) Schematic illustration of the synthesis of MXene sheets from compacted MAX phases. Reproduced with permission from Ref. [37].

altering the electrophilicity of catalysts in those systems [31]. The potential of MXenes, in particular titanium carbide (Ti_3C_2), to catalyze the oxygen reduction reaction and other fuel cell-related reactions has been studied. MXenes are a fantastic choice to raise anodic conductance because they are excellent conducting transporters that improve electron transmission [32]. MXenes' unique spatial organization and electrophilicity make them an ideal anodic nanomaterial for the creation of microbial fuel cells. With over 30 stoichiometric compositions that have been synthesized experimentally and numerous others that have been determined theoretically, the MXene family offers a plethora of opportunities to enhance particular applications due to their unique electronic, physical, and (electro)chemical properties [33]. The performance of ORR could be enhanced by the effective improvement of composite materials' electrochemical characteristics by the integration of MXenes. Because transition metal nanoparticles often function as active catalysts for ORR in alkaline conditions, MXenes may operate as supports or scaffolds because of their large number of embedding sites. MXene–Ag composites were created by Peng and Zhang's group by reducing $AgNO_3$ in situ on intercalated $Ti_3C_2(OH/ONa)_2$ MXenes [34] These bimetallic nanowire composites, MXene-$Ag_{0.9}Ti_{0.1}$, shaped like urchins, showed encouraging activity for ORR. A systematic screening of single-atom Pt ORR/OER catalysts supported by MXene materials was reported by Wei and colleagues. The results showed that MXenes based on vanadium (V), titanium (Ti), niobium (Nb), and chromium (Cr) are promising supports for single-atom Pt [35].

It has also been shown that Pristine MXenes are capable of actively catalyzing the ORR in metal-oxygen batteries. A consistent O-terminated Nb_2C MXene cathode material for lithium-oxygen batteries was created by Zhang, Han, and Dang's group. Using cutting-edge characterization techniques, a greater knowledge of the physical and electrical features of MXenes has led to significant advancements. It may be possible to maximize the electrocatalytic effectiveness of MXene-based composites by using their basal plane as a platform to collect the active phase. Almost all of the current works primarily focus on Ti_3C_2 materials, despite the fact that there are various types of MXenes now available. Therefore, developing simple and affordable processes to consistently create various kinds of carbide and nitride MXenes should be the focus of future research. Nitride MXenes, for instance, exhibit considerable promise for a variety of uses, such as electrocatalysis because they are more electrically conductive than carbide MXenes. To advance these technologies steadily, more research into the ORR process based on the special qualities of MXenes is essential.

14.4 2D NOBLE METALS

Long-term, hydrogen technologies are expected to be the foundations of sustainable and renewable energy sources and will eventually replace existing fossil fuels [38]. The HER is a critical process in electrochemical water splitting for clean hydrogen production. A 3D-to-2D phase transformation can help with this to some extent. This will allow not only the manufacturing of low-cost and mass-efficient large-area metal-based catalysts, which is critical for overall catalytic efficacy against HERs. Indeed, the low coordination numbers of the surficial atoms of 2D metal nanosheets can enable hydrogen proton adsorption for the efficient HER.

This is due to the fact that practically all of the metal atoms in the 2D catalyst will be engaged in the catalytic reaction.

14.4.1 2D Metal Nanosheets

Nanosheets (NSs) are a type of two-dimensional (2D) nanostructure with a thickness ranging from 1 to 100 nm. Graphene is the thinnest nanosheet—0.34 nm—and the most widely utilized one.

Noble metal-based 2D NSs have recently attracted a lot of attention due to their unique and adaptable characteristics, which boost their catalytic efficacy [39]. Examples of these characteristics include a larger surface area and a large number of uncovered unsaturated atoms. With tremella-like morphology, the NSs of Pd–Ag, Pd–Pb, and Pd–Au demonstrated improved resistance to CO poisoning and better electronic effects when it came to ethanol oxidation and ethylene glycol oxidation reactions. PdAuBiTe-based 2D ultrathin NSs were created by Zhao et al. [40] utilizing a unique visible-light-induced template technique. With a mass activity of 2.48 A mg $Pd^{\circ 1}$ towards ORR, the manufactured catalyst exhibits a greater activity than both industrial Pd/C and commercial Pt/C, by a factor of 17.7. After 10,000 cycles, there was no discernible decline in ORR performance, and it exhibited a high tolerance to CO and methanol poisoning. These days, ultrathin 2D NSs are receiving a lot of attention because of their amazing surface shape and electrical properties, which improved the fuel cell oxidation reaction's electrochemical activity. Numerous research studies have come to the conclusion that the physicochemical and electrical behavior of 2D NSs is unaffected by their more faulty structure when lowered to the atomic scale. As a result, they significantly improved the MOR's performance. In terms of choosing noble metals, Pd, which is very inexpensive, has been suggested as a good replacement for Pt. The primary disadvantage of employing FC methods is the expense of Pt-based catalysts, as the catalyst only projects the acquisition cost to be 54% of the entire stack cost [41]. It has been noted that adding an oxygen-loving metal to the electrode configuration can improve Pd's electrocatalytic activity. The bi-metal-based homogeneous mixture of Pd with an initial transition metal, such as Pd–Co [42], Pd–Cu [43], Pd–Ni [44], Pd–Ag [45], etc., is used for FC purposes. Meanwhile, in lower-temperature FCs, carbon-based Pt is typically used as an anode and cathode electrocatalyst. Research has been done on a variety of Pt-based electrode catalyst topologies, including three-dimensionally structured macroporous structures and nanoplates, nanowires, and nanotubes [46]. The best way to increase catalytic accomplishment, including durability and performance, is to use foreign metal. Additionally, it reduces Pt-metal usage.

14.4.2 2D Metals for Hydrogen Evolution

The HER is a crucial step in the electrochemical water-splitting process as it produces pure hydrogen. Platinum group metals (PGMs) are thought to be better catalysts for the HER. Most two-dimensional metals are projected to be most suited from a theoretical standpoint, at least when they have hexagonal and honeycomb structures. However, the great inclination of 2D noble metals to clump together to form 3D densely packed formations means that growing them is a difficult undertaking. There are some 2D metals that can be synthesized using different methods are ruthenium (Ru), iridium (Ir), rhodium (Rh), platinum (Pt), palladium (Pd), gold (Au), and silver (Ag). Chen et al. [47] reported the production of two-dimensional Ru nanocrystals utilizing a solution-based colloidal technique that is mediated by O_2. Rh nanosheets are generally synthesized in the presence of surface-capping chemicals, just as other 2D noble metals. Additionally, several attempts were made to create ultrathin Ru films using atomic layer deposition (ALD), electrochemical deposition, and dc-magnetron sputtering on glass, Au (111), and Si (100) substrates, respectively. Platinum has always drawn significant interest from the scientific community because of its exceptional catalytic ability. Nevertheless, the high price of bulk platinum makes it difficult to produce Pt-based catalysts on a large scale and encourages the creation of new technologies that assume a decrease in platinum consumption and an increase in active surface area at the same time. Using 2D Pt-based technologies is a dependable approach in this situation to address the aforementioned difficulty.

The literature review indicates that there is still a dearth of research on the hydrogen evolution reaction (HER) performance of 2D noble metal-based catalysts, despite their enormous potential. Chen et al. [47] discovered that, among various electrodes such as Ru nanowires, Ru nanoparticles, and commercial Ru electrodes, 2D Ru nanosheets synthesized by the O_2-mediated solution-based colloidal method have the highest HER catalytic activity (with a Tafel slope of 58.6 mV dec^{-1}) under acidic conditions (H_2-saturated 0.1 M $HClO_4$ aqueous solution). Moving on to other noble metals, electrochemical electrodes based on partially hydroxylated 2D iridium nanosheets exhibit outstanding HER catalytic performance under both acidic (0.5 M H_2SO_4) and alkaline conditions (in 1 M KOH), as evidenced by a 50 mV improvement in over potential (at 10 mA cm^{-2}) compared to commercial Pt/C and Ir/C electrodes [48]. In conclusion, the investigation of 2D metals for HER is an intriguing area of research that has the potential to improve the efficiency and sustainability of hydrogen production for a variety of applications (Figure 14.5). Ongoing research strives to address difficulties and optimize these materials' qualities for practical use.

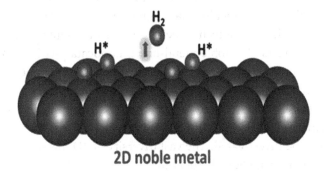

FIGURE 14.5 2D noble metals design for designing electrochemical electrodes for hydrogen production systems.

Pt has been the most extensively researched HER electrocatalysts with superior activity over a wide pH range, despite its reasonable price and somewhat low stability. Ru is investigated as a great HER electrocatalyst to produce hydrogen in both acidic and alkaline settings, specifically when it is on the N-doped porous carbon matrix. Ru has the lowest price and the highest stability.

14.4.3 2D Metals for Oxygen Evolution

The lowering of greenhouse gases (CO_2) and the process of water splitting both depend on the oxygen evolution reaction (OER). A highly efficient OER response is now needed due to growing concerns about renewable energy sources and environmental issues. But the slow kinetics has been restricting OER's reaction rate. A lot of work has gone into creating or developing OER catalysts that are more effective [49]. 3D Ir superstructures with enormous surface area, appropriate layer spacing, and 3D accessible active sites were created by Huang et al. [50] using ultrathin Ir nanosheets. Better than Ir nanoparticles, the resulting 3D Ir superstructures exhibit ordinary OER catalytic capabilities in both alkaline and acidic environments. Conversely, the primary emphasis in the development of CO_2 reduction catalysts is on the selectivity and Faradaic efficiency with respect to value-added carbon molecules. The partially oxidized Co nanosheets outperformed the bulk Co metal, partially oxidized bulk Co, and pure Co nanosheets in terms of OER performance.

14.5 CONCLUSION AND OUTLOOK

The important advancements in noble 2D metals for fuel cells that have been made are illustrated in this chapter. The development of Pt- and Pd-based nanocomposites for electrochemical reactions in fuel cells has advanced significantly during the last ten years. High electrochemical activity and enhanced permeation performance were offered by membranes and nanosheets based on the high specific areas of 2D materials from the ultrathin layers. MXenes or MXene composites with conducting materials have previously been employed as catalytic supports for noble metal nanoparticles or alloys in direct alcohol fuel cells. As a support, MXene promotes hydrogen bonding between alcohol molecules and hydroxyl groups on its surface, boosting catalytic activity. Further research should be done on the 2D noble metals for the hydrogen evolution reaction. Although anisotropic development of 2D metals is a technically challenging process due to the high surface energy of most metals, a kinetic control technique offers a viable route to synthesize exceedingly thin 2D metal nanosheets. 2D noble metal-based nanocatalysts have shown tremendous promise for improving electrocatalytic characteristics in various fuel-cell-related electrocatalysis. The 2D materials are generally employed as catalyst supports and have advanced significantly; however, scalability and synthetic processes must be improved. As non-noble electrocatalysts, more unique 2D architectures and nanocomposites with other nanostructured materials can be investigated. The vast diversity of 2D materials opens up enormous possibilities for constructing flexible fuel cells.

REFERENCES

1. C.-J. Tseng, B. T. Tsai, Z.-S. Liu, T.-C. Cheng, W.-C. Chang, S.-K. Lo, A PEM fuel cell with metal foam as flow distributor, *Energy Convers. Manag.* 62 (2012) 14–21.

2. T. J. Bvumbe, P. Bujlo, I. Tolj, K. Mouton, G. Swart, S. Pasupathi, B. G. Pollet, Review on management, mechanisms and modelling of thermal processes in PEMFC, *Hydrogen Fuel Cells* 1(1) (2016) 1–20.

3. O. Erdinc, M. Uzunoglu, Recent trends in PEM fuel cell-powered hybrid systems: Investigation of application areas, design architectures and energy management approaches, *Renew. Sust. Energ. Rev.* 14(9) (2010) 2874–2884.

4. D. Larcher, J.-M. Tarascon, Towards greener and more sustainable batteries for electrical energy storage, *Nat. Chem.* 7(1) (2014) 19–29.

5. H. Zhang, Ultrathin two-dimensional nanomaterials, *ACS Nano* 9(10) (2015) 9451–9469.

6. M. Fan, J. Wu, J. Yuan, L. Deng, N. Zhong, L. He, J. Cui, Z. Wang, S. K. Behera, C. Zhang, J. Lai, B. I. Jawdat, R. Vajtai, P. Deb, Y. Huang, J. Qian, J. Yang, J. M. Tour, J. Lou, C. Chu, D. Sun, P. M. Ajayan, Doping nanoscale graphene domains improves magnetism in hexagonal boron nitride, *Adv. Mat.* 31(12) (2019) 1805778.

7. S. K. Behera, P. Deb, A. Ghosh, Mechanistic study on electrocatalytic hydrogen evolution by high efficiency graphene/MoS2 heterostructure, *Chem. Select* 2(13) (2017) 3657–3667.

8. F. Fina, S. K. Callear, G. M. Carins, J. T. S. Irvine, Structural investigation of graphitic carbon nitride via XRD and neutron diffraction, *Chem. Mat.* 27(7) (2015) 2612–2618.

9. M. Naguib, M. Kurtoglu, V. Presser, J. Lu, J. Niu, M. Heon, L. Hultman, Y. Gogotsi, M. W. Barsoum, Two-dimensional nanocrystals produced by exfoliation of Ti3AlC2, *Adv. Mat.* 23(37) (2011) 4248–4253.

10. B. Singh, G. Guest, R. M. Bright, A. H. Strømman, Life cycle assessment of electric and fuel cell vehicle transport based on forest biomass. *J. Ind. Ecol.* 18(2) (2014) 176–186.

11. B. Sundén, Academic Press, Hydrogen, Batteries and Fuel Cells, Chapter 8, (2019) Amsterdam Netherlands Elsiever.

12. M. Singh, D. Zappa, E. Comini, Solid oxide fuel cell: Decade of progress, future perspectives and challenges, *Int. J. Hydrog. Energy* 46(54) (2021) 27643–27674.

13. D. Young, A. M. Sukeshini, R. Cummins, H. Xiao, M. Rottmayer, T. Reitz, Ink-jet printing of electrolyte and anode functional layer for solid oxide fuel cells, *J. Power Sources* 184(1) (2008) 191–196.

14. I. Jang, G. H. Kelsall, Fabrication of 3D NiO-YSZ structures for enhanced performance of solid oxide fuel cells and electrolysers, *Electroch. Commun.* 137 (2022) 107260.

15. K. S. Novoselov, A. K. Geim, S. V. Morozov, D. Jiang, Y. Zhang, S. V. Dubonos, I. V. Grigorieva, A. A. Firsov, Electric field effect in atomically thin carbon films, *J. Sci.* 306(5696) (2005) 666–669.

16. D. S. Hecht, L. Hu, G. Irvin, Emerging transparent electrodes based on thin films of carbon nanotubes, graphene, and metallic nanostructures, *J. Adv. Mater.* 23(13) (2011) 1482–1513.

17. A. Ali, P. K. Shen, Recent advances in graphene-based platinum and palladium electrocatalysts for the methanol oxidation reaction, *J. Mater. Chem. A* 7(39) (2019) 22189–22217.

18. S. Cui, S. Mao, G. Lu, J. Chen, Graphene coupled with nanocrystals: opportunities and challenges for energy and sensing applications, *J. Phys. Chem. Lett.* 4(15) (2013) 2441–2454.

19. M. Lei, Z. B. Wang, J. S. Li, H. L. Tang, W. J. Liu, Y. G. Wang, CeO2 nanocubes-graphene oxide as durable and highly active catalyst support for proton exchange membrane fuel cell, *Sci. Rep.* 4(1) (2014) 7415.

20. H. Huang, S. Yang, R. Vajtai, X. Wang, P. M. Ajayan, Pt-Decorated 3D architectures built from graphene and graphitic carbon nitride nanosheets as efficient methanol oxidation catalysts, *Adv. Mater.* 26(30) (2014) 5160–5165.

21. C.-H. A. Tsang, K. N. Hui, K. S. Hui, L. Ren, Deposition of Pd/graphene aerogel on nickel foam as a binder-free electrode for direct electro-oxidation of methanol and ethanol, *J. Mater. Chem. A* 2(42) (2014) 17986–17993.
22. M. Z. Iqbal, A.-U. Rehman, S. Siddique, Prospects and challenges of graphene-based fuel cells, *J. Phys. Chem. C* 39 (2019) 217–234.
23. M. Li, Q. Jiang, M. Yan, Y. Wei, J. Zong, J. Zhang, Y. Wu, H. Huang, Three-dimensional boron- and nitrogen-codoped graphene aerogel-supported Pt nanoparticles as highly active electro-catalysts for methanol oxidation reaction, *ACS Sustain. Chem. Eng.* 6(5) (2018) 6644–6653.
24. B. P. Vinayan, R. Nagar, N. Rajalakshmi, S. Ramaprabhu, Novel platinum–cobalt alloy nanoparticles dispersed on nitrogen-doped graphene as a cathode electrocatalyst for PEMFC applications, *Adv. Funct. Mater.* 22(16) (2012) 3519–3526.
25. Y. Liang, Y. Li, H. Wang, J. Zhou, J. Wang, T. Regier, H. Dai, Co3O4 nanocrystals on graphene as a synergistic catalyst for oxygen reduction reaction, *Nat. Mater.* 10(10) (2011) 780–786.
26. Y. Zhou, C. H. Yen, Y. H. Hu, C. Wang, X. Cheng, C. M. Wai, J. Yang, Y. Lin, Making ultra-fine and highly-dispersive multimetallic nanoparticles in three-dimensional graphene with supercritical fluid as excellent electrocatalyst for oxygen reduction reaction, *J. Mater. Chem. A* 4(47) (2016) 18628–18638.
27. Y. Li, Y. Zhou, C. Zhu, Y. H. Hu, S. Gao, Q. Liu, X. Cheng, L. Zhang, J. Yang, Y. Lin, Porous graphene doped with Fe/N/S and incorporating Fe_3O_4 nanoparticles for efficient oxygen reduction, *Catal. Sci. Technol.* 8(20) (2018) 5325–5333.
28. D. Chen, H. Feng, J. Li, Graphene oxide: Preparation, functionalization, and electrochemical applications, *Chem Rev.* 112(11) (2012) 6027–6053.
29. M. Perez-Page, M. Sahoo, S. M. Holmes, Single layer 2D crystals for electrochemical applications of ion exchange membranes and hydrogen evolution catalysts, *Adv. Mater. Interfaces* 6(7) (2019) 1801838.
30. S. Navalon, A. Dhakshinamoorthy, M. Alvaro, H. Garcia, Carbocatalysis by graphene-based materials, *Chem Rev.* 114(12) (2014) 6179–6212.
31. L. Zhao, B. Dong, S. Li, L. Zhou, L. Lai, Z. Wang, S. Zhao, M. Han, K. Gao, M. Lu, X. Xie, B. Chen, Z. Liu, X. Wang, H. Zhang, H. Li, J. Liu, H. Zhang, X. Huang, W. Huang, Interdiffusion reaction-assisted hybridization of two-dimensional metal–organic frameworks and Ti3C2Tx nanosheets for electrocatalytic oxygen evolution, *ACS Nano* 11(6) (2017) 5800–5807.
32. J. Qiao, L. Kong, S. Xu, K. Lin, W. He, M. Ni, Q. Ruan, P. Zhang, Y. Liu, W. Zhang, L. Pan, Z. Sun, Research progress of MXene-based catalysts for electrochemical water-splitting and metal-air batteries, *Energy Storage Mater.* 43 (2021) 509–530.
33. A. D. Handoko, S. N. Steinmann, F. Wei, Z. W. Seh, Correction: Theory-guided materials design: two-dimensional MXenes in electro- and photocatalysis, *Nanoscale Horiz.* 4(4) (2019) 1014.
34. Z. Zhang, H. Li, G. Zou, C. Fernandez, B. Liu, Q. Zhang, J. Hu, Q. Peng, Self-reduction synthesis of new MXene/Ag composites with unexpected electrocatalytic activity, *ACS Sustain. Chem. Eng.* 4(12) (2016) 6763–6771.
35. D. Kan, R. Lian, D. Wang, X. Zhang, J. Xu, X. Gao, Y. Yu, G. Chen, Y. Wei, Screening effective single-atom ORR and OER electrocatalysts from Pt decorated MXenes by first-principles calculations, *J. Mater. Chem. A* 8(33) (2020) 17065–17077.
36. M. Naguib, V. N. Mochalin, M. W. Barsoum, Y. Gogotsi, 25th Anniversary Article: MXenes: A new family of two-dimensional materials, *Adv. Mater.* 26(7) (2013) 992–1005.
37. M. Naguib, O. Mashtalir, J. Carle, V. Presser, J. Lu, L. Hultman, Y. Gogotsi, M. W. Barsoum, Two-dimensional transition metal carbides, *ACS Nano* 6(2) (2012) 1322–1331.
38. S. van Renssen, The hydrogen solution?, *Nat. Clim. Chang.* 10(9) (2020) 799–801.
39. X. Peng, D. Lu, Y. Qin, M. Li, Y. Guo, S. Guo, Pt-on-Pd dendritic nanosheets with enhanced bifunctional fuel cell catalytic performance, *ACS Appl. Mater. Interfaces* 12(27) (2020) 30336–30342.

40. F. Zhao, L. Zheng, Q. Yuan, X. Yang, Q. Zhang, H. Xu, Y. Guo, S. Yang, Z. Zhou, L. Gu, X. Wang, Ultrathin PdAuBiTe nanosheets as high-performance oxygen reduction catalysts for a direct methanol fuel cell device, *Adv. Mater.* 33(42) (2021) 2103383.

41. L. F. Brown, A comparative study of fuels for on-board hydrogen production for fuel-cell-powered automobiles, *Int. J. Hydrog. Energy* 26(4) (2001) 381–397.

42. Y. Wang, X. Wang, C.M. Li, Electrocatalysis of Pd–Co supported on carbon black or ball-milled carbon nanotubes towards methanol oxidation in alkaline media, *Catal. B: Environ.* 99 (2010) 229–234.

43. Y. Ren, S. Zhang, R. Lin, X. Wei, Electro-catalytic performance of Pd decorated Cu nanowires catalyst for the methanol oxidation, *Int. J. Hydrog. Energy* 40(6) (2015) 2621–2630.

44. M. G. Hosseini, M. Abdolmaleki, and S. Ashrafpoor, Methanol electro-oxidation on a porous nanostructured Ni/Pd-Ni electrode in alkaline media, *Chin. J. Catal.* 34(9) (2013) 1712–1719.

45. Z. Yin, Y. Zhang, K. Chen, J. Li, W. Li, P. Tang, H. Zhao, Q. Zhu, X. Bao, D. Ma, Monodispersed bimetallic PdAg nanoparticles with twinned structures: Formation and enhancement for the methanol oxidation, *Sci. Rep.* 4(1) (2014) 4288.

46. Z. Yan, B. Li, D. Yang, and J. Ma, Pt nanowire electrocatalysts for proton exchange membrane fuel cells, *Chin. J. Catal.* 34 (2013) 1471–1481.

47. L. Chen, Y. Li, X. Liang, Ar/H_2/O_2-controlled growth thermodynamics and kinetics to create zero-, one-, and two-dimensional ruthenium nanocrystals towards acidic overall water splitting, *Adv. Funct. Mater.* 31(31) (2021) 2007344.

48. Z. Cheng, B. Huang, Y. Pi, L. Li, Q. Shao, X. Huang, Partially hydroxylated ultrathin iridium nanosheets as efficient electrocatalysts for water splitting. *Natl. Sci.* 7(8) (2020) 1340–1348.

49. Z. Ding, J. Bian, S. Shuang, X. Liu, Y. Hu, C. Sun, Y. Yang, High entropy intermetallic–oxide core–shell nanostructure as superb oxygen evolution reaction catalyst, *Adv. Sustain. Syst.* 4 (2020) 1900105.

50. Y. Pi, N. Zhang, S. Guo, J. Guo, X. Huang, Ultrathin laminar Ir superstructure as highly efficient oxygen evolution electrocatalyst in broad pH range, *Nano Lett.* 16(7) (2016) 4424–4430.

CHAPTER 15

2D Metals for Methanol/ Ethanol Oxidation

Yuli Ma and Junyu Lang

15.1 INTRODUCTION OF METHANOL/ETHANOL OXIDATION

15.1.1 The Developmental History of the Oxidation of Methanol/Ethanol

Methanol (CH_3OH) is regarded as a renewable alcohol due to its production from sustainable sources, effectively diminishing the dependence on fossil fuels and reducing CO_2 emissions [1]. This is achieved by synthesizing methanol from carbon dioxide and hydrogen, both obtainable using renewable sources. Similarly, ethanol (C_2H_5OH) can also be derived from renewable resources, and in recent years, it has been predominantly produced from agricultural feedstocks like corn or sugarcane. Furthermore, ethanol can be sourced from biomass derived from lignocellulosic feedstock, offering clear advantages, including widespread availability and cost-effectiveness of the feedstock [2].

Research on the oxidation of methanol and ethanol has a long history dating back to the late 18th century when scientists first began to investigate alcohol properties. These early experiments primarily focused on combustion phenomena, observing that alcohols, when ignited, produced flames and light. As advancements in chemical analysis techniques occurred, a more in-depth exploration of alcohol oxidation began. In 1867, German professor Hofmann achieved a significant milestone by successfully synthesizing formaldehyde using a platinum spiral catalyst, which is the oxidation product of methanol [3]. This breakthrough marked a crucial development in the field. In the early 20th century, scientists furthered their investigations and gradually realized that ethanol could be oxidized to produce acetic acid and carbon dioxide, a more complex oxidation reaction. This discovery held considerable industrial significance, particularly in the production of acetic acid. As the mid-20th century approached, research into alcohol oxidation intensified, with scientists exploring different catalysts and reaction conditions for both methanol and

218

DOI: 10.1201/9781032645001-15

ethanol. These studies laid the foundation for the industrial applications of methanol and ethanol oxidation, especially in the realm of organic synthesis. With increasing environmental concerns and the rise of green chemistry and clean energy, research into alcohol oxidation shifted toward more efficient and environmentally friendly approaches. This evolution in research supported the development of technologies like direct methanol fuel cells (DMFCs), contributing to the advancement of cleaner and more sustainable energy solutions [4].

15.1.2 Introduction of Alcohol Fuel Cells

Fuel cells play a crucial role in tackling the urgent energy and environmental issues brought about by our heavy reliance on fossil fuels as the main source of energy. Within the broad spectrum of fuel cell technologies, direct alcohol fuel cells (DAFCs) stand out as a highly promising solution for producing clean energy. They excel in efficiently converting the chemical energy stored in alcohols such as methanol and ethanol into electrical energy. Importantly, this process helps reduce environmental impacts, with methanol offering an energy density of 6.07 kWh per kilogram and ethanol providing 8.01 kWh per kilogram, making them environmentally friendly choices for power generation [5].

The oxidation process of alcohol is of paramount significance in the context of alcohol fuel cells and serves as a fundamental principle for their operation. Alcohol fuel cells encompass two main variants: DMFCs and direct ethanol fuel cells (DEFCs). These devices hold immense potential for energy conversion, with the oxidation reaction catalyzed to enhance reaction kinetics. In this regard, metal catalysts have emerged as vital components due to their high activity and tunable properties. Take methanol fuel cells as an example: methanol undergoes electrochemical oxidation at the anode electrocatalyst, generating electrons that travel through the external circuit to the cathode electrocatalyst. At the cathode, these electrons combine with oxygen in a reduction reaction. The continuity of the circuit within the cell is maintained by the conduction of protons in the electrolyte. In modern fuel cells, proton-conducting polymer electrolyte membranes, such as Nafion™, are frequently employed. These membranes enable convenient cell operation, even at high temperatures and pressures. A schematic depiction of the components in a DMFC is provided in Figure 15.1.

Alcohol fuel cells have a wide range of practical applications, from powering portable electronic devices to acting as backup power sources. Moreover, there is a growing interest in their potential use in electric vehicles. However, the current state of the art in alcohol fuel cells involves ongoing efforts to enhance their efficiency, durability, and cost-effectiveness. These efforts aim to make alcohol fuel cells more practical and accessible for real-world deployment. One crucial aspect of alcohol fuel cell functionality is the principle of methanol and ethanol oxidation, which provides a clean and efficient pathway for energy conversion. The use of metal catalysts plays a pivotal role in advancing the state of these fuel cells, contributing to their potential as a transformative clean energy technology.

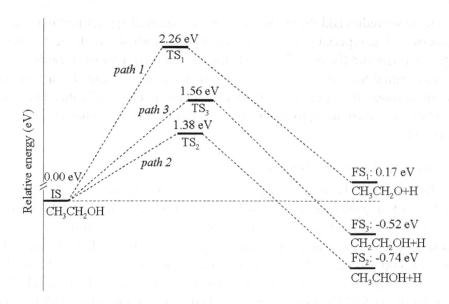

FIGURE 15.1 A schematic depiction of the components in a DMFC.

15.1.3 TRADITIONAL METAL CATALYSTS FOR THE OXIDATION OF ETHANOL/METHANOL

Traditional metal catalysts such as Pt, Pt–Ru, Pt–Sn, and Pd nanoparticles supported on high-surface-area carbon supports are well known for their high catalytic activity in reactions like methanol oxidation reaction (MOR) and ethanol oxidation reaction (EOR). These catalysts, however, come with their own unique challenges and considerations:

(1) Pt-based catalysts, while highly effective, are expensive and sensitive to carbon monoxide (CO) poisoning, limiting their use in large-scale applications. To address these challenges, researchers are actively working on novel catalyst designs and configurations, including specific geometric layouts, shapes, and compositions, to maximize the efficiency and longevity of Pt-based catalysts.

(2) Cu catalysts play a crucial role in ethanol oxidation, particularly in the production of valuable products like acetaldehyde and acetic acid. Yet, they come with selectivity issues, stability concerns, and potential toxicity due to trace copper ions. Ongoing research focuses on alloying and structural optimization to enhance their efficiency and reliability.

(3) Pb-based Jones catalyst has been used for methanol oxidation, but its usage is becoming restricted due to lead's toxicity. As a result, researchers are exploring alternative, more environmentally friendly catalysts.

(4) Pd provides a good balance between cost and catalytic activity in ethanol oxidation. Still, concerns about long-term stability, especially at high temperatures and potentials, need to be addressed.

(5) Ru is often combined with other metal catalysts to enhance catalytic activity, particularly in methanol oxidation, but its relative expense is a consideration.

(6) Au catalysts offer unique performance based on nanoparticle size and shape, requiring precise control for specific niche applications.

(7) Fe and iron compounds are employed as traditional catalysts for ethanol and methanol oxidation in laboratory research settings due to their affordability and abundance. While they may not always match the performance of noble metals in commercial applications, ongoing efforts aim to improve their efficiency and stability for wider applications.

These traditional metal catalysts remain crucial for various applications, including fuel cells and organic synthesis. Researchers are continuously exploring and refining these catalysts to enhance their performance, cost-effectiveness, and sustainability. The ongoing efforts to address their limitations demonstrate the commitment to making these catalysts more practical and environmentally friendly for a wide range of applications.

15.2 PERFORMANCE OF 2D METAL CATALYSTS IN METHANOL/ETHANOL OXIDATION

The catalytic oxidation of methanol and ethanol is of significant interest due to their widespread applications in fuel cells, chemical synthesis, and energy conversion processes. Traditional metal catalysts have been extensively studied for these reactions, but recent advancements in materials science have led to the development of 2D metal catalysts with unique properties that offer promising potential for enhancing the performance of these oxidation reactions. This section explores the performance of 2D metal catalysts in the oxidation of methanol and ethanol, highlighting their advantages and limitations.

15.2.1 2D Pure Metals

The emergence of 2D nanostructures of noble metals in the field of electrocatalysis represents a remarkable advancement with profound implications. These materials have gained significant attention due to their extraordinary catalytic potential, particularly in the efficient oxidation of alcohols.

One striking example is the work by Zhu and colleagues [6], who introduced a novel method for creating ultrathin, free-floating Pt nanoplates using the T7 peptide. The assembly of small nanocrystals into 2D anisotropic Pt nanosheets is not only innovative but also provides a fresh perspective on electrocatalyst design. What makes this development truly intriguing is the presence of abundant grain boundaries in the Pt nanosheets. These grain boundaries are believed to be a key factor in enhancing the catalytic activity of Pt nanosheets when compared to conventional Pt nanoparticles. In another noteworthy study, Slabon and colleagues tackled the challenges of alcohol oxidation by synthesizing 2D arrays of uniform Pd nanocrystals through the reduction of $Pd(acac)_2$ using oleylamine (OAm) and borane tributylamine (BTB) at 90°C for 1 hour [7]. The method they employed was unique in its ability to prevent nanoparticle agglomeration, a common issue in traditional catalysts. As a result, these 2D Pd nanocrystal arrays exhibited exceptional peak current densities in electrochemical tests for both methanol (1040 A g^{-1} (6.7 mA cm^{-2})) and ethanol oxidation (5500 A g^{-1} (35.7 mA cm^{-2})). This outstanding performance can be attributed to the capability of 2D arrays to maintain the dispersion of active sites and to minimize the loss of the active surface area. The exploration of 2D arrays of gold nanocrystals, each with a diameter of 8.0 ± 1.0 nm, through a one-pot reduction process involving gold(III) acetate, oleylamine (OAm), and 1,2-hexadecanediol at 30°C for 24 hours, further underscores the versatility and potential of these advanced materials [8]. The electrocatalytic activity of these nanocrystals for both MOR and EOR highlights their adaptability in diverse electrochemical applications. Their performance is particularly intriguing due to their dependence on the size and shape of the nanoparticles. Moreover, the synthesis of ultrathin 2D hierarchical porous Rh nanosheets mediated by 1-hydroxyethylidene and 1,1-diphosphonic acid (HEDP) represents a significant advancement in coordination chemistry and materials design [9]. These Rh nanosheets, with a thickness of approximately 1.7 nm, exhibit a wealth of pores, edge atoms, and grain boundary atoms. This unique architectural structure greatly facilitates electron flow, mass transport, and access to active sites for the MOR. As compared to conventional Rh black catalysts, these porous Rh nanosheets offer an expanded electrochemically active surface area, reduced charge transfer resistance, and elevated specific activity. These findings underscore the potential of advanced materials in electrocatalysis, showing how precise control of structure and composition can lead to enhanced performance and efficiency.

2D pure metals, despite their potential, come with inherent limitations, such as susceptibility to corrosion, mechanical fragility, and suboptimal electronic properties. Alloying, the introduction of additional elements into the structure, is a key strategy to address these limitations. It enhances corrosion resistance, improves mechanical strength, and fine-tunes electronic properties, making the materials more versatile and suitable for various applications.

15.2.2 2D Binary Metals

15.2.2.1 2D Pt-Based Binary Alloy

Platinum stands as a standout metal catalyst for electro-oxidation reactions. However, its effectiveness is hampered by unwanted carbonaceous species, particularly carbon monoxide, which is generated during the fuel oxidation process. To counter this poisoning issue, Pt is often alloyed with an oxophilic metal (M) to create PtM catalysts. This alloying promotes the generation of oxygenated species on the adjacent M, facilitating the oxidative removal of CO through a bifunctional mechanism. Among PtM catalysts, PtRu and PtSn have gained significant attention for their applications in MOR and EOR [10]. For instance, Tuan's group successfully prepared 2D PtSn nanosheets with a thickness of less than 1 nm using a simple colloidal method [11]. These nanosheets have proven to be highly effective electrocatalysts for both MOR and EOR. PtSn nanosheets exhibit remarkable mass activity in MOR, boasting 871.6 mA/mg Pt, which surpasses commercial Pt/C (371 mA/mg Pt) and Pt black (86.1 mA/mg) by 2.3 and 10.1 times, respectively. In the case of EOR, PtSn nanosheets display a mass activity of 673.6 mA/mg Pt, which is 5.3 times higher than commercial Pt black (127.7 mA/mg Pt) and 2.3 times higher than commercial Pt/C catalyst (295 mA/mg Pt). The exceptional performance of PtSn nanosheets can be attributed to the highly reactive exposed (1 1 1) facet sites that result from their sub-1 nm 2D sheet-like morphology. This development marks a significant stride in the quest for more efficient and sustainable electrocatalysts for fuel oxidation reactions.

Other 2D PtM alloys, such as those with M being Cu, Fe, Co, Ni, and Bi, have also been investigated as catalysts for the oxidation reactions [12, 13]. While these alloys hold promise, their catalytic performance is often hindered by specific challenges, necessitating further investigation and improvement. One major challenge encountered with these PtM alloys is the uncontrolled leaching of the M component from the PtM structure. This leaching can significantly impact the stability and longevity of the catalyst, potentially rendering it less effective over time. Developing methods to prevent or mitigate this leaching is a critical area of focus in catalyst research. Furthermore, in the case of ethanol oxidation reaction, these PtM alloys face difficulties in achieving complete fuel oxidation. This challenge arises from the formidable task of breaking the strong C–C bond present in ethanol molecules. The incomplete oxidation of fuel can result in reduced efficiency and incomplete energy extraction. Researchers are actively working to address these challenges through various approaches, including alloy composition optimization, nanostructuring, and the development of novel support materials. Additionally, advanced characterization techniques and in situ studies are being employed to gain a deeper understanding of the catalytic mechanisms and identify ways to enhance the catalytic performance of PtM alloys.

15.2.2.2 2D Pd-Based Binary Alloy

The use of monometallic Pd as a catalyst also often results in the adsorption of CO^* species on its surface, leading to a detrimental impact on electrocatalytic activity. To address this issue effectively, alloying Pd with other metals like Ag, Co, Cu, Ru, or Sn has proven to be a promising strategy. This alloying process serves to modulate Pd's electronic structure and introduces OH_{ads} species, facilitating the oxidation of adsorbed CO^* into CO_2. The introduction of a second element into the alloy induces a shift in the d-band center of Pd, leading to alterations in its electronic structure. Furthermore, this incorporation of another element creates a ligand effect, influencing the efficient transfer of electrons between different atoms within the alloy. Therefore, the construction of 2D Pd-based nanostructures presents distinct advantages for fuel cell reactions, offering desirable electrochemical properties. This approach represents a sophisticated and effective means of optimizing catalysts for clean energy applications.

Pd-based bimetallic catalyst nanosheets are known for their notable characteristics, including a substantial specific surface area and an abundance of active sites. For instance, Zhang and his research team [14] successfully synthesized ultrathin PdZn and PdCd nanosheets with a thickness of less than 5 nm. They achieved this through a wet chemistry method involving $Mo(CO)_6$, which guided the growth of metallic palladium along specific crystal surfaces. Similarly, Luo et al. [15] delved into PdMo bimetallene, revealing its outstanding oxygen regeneration capabilities. Zhang and his colleagues [16] took an innovative approach by synthesizing PdCu nanosheets using CO gas, underscoring the pivotal role of CO* in the nanosheet synthesis process. In a separate study, 2D PdAg nano-dendrites were crafted through the simultaneous reduction of H_2PdCl_4 and $AgNO_3$ in an aqueous solution with the assistance of OTAC at 20°C [17]. Furthermore, researchers have explored the synthesis of PdPt alloy ultrathin-assembled nanosheets by carefully regulating the growth behavior of PdPt nanostructures [18]. These nanosheets demonstrated a remarkable enhancement in catalytic activity and stability when applied to the ethanol oxidation reaction, outperforming irregularly shaped PdPt alloy nanocrystals, commercial Pd/C, and commercial Pt/C catalysts. This achievement underscores the importance of the PdPt alloy composition and the ultrathin 2D morphology, which provide high accessibility to active sites and an advantageous electronic structure for enhancing electrocatalytic activity. In addition, strained 2D PdPb nanosheets [19] with enhanced catalytic activity for ethanol oxidation have been successfully prepared. These nanosheets offer a promising approach for constructing efficient low-dimensional nanocatalysts with abundant active sites and controllable strain levels in fuel cell applications.

15.2.2.3 Other 2D Alloys

While platinum-based and palladium-based binary alloys have been prominent choices, there exists a broader spectrum of 2D binary alloy materials that hold great potential for catalysis. These materials, including gold-based alloys, silver-based alloys, copper-based alloys, and indium-based alloys, offer a diverse array of options to address the challenges and requirements of alcohol oxidation reactions.

15.2.3 2D Ternary Metals

Ternary metals in a 2D configuration, comprising three distinct elements, belong to a versatile class of materials that hold promise for various applications, particularly in alcohol oxidation reactions. Contrasted with binary alloys, ternary metals bring forth a broader spectrum of properties by virtue of the incorporation of a third element. This expanded compositional diversity allows scientists to precisely adjust the electronic and structural attributes of the 2D material, a pivotal factor in augmenting its catalytic efficacy in alcohol oxidation reactions. The choice of precise ternary metal compositions can be customized to achieve the desired catalytic performance. For example, through thoughtful selection of the constituent elements, scientists can exert control over surface reactivity, binding energy of reactants, and the overall electrocatalytic efficiency of the material. Consequently, the catalytic characteristics of distinct 2D ternary metals may exhibit notable variations, and the ideal composition hinges on the particular demands of the alcohol oxidation reaction at hand.

Ternary Pt-based nanomaterials have garnered significant attention for their exceptional electrocatalytic performance in fuel cell reactions while simultaneously reducing the utilization of precious Pt atoms. Ultrathin and highly flexible PtTeCu nanosheets [20] were successfully prepared using a convenient one-pot polyol method. Additionally, a hydrothermal synthesis method was developed to produce ternary ultrathin PdPtNi nanosheets. By adjusting the ratio of mixed solvents, it is possible to modify the pore structure on the surface of the nanosheets. Remarkably, thanks to their high atom utilization efficiency and rapid electron transfer capabilities, the PdPtNi porous nanosheets [21] exhibited remarkable activity in MOR and EOR. The mass activities for MOR and EOR were measured at 6.21 A/mg and 5.12 A/mg, respectively, significantly surpassing those of commercial Pt/C and Pd/C catalysts. Moreover, the three-element alloy nanosheets PtCoFe [22], PdPtAg [23], PdIrCu [24], and PdCuM (M = Ru, Rh, Ir) [25] have also been successfully synthesized.

In the search for the most suitable 2D alloy materials, researchers conduct systematic evaluations of their catalytic activity, stability, and availability. Electrochemical testing and characterization techniques help in assessing how these alloys perform under specific reaction conditions. The design and fabrication of nanomaterials are essential in optimizing alloy properties. Researchers use advanced techniques to precisely control the composition, structure, and morphology of 2D alloys.

15.3 MECHANISTIC INSIGHTS INTO METHANOL/ETHANOL OXIDATION ON 2D METAL CATALYSTS

15.3.1 In Situ and Operando Characterization Techniques

In situ and operando characterization techniques play a crucial role in the comprehensive study of methanol and ethanol oxidation reactions, which are of significant importance in various applications like fuel cells and catalysis. These techniques involve real-time monitoring and analysis of the reactions under working conditions, providing valuable insights into the underlying mechanisms and catalyst performance. Several advanced in situ testing techniques are listed below:

(1) In situ Fourier transform infrared spectroscopy (FTIR): In situ FTIR allows researchers to track the adsorption and desorption of reactants and intermediates on the catalyst surface during the oxidation reactions. This technique provides information about the nature of surface species and the reaction pathways, helping to elucidate the catalytic mechanism.

(2) In situ Raman spectroscopy: In situ Raman spectroscopy is employed to study the surface chemistry of catalysts during the reactions. It provides information about adsorbed species and their vibrational modes, aiding in the analysis of reaction mechanisms.

(3) Operando X-ray diffraction (XRD): Operando XRD is used to study structural changes in catalyst materials as they function during methanol and ethanol oxidation. It reveals information about crystal phase transformations, lattice strain, and catalyst stability, contributing to a better understanding of catalyst performance.

(4) Operando scanning electrochemical microscopy (SECM): Operando SECM enables researchers to map the electrochemical activity of catalyst surfaces as they function during the reactions. This method helps in identifying local reactivity and electrochemical corrosion behavior, shedding light on the catalyst's performance.

226 ■ 2D Metals

(5) Operando differential electrochemical mass spectrometry (DEMS): Operando DEMS is used to analyze gas-phase products generated during the reactions. It provides valuable information about product distribution and reaction kinetics, offering crucial data for reaction mechanism elucidation.

These in situ and operando techniques collectively allow researchers to gain a comprehensive understanding of the intricate processes, reaction pathways, and catalyst behavior during methanol and ethanol oxidation. This knowledge is essential for optimizing reaction efficiency, selectivity, and catalyst design, ultimately advancing the practical applications of these reactions, such as in clean energy technologies like fuel cells.

15.3.2 Reaction Mechanism of MOR and EOR

Extensive research efforts have focused on understanding the complex nature of intermediates and final products in Pt- or Pd-based catalysts for MOR and EOR. The oxidation of methanol can proceed via the intermediacy of either formate ($HCOO^-$) or CO [26], as shown in Figure 15.2a. Before the formation of CO_{ads} (adsorbed carbon monoxide), various adsorbates have been spectroscopically detected, including CH_xOHads (adsorbed methanol species), $-COH_{ads}$ (hydroxycarbonyl species), $-HCO_{ads}$ (adsorbed formate), $-COOH_{ads}$ (adsorbed carboxylic acid), and even $(HCOOH)_{2ads}$ (adsorbed formic acid dimers) [27]. For EOR, the research has resulted in the development of a widely accepted dual-pathway model on Pt- or Pd-based catalysts, as depicted in Figure 15.2b [28]. The C1 pathway involves the comprehensive conversion of ethanol into CO_2 or carbonates through the CO_{ads} intermediate, requiring the transfer of 12 electrons. In contrast, the C2 pathway represents the partial oxidation of ethanol to either acetate with a four-electron transfer or acetaldehyde with a two-electron transfer, all while preserving the C–C bond intact. EOR has been shown to occur via a series of complex reactions involving a number of sequential and parallel

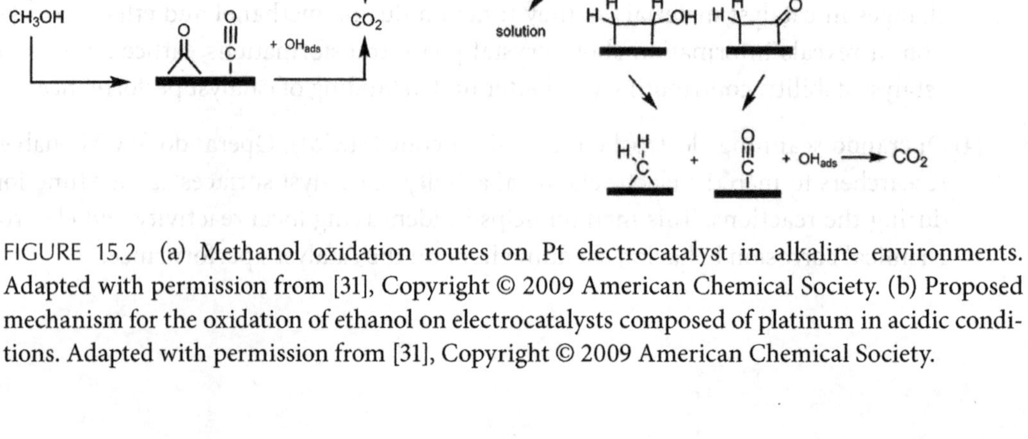

FIGURE 15.2 (a) Methanol oxidation routes on Pt electrocatalyst in alkaline environments. Adapted with permission from [31], Copyright © 2009 American Chemical Society. (b) Proposed mechanism for the oxidation of ethanol on electrocatalysts composed of platinum in acidic conditions. Adapted with permission from [31], Copyright © 2009 American Chemical Society.

reaction steps, thus resulting in more than 40 possible volatile and adsorbed intermediates or oxidative derivatives [29]. These complex reaction mechanisms are highly dependent on adsorption, oxidation, and the generation of intermediate products. Numerous comprehensive reviews [28, 30] have excellently summarized the mechanism, and, therefore, detailed elaboration on this topic will not be provided here.

The mechanism is influenced by various factors, including reaction conditions and catalyst characteristics. The diversity in reaction pathways and intermediate species underscores the complexity of these catalytic processes and the need for a profound understanding of the underlying mechanisms. A comprehensive grasp of these mechanisms can be harnessed to design and synthesize more efficient catalysts. By tailoring catalyst characteristics and reaction conditions, researchers can optimize the efficiency and selectivity of ethanol and methanol oxidation reactions, ultimately contributing to the development of more sustainable and effective catalytic systems.

15.3.3 Influencing Factors for the Kinetics of Methanol and Ethanol Oxidation

In the realm of methanol and ethanol oxidation on 2D metal catalysts, various factors exert influence on the kinetics of these reactions, contributing to their overall efficiency and performance:

(1) Catalyst composition: The choice of metal(s) in the catalyst is a pivotal factor that significantly impacts reaction kinetics. For instance, platinum-based catalysts are well-known for their high catalytic activity in alcohol oxidation. However, the choice of other metals, especially in bimetallic or ternary catalysts, can bring about unique catalytic properties.

(2) Temperature: Reaction kinetics are temperature-dependent and can be described using the Arrhenius equation. Elevated temperatures generally result in increased reaction rates. This phenomenon occurs because higher temperatures enhance reactant mobility, leading to more frequent collisions and faster reactions.

(3) Electrode potential: In the context of electrochemical oxidation, the electrode potential plays a critical role in influencing reaction kinetics. Raising the electrode potential facilitates electron transfer at the catalyst–electrolyte interface, thereby accelerating the reaction. This is particularly important in electrocatalytic processes.

(4) Surface coverage: The availability of active sites on the catalyst surface is a dynamic factor that can change as reactants and intermediates adsorb onto the catalyst. Surface coverage by reaction intermediates can significantly affect reaction rates. Catalyst poisoning, where the active sites are obstructed or blocked by reaction intermediates, can slow down the reaction and influence selectivity.

(5) Reaction intermediates: The presence and concentration of reaction intermediates, such as surface-bound species like methoxy and ethoxy groups, have a direct impact on reaction kinetics. The removal or further transformation of these intermediates can influence the overall reaction rate and selectivity.

(6) Electron transfer kinetics: The rate at which electrons are transferred from the reactants to the catalyst surface and vice versa can be a limiting factor in the overall reaction kinetics. This aspect is particularly relevant in electrochemical reactions where efficient electron transfer is crucial for catalytic activity.

Elucidating the kinetics of these complex reactions involves a combination of experimental techniques and theoretical methods: Experimental techniques such as chronoamperometry, cyclic voltammetry, and electrochemical impedance spectroscopy provide valuable insights into current–time profiles, electrochemical features, and charge transfer resistances. These experiments offer a practical means to assess reaction rates and understand the role of different factors. Computational methods, including density functional theory calculations (DFT) and kinetic modeling [32], offer theoretical insights into the energy barriers and reaction mechanisms involved. DFT can provide information on the electronic structure of catalysts and intermediates, while kinetic modeling allows researchers to simulate and predict reaction pathways (Figure 15.3) and rates under various conditions.

The integration of both experimental and computational approaches, along with the utilization of advanced in situ characterization techniques, empowers researchers to attain

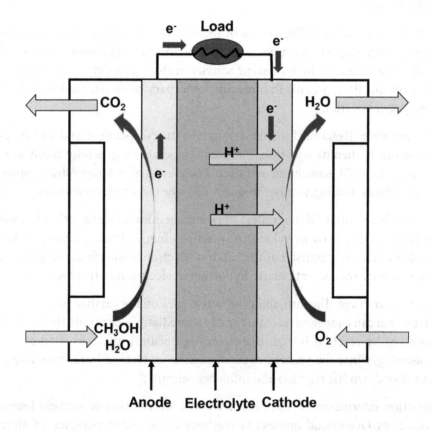

FIGURE 15.3 Schematic diagram of potential energy change during alcohol dehydrogenation on the surface of Pd(111). Adapted with permission from [33], Copyright © 2010 American Chemical Society.

a comprehensive understanding of the kinetics and reaction mechanisms governing methanol and ethanol oxidation on 2D metal catalysts. This holistic knowledge is of paramount importance for the design and optimization of catalysts, ultimately enhancing their efficiency and selectivity.

15.4 OPTIMIZATION AND DESIGN STRATEGIES FOR ENHANCED CATALYTIC PERFORMANCE

Rational design of active sites and interfaces is a key strategy for optimizing the catalytic performance of 2D metal catalysts in methanol and ethanol oxidation reactions. Here are specific approaches:

(1) Alloy formation: As detailed in Section 15.2, alloying metals in a catalyst creates unique surface properties that differ from those of individual metals. Researchers can control the alloy composition to fine-tune catalytic properties.

(2) Surface functionalization: Modifying the catalyst surface with functional groups, such as oxides, hydroxides, or nitrogen species, can alter the electronic structure and reactivity. Surface functionalization can promote the adsorption of reactants, facilitate electron transfer, and stabilize reaction intermediates. Nitrogen-doped graphene-supported catalysts have shown improved performance due to altered electronic properties [34].

(3) Controlling particle size and morphology: Catalyst particle size and morphology play a critical role in reaction kinetics. Smaller particle sizes and high-surface-area structures, such as nanoparticles or nanosheets, provide more active sites and shorter diffusion pathways for reactants. These structures promote better catalytic performance.

(4) Surface defect engineering: Creating controlled defects or vacancies on the catalyst surface can introduce highly active sites. Surface defects can serve as active sites for adsorption and reaction, enhancing catalytic performance. Careful control of defect density is essential for optimizing performance [35].

(5) Catalyst support: Choosing an appropriate support material can influence catalyst stability and reactivity. A suitable support material can prevent catalyst aggregation and enhance dispersion. Carbon-based supports (e.g., graphene and carbon nanotubes) are commonly used due to their high surface area and excellent electron conductivity.

(6) Interface engineering: Designing catalysts with heterostructures, where different materials or phases are in close proximity, can create synergistic effects. These interfaces promote charge transfer and facilitate reaction pathways. Core–shell catalysts, involving a core material encapsulated by a shell with different properties, enable precise control over catalytic activity and selectivity. For instance, a core of platinum surrounded by a less reactive shell can promote selective alcohol oxidation.

By carefully tailoring the composition and structure of 2D metal catalysts using these strategies, researchers can achieve enhanced catalytic performance in methanol and ethanol oxidation reactions. These optimized catalysts hold great potential for applications in energy conversion devices like fuel cells and for promoting more sustainable chemical synthesis processes.

15.5 PROSPECTS AND FUTURE DIRECTIONS

The development of 2D metal catalysts for the oxidation of methanol and ethanol is an area of significant interest in the field of catalysis due to its potential applications in fuel cells, energy storage, and renewable energy technologies. Here are some prospects and future directions in this research area:

(1) Enhanced catalytic activity: Researchers are continually striving to improve the catalytic activity of 2D metal catalysts for methanol and ethanol oxidation. This involves exploring new materials, optimizing the structure of existing catalysts, and enhancing their electrochemical performance. Future developments may focus on identifying catalysts with even higher activity and selectivity.

(2) Durability and stability: One of the major challenges in using 2D metal catalysts is their long-term stability and durability. Catalyst degradation over time can limit their practical applications. Future research may involve the development of stable and robust 2D metal catalysts through the design of protective coatings, doping with other elements, or novel synthesis methods.

(3) Understanding reaction mechanisms: A deeper understanding of the reaction mechanisms involved in methanol and ethanol oxidation on 2D metal catalysts is crucial. Future directions may include advanced characterization techniques such as in situ spectroscopy and microscopy to gain insights into the active sites and reaction pathways, allowing for more rational catalyst design.

(4) Scalability and cost-efficiency: To make 2D metal catalysts practical for large-scale applications, researchers will need to find scalable synthesis methods that are cost-effective. Future work may involve developing scalable production techniques and using earth-abundant materials to reduce costs.

(5) Catalyst engineering: Tailoring the properties of 2D metal catalysts for specific applications is an exciting avenue. Researchers may explore the engineering of catalysts with specific surface structures, electronic properties, and active sites to optimize their performance for different reactions or conditions.

(6) Multifunctional catalysts: Developing catalysts that can perform multiple reactions simultaneously or sequentially can be highly beneficial. Future research may focus on creating multifunctional 2D metal catalysts that can efficiently catalyze both methanol and ethanol oxidation as well as other relevant reactions.

(7) Integration into energy technologies: The integration of 2D metal catalysts into practical energy conversion and storage devices is a promising future direction. This includes their incorporation into fuel cells, supercapacitors, and electrochemical cells for energy generation and storage.

In summary, the prospects for 2D metal catalysts for methanol and ethanol oxidation are promising, with ongoing research focusing on improving catalytic activity, stability, scalability, and integration into practical applications. These catalysts have the potential to play a crucial role in the transition to cleaner and more sustainable energy technologies.

15.6 CONCLUSION

This chapter serves as a comprehensive exploration of the performance, mechanisms, and optimization strategies of 2D metal catalysts in the context of methanol and ethanol oxidation. This chapter begins with a historical perspective on the development of methanol and ethanol oxidation and introduces the relevance of alcohol fuel cells, followed by an examination of traditional metal catalysts. The focus then shifts to the assessment of 2D metal catalysts, encompassing pure metals, binary metals, and ternary metals, emphasizing their applications in methanol and ethanol oxidation. The pivotal role of in situ and operando characterization techniques in unraveling reaction mechanisms is underscored. Lastly, the chapter expounds on optimization and design strategies aimed at enhancing catalytic performance, culminating in a discussion on prospects and future research directions. The profound insights and knowledge presented in this chapter are instrumental in advancing the development of efficient and sustainable 2D metal catalysts. These catalysts are pivotal components in the ongoing quest to address global energy and environmental challenges and usher in a cleaner and more sustainable future.

REFERENCES

1. M. Pérez-Fortes, J.C. Schöneberger, A. Boulamanti, E. Tzimas, Methanol synthesis using captured CO_2 as raw material: techno-economic and environmental assessment, *Applied Energy* 161 (2016) 718–732.
2. M. Balat, H. Balat, C. Öz, Progress in bioethanol processing, *Progress in Energy and Combustion Science* 34(5) (2008) 551–573.
3. R.N. Hader, R.D. Wallace, R.W. McKinney, Formaldehyde from methanol, *Industrial & Engineering Chemistry* 44(7) (1952) 1508–1518.
4. C. Lamy, A. Lima, V. LeRhun, F. Delime, C. Coutanceau, J.-M. Léger, Recent advances in the development of direct alcohol fuel cells (DAFC), *Journal of Power Sources* 105(2) (2002) 283–296.
5. H. Peng, J. Ren, Y. Wang, Y. Xiong, Q. Wang, Q. Li, X. Zhao, L. Zhan, L. Zheng, Y. Tang, One-stone, two birds: alloying effect and surface defects induced by Pt on $Cu_{2-x}Se$ nanowires to boost C-C bond cleavage for electrocatalytic ethanol oxidation, *Nano Energy* 88 (2021) 106307.
6. E. Zhu, X. Yan, S. Wang, M. Xu, C. Wang, H. Liu, J. Huang, W. Xue, J. Cai, H. Heinz, Y. Li, Y. Huang, Peptide-assisted 2-D assembly toward free-floating ultrathin platinum nanoplates as effective electrocatalysts, *Nano Letters* 19(6) (2019) 3730–3736.

7. M. Davi, D. Keßler, A. Slabon, Electrochemical oxidation of methanol and ethanol on two-dimensional self-assembled palladium nanocrystal arrays, *Thin Solid Films* 615 (2016) 221–225.

8. M. Davi, T. Schultze, D. Kleinschmidt, F. Schiefer, B. Hahn, A. Slabon, Gold nanocrystal arrays as electrocatalysts for the oxidation of methanol and ethanol, *Zeitschrift für Naturforschung B* 71(7) (2016) 821–825.

9. J.-Y. Zhu, S. Chen, Q. Xue, F.-M. Li, H.-C. Yao, L. Xu, Y. Chen, Hierarchical porous Rh nanosheets for methanol oxidation reaction, *Applied Catalysis B: Environmental* 264 (2020) 118520.

10. W. Sugimoto, D. Takimoto, Platinum group metal-based nanosheets: synthesis and application towards electrochemical energy storage and conversion, *Chemistry Letters* 50(6) (2021) 1304–1312.

11. J.-Y. Chen, S.-C. Lim, C.-H. Kuo, H.-Y. Tuan, Sub-1 nm PtSn ultrathin sheet as an extraordinary electrocatalyst for methanol and ethanol oxidation reactions, *Journal of Colloid and Interface Science* 545 (2019) 54–62.

12. J. Wang, J. Zhang, G. Liu, C. Ling, B. Chen, J. Huang, X. Liu, B. Li, A.-L. Wang, Z. Hu, M. Zhou, Y. Chen, H. Cheng, J. Liu, Z. Fan, N. Yang, C. Tan, L. Gu, J. Wang, H. Zhang, Crystal phase-controlled growth of PtCu and PtCo alloys on 4H Au nanoribbons for electrocatalytic ethanol oxidation reaction, *Nano Research* 13(7) (2020) 1970–1975.

13. M.A.Z.G. Sial, M.A.U. Din, X. Wang, Multimetallic nanosheets: synthesis and applications in fuel cells, *Chemical Society Reviews* 47(16) (2018) 6175–6200.

14. Q. Yun, Q. Lu, C. Li, B. Chen, Q. Zhang, Q. He, Z. Hu, Z. Zhang, Y. Ge, N. Yang, Synthesis of PdM (M= Zn, Cd, ZnCd) nanosheets with an unconventional face-centered tetragonal phase as highly efficient electrocatalysts for ethanol oxidation, *ACS Nano* 13(12) (2019) 14329–14336.

15. M. Luo, Z. Zhao, Y. Zhang, Y. Sun, Y. Xing, F. Lv, Y. Yang, X. Zhang, S. Hwang, Y. Qin, J.-Y. Ma, F. Lin, D. Su, G. Lu, S. Guo, PdMo bimetallene for oxygen reduction catalysis, *Nature* 574(7776) (2019) 81–85.

16. T. Zeng, X. Meng, H. Huang, L. Zheng, H. Chen, Y. Zhang, W. Yuan, L.Y. Zhang, Controllable synthesis of web-footed PdCu nanosheets and their electrocatalytic applications, *Small* 18(14) (2022) 2107623.

17. W. Huang, X. Kang, C. Xu, J. Zhou, J. Deng, Y. Li, S. Cheng, 2D PdAg alloy nanodendrites for enhanced ethanol electrooxidation, *Advanced Materials* 30(11) (2018) 1706962.

18. Y. Han, J. Kim, S.U. Lee, S.I. Choi, J.W. Hong, Synthesis of Pd-Pt ultrathin assembled nanosheets as highly efficient electrocatalysts for ethanol oxidation, *Chemistry–An Asian Journal* 15(8) (2020) 1324–1329.

19. L. Xu, B. Fu, F. Gao, J.-W. Ma, H. Gao, P. Guo, Strain engineering of face-centered cubic Pd–Pb nanosheets boosts electrocatalytic ethanol oxidation, *ACS Applied Energy Materials* 6(4) (2023) 2471–2478.

20. K. Dong, H. Dai, H. Pu, T. Zhang, Y. Wang, Y. Deng, Constructing efficient ternary PtTeCu nano-catalysts with 2D ultrathin-sheet structures for oxidation reaction of alcohols, *Applied Surface Science* 609 (2023) 155301.

21. D. Wang, Y. Zhang, K. Zhang, X. Wang, C. Wang, Z. Li, F. Gao, Y. Du, Rapid synthesis of Palladium-Platinum-Nickel ultrathin porous nanosheets with high catalytic performance for alcohol electrooxidation, *Journal of Colloid and Interface Science* 650 (2023) 350–357.

22. M.A.Z.G. Sial, H. Lin, M. Zulfiqar, S. Ullah, B. Ni, X. Wang, Trimetallic PtCoFe alloy monolayer superlattices as bifunctional oxygen-reduction and ethanol-oxidation electrocatalysts, *Small* 13(24) (2017) 1700250.

23. J.W. Hong, Y. Kim, D.H. Wi, S. Lee, S.-U. Lee, Y.W. Lee, S.-I. Choi, S.W. Han, Ultrathin free-standing ternary-alloy nanosheets, *Angewandte Chemie International Edition* 55(8) (2016) 2753–2758.

24. H.M. An, Z.L. Zhao, L.Y. Zhang, Y. Chen, Y.Y. Chang, C.M. Li, Ir-alloyed ultrathin ternary PdIrCu nanosheet-constructed flower with greatly enhanced catalytic performance toward formic acid electrooxidation, *ACS Applied Materials & Interfaces* 10(48) (2018) 41293–41298.
25. L. Jin, H. Xu, C. Chen, H. Shang, Y. Wang, C. Wang, Y. Du, Three-dimensional PdCuM (M = Ru, Rh, Ir) Trimetallic alloy nanosheets for enhancing methanol oxidation electrocatalysis, *ACS Applied Materials & Interfaces* 11(45) (2019) 42123–42130.
26. L. Gong, Z. Yang, K. Li, W. Xing, C. Liu, J. Ge, Recent development of methanol electro-oxidation catalysts for direct methanol fuel cell, *Journal of Energy Chemistry* 27(6) (2018) 1618–1628.
27. H. Tian, Y. Yu, Q. Wang, J. Li, P. Rao, R. Li, Y. Du, C. Jia, J. Luo, P. Deng, Y. Shen, X. Tian, Recent advances in two-dimensional Pt based electrocatalysts for methanol oxidation reaction, *International Journal of Hydrogen Energy* 46(61) (2021) 31202–31215.
28. Y. Wang, S. Zou, W.-B. Cai, Recent advances on electro-oxidation of ethanol on Pt- and Pd-based catalysts: from reaction mechanisms to catalytic materials, *Catalysts* 5(3) (2015) 1507–1534.
29. H.-F. Wang, Z.-P. Liu, Comprehensive mechanism and structure-sensitivity of ethanol oxidation on platinum: new transition-state searching method for resolving the complex reaction network, *Journal of the American Chemical Society* 130(33) (2008) 10996–11004.
30. S. Wasmus, A. Küver, Methanol oxidation and direct methanol fuel cells: a selective review, *Journal of Electroanalytical Chemistry* 461(1) (1999) 14–31.
31. C. Bianchini, P.K. Shen, Palladium-based electrocatalysts for alcohol oxidation in half cells and in direct alcohol fuel cells, *Chemical Reviews* 109(9) (2009) 4183–4206.
32. R. Wu, L. Wang, A density functional theory study on the mechanism of complete ethanol oxidation on Ir (100): surface diffusion-controlled C–C bond cleavage, *The Journal of Physical Chemistry C* 124(49) (2020) 26953–26964.
33. E.D. Wang, J.B. Xu, T.S. Zhao, Density functional theory studies of the structure sensitivity of ethanol oxidation on palladium surfaces, *The Journal of Physical Chemistry C* 114(23) (2010) 10489–10497.
34. Y. Wang, L. Jin, C. Wang, Y. Du, Nitrogen-doped graphene nanosheets supported assembled Pd nanoflowers for efficient ethanol electrooxidation, *Colloids and Surfaces A: Physicochemical and Engineering Aspects* 587 (2020) 124257.
35. X. Meng, Y. Ouyang, H. Wu, H. Huang, F. Wang, S. Wang, M. Jiang, L.Y. Zhang, Hierarchical defective palladium-silver alloy nanosheets for ethanol electrooxidation, *Journal of Colloid and Interface Science* 586 (2021) 200–207.

CHAPTER 16

2D Metals for Energy Storage Applications

Anit Joseph, Sudha Priyanga, and Tiju Thomas

16.1 INTRODUCTION

Since the first discovery of graphene, low-dimensional materials have received a lot of interest [1]. The family of monolayer or two-dimensional (2D) materials has seen accelerated growth due to their unique physical and chemical characteristics. The morphology of metals and alloys is typically seen as in three dimensions (3D) rather than in two dimensions (2D). It is important to note that the definition of "two-dimensional" is broadened to incorporate freestanding ultrathin metallic nanosheets. Recent research roadmaps show that 2D nanomaterials, such as graphene, layered double hydroxides (LDHs), transition metal dichalcogenides (TMDCs), graphitic carbon nitride (g-C_3N_4), 2D carbon nanosheets, 2D transition metal carbides, nitrides or carbonitrides (MXenes), 2D transition metal oxides (TMOs), 2D metal-organic frameworks (MOFs), and phosphorene, have become widely used as electrode material for electrochemical energy storage (EES) devices.

The following characteristics explain why well-designed 2D nanomaterials are appropriate as cutting-edge electrode materials for EES devices. (i) A larger contact area between the active materials and the electrolytes, leading to improved charge distribution. It can increase the faradaic reaction rate of electrode materials used in batteries and capacitors. (ii) The rate capability of battery-type electrodes is enhanced due to the decreased ion diffusion lengths in 2D nanomaterials. (iii) 2D nanomaterials have interlayer structures and/ or tuned surfaces. The electrical conductivity of the materials is improved by the presence of several surface heteroatoms or functional groups. Through interlayer engineering, it is possible to increase the interlayer spacing, providing rapid ion transport routes and potential reactive sites. (iv) The unique 2D characteristics may efficiently reduce or control the volume fluctuation when the EES devices are repeatedly charged and discharged.

234

DOI: 10.1201/9781032645001-16

Energy storage alternatives come in a wide variety of forms, and novel concepts are constantly being developed.

16.1.1 Forms of Energy Storage: Energy Storage Devices

The five basic types of energy storage technologies are mechanical, electrochemical (or battery), thermal, electrical, and hydrogen (chemical) storage methods (Figure 16.1). Advanced energy storage systems can provide power backup that lasts a few minutes to many hours. It is crucial to select the right technology according to the needs and limitations of the application.

16.1.1.1 Pumped Hydro Storage (PHS)

Hydroelectric energy storage includes PHS. It is an arrangement of two water reservoirs at various elevations that can produce power as water flows through a turbine and descends from one to the other (discharge). PHS functions similarly to a giant battery.

16.1.1.2 Compressed Air Energy Storage (CAES)

In CAES systems, additional energy is mechanically retained by compressing the air in artificial or naturally occurring caverns. The air from the atmosphere is utilized in the most common designs. The compressor keeps the air at high pressure in a sealed container when there is extra energy. A CAES plant's standard power output falls between 100 and 300 MW.

16.1.1.3 Flywheel

A flywheel is a mechanical device that stores energy in the form of rotational momentum (Figure 16.2). Torque can be applied to a flywheel to cause it to spin, which boosts

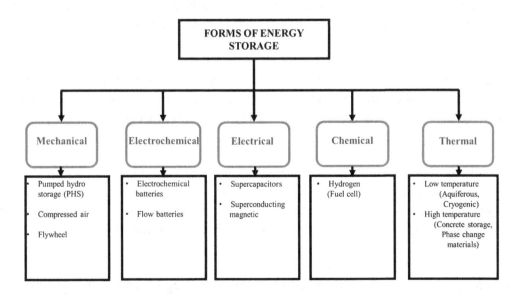

FIGURE 16.1 Energy storage technologies classification.

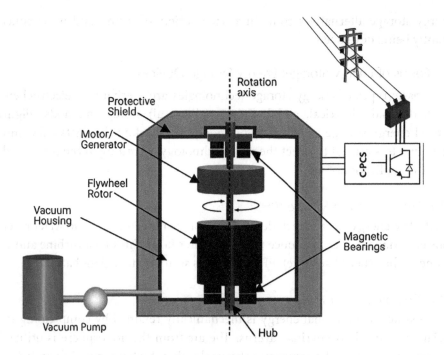

FIGURE 16.2 Schematic representation of a flywheel. Adapted with permission from [2], Copyright (2018), Elsevier.

its rotational speed. This stored momentum can apply torque to any rotating object. The energy stored inside a flywheel is related to its rotation speed and moment of inertia.

16.1.1.4 Batteries
The most well-established and widely used energy storage is the battery. They fall under the category of long-term energy storage systems. Electrochemical batteries can store energy by employing chemical interactions between positive and negative plates.

16.1.1.5 Fuel Cell
A fuel cell is an electrochemical device that uses fuel to produce electricity. On either side of an electrolyte, there are two electrodes that make up the cell. The system functions as an electrolyzer when a current is provided, splitting water into hydrogen and oxygen.

16.1.1.6 Aquiferous Thermal Energy Storage (TES)
In aquiferous low-temperature TES, wherein during off-peak hours, water is chilled or iced by a refrigerator and then stored for consumption during peak hours. The temperature difference between the chilled or iced water kept in the tank and the warm water returned from the heat exchanger determines the quantity of stored cooling energy [3].

16.1.1.7 Cryogenic Energy Storage (CES)
A cutting-edge technique for storing grid electricity is cryogenic energy storage. The concept is to liquefy air using off-peak or inexpensive electricity. When power is needed, cold

liquid air is pumped to boost pressure and then expanded through a turbine to produce useful electricity [3].

16.1.1.8 Concrete Storage

This technology uses castable ceramics or concrete to store energy at high temperatures for parabolic trough power plants. Synthetic oil may be used as the heat transfer fluid.

16.1.1.9 Phase Change Materials (PCM)

A substance continuously absorbing and releasing thermal energy is known as a phase change material (PCM). PCMs store and release heat through the melting and hardening stages. Usually, these materials have huge latent heat capacities.

16.1.1.10 Superconducting Magnetic Energy Storage (SMES)

To store electrical energy in a magnetic field, SMES, a relatively recent technology method, first generates DC current into a coil of a superconducting wire (Figure 16.3). There are no resistive losses, and no energy conversion into alternative forms is required. Typically, SMES has a higher power density than other devices.

16.1.1.11 Supercapacitors

Supercapacitors, commonly referred to as ultracapacitors or electrochemical double-layer capacitors, are relatively recent energy storage technologies. An electric field created between two electrodes serves as energy storage without a chemical reaction. Supercapacitors have a much higher energy density than ordinary capacitors. They both operate on the same principle, except that supercapacitors transfer ions over a conducting electrode with a large specific surface [2].

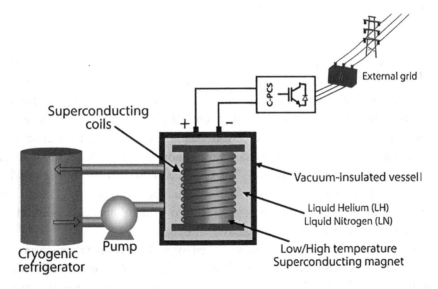

FIGURE 16.3 Schematic representation of the SMES system. Adapted with permission from [2], Copyright (2018), Elsevier.

16.2 2D METALS

Due to their unique physical and chemical characteristics, two-dimensional (2D) materials like graphene and MoS_2 have recently drawn much attention. These materials hold enormous potential for broad applications, from sensing to separation. It is possible to create single-layered 2D metals, although these materials are typically only stable at lateral sizes under 2 nm. Several top-down or bottom-up strategies can be used to synthesize thicker metal films, referred to as "two-dimensional."

16.2.1 Thermodynamic Stability

The chemical bonding within a layer is covalent in bulk materials having layered structures, such as graphite, hexagonal boron nitride, transition metal oxides, and transition metal dichalcogenides. In contrast, the interlayer bonding is of a weak van der Waals contact. As a result, it is simple to separate the various atomic layers [4]. In contrast, the strong non-directional metallic bonding in 3D for bulk metals with a non-layered atomic structure makes it challenging to form 2D metals. In a freestanding geometry with too many dangling bonds and was, therefore, previously thought to be thermodynamically unstable. However, Nevalaita and Koskinen [5] used the liquid-drop model to thoroughly explore the energy stability of 45 elemental 2D metal patches. They discovered that a few metals demonstrate intrinsic 2D stability despite their small sizes and only a few atoms.

Most transition metals, including Au, Zn, and Cd, tend to form small 2D structures. Numerous theoretical research [6,7] supported the conclusion. These ultrathin 2D metallic materials are thermodynamically unfavorable because of the high surface energies and potential strain energy induced by defects [8].

Thus, several approaches in terms of thermodynamics and kinetics are put out to synthesize 2D metallic materials, including surface-capping [8, 9] or chemical processes [10, 11, 12] or template-assisted growth [13, 14].

16.2.2. Synthesis Methods

Synthesis methods can be broadly divided into bottom-up and top-down approaches (Figure 16.4). Low-dimensional metals have expanded significantly since the first report of the successful production of Au nanoparticles [15]. Chemical syntheses are the main process used to create 2D metals today. These chemical processes have a "bottom-up" design that allows them to develop 2D metals. More recently, templates for the confined growth of 2D metals were also made from monolayer 2D non-metals such as graphene, graphene oxides (GOs), graphite, and MoS_2 [16]. The chemical methods do have some drawbacks such as low synthesis efficiency and restrict the lateral size of the 2D metals produced. Additionally, the synthetic chemical processes used in these chemical methods are typically only suitable for one or two elemental metals.

Comparatively, the "top-down" processes used to produce 2D metals, such as mechanical compression [17] and polymer surface buckling-enabled exfoliation (PSBEE) [18], typically result in large lateral sizes. The traditional chemical techniques, such as ligand-assisted growth and small-molecule-mediated growth, are used to create 2D metals, typically 1 µm

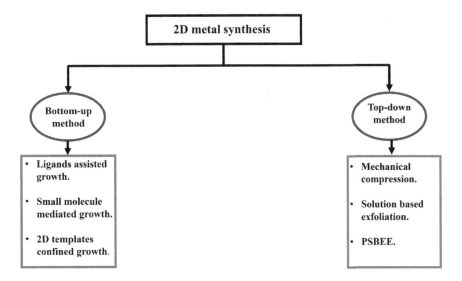

FIGURE 16.4 Schematic representation of the 2D metal synthesis method.

wide, and have an aspect ratio between 10 and 10^2. In comparison, the 2D metals created using top-down methods have an aspect ratio between 10^2 and 10^6 [19].

16.3 PROPERTIES OF 2D METALS

Metals in two-dimensional (2D) form, often referred to as 2D metals or 2D materials with metallic properties, have gained significant attention in materials science due to their unique electronic, mechanical, and thermal characteristics. Graphene, a single layer of carbon atoms arranged in a hexagonal lattice, is one of the most well-known 2D metals. Here are some properties and characteristics of 2D metals (Figure 16.5).

The properties of 2D metals can be broadly classified into electronic, structural, thermal, mechanical, optical, and surface properties. These are explained in detail below in the context of 2D metals.

16.3.1 Electronic Properties

The electronic structure of 2D metals is characterized by unique features due to their two-dimensional nature. In 2D metals, electrons are confined to move within a single atomic layer, which can lead to quantization effects and distinctive electronic properties. Let's explore the electronic structure of 2D metals with examples.

16.3.1.1 Graphene
16.3.1.1.1 Electronic Band Structure
Graphene is a single layer of carbon atoms arranged in a hexagonal lattice. It exhibits a linear electronic band structure near the Fermi level, forming a "Dirac cone" with two degenerate bands. The electronic band structure of a material describes how the energy levels (bands) of its electrons vary with momentum within the crystal's Brillouin zone. In the case of graphene, which consists of a hexagonal lattice of carbon atoms, the electronic

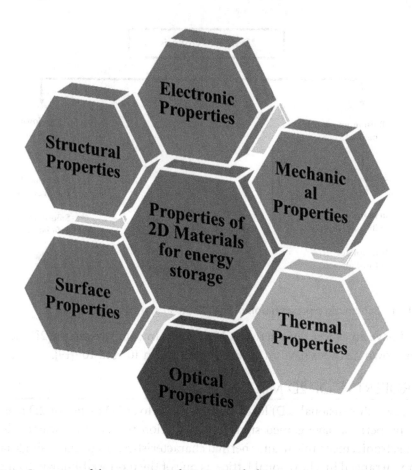

FIGURE 16.5 Summary of the properties of 2D materials for energy storage.

band structure exhibits some unique features: The most distinctive feature of graphene's electronic band structure is a linear dispersion relationship near the Fermi level (E_F). This linear relationship forms what is known as a "Dirac cone." In this cone, the energy (E) of electrons varies linearly with momentum (k), and it is given by the following equation:

$$E(k) = \hbar v_f |k|$$

where $E(k)$ is the energy, $\hbar$ is the reduced Planck's constant, v_f is the Fermi velocity (approximately 10^6 m/s), and k is the momentum [20–22].

Another critical feature of graphene's band structure is the absence of a band gap [20]. The valence and conduction bands meet at the Dirac point without an energy gap. This feature makes graphene a semi-metal or *zero-gap semiconductor*. The electronic properties of graphene are strongly influenced by its chirality, which refers to the orientation of the hexagonal lattice. Zigzag and armchair edges have distinct electronic behaviors.

16.3.1.1.2 Density of States (DOS)

The density of states (DOS) represents the distribution of electron energy levels as a function of energy within a material. It provides information about the number of electronic states available at different energy levels [21]. In the case of graphene, the DOS is zero at the Dirac point ($E = 0$), where the valence and conduction bands meet. This means there are no available electronic states at $E = 0$, consistent with the absence of an energy gap [21]. Away from the Dirac point, the DOS in graphene increases linearly with energy.

16.3.1.2 Transition Metal Dichalcogenides (TMDs) (e.g., MoS$_2$, WSe$_2$)

Unlike graphene, many TMDs have a finite band gap in their electronic structure, making them semiconductors in their 2D form. TMDs typically have two types of valence bands (heavy hole and light hole) and conduction bands. The energy gap between these bands determines the semiconductor properties. Monolayer TMDs are direct band gap semiconductors, while few-layer or bulk TMDs may have indirect band gaps. The electronic band structure and density of states (DOS) of TMDs can vary depending on the specific material and its layer thickness. Let's explore with detailed explanations. TMDs, such as molybdenum disulfide (MoS_2) and tungsten diselenide (WSe_2), exhibit distinct electronic band structures. Monolayer MoS_2 has a direct band gap of about 1.8 eV [22]. As the number of layers increases in TMDs, the band gap decreases, and they may exhibit an indirect band gap [23]. Spin–orbit coupling (SOC) is significant in TMDs and can lead to splitting energy bands near the band edges. In monolayer TMDs, there is a gap in the DOS at the band gap energy. This gap represents the absence of electronic states within the energy range of the band gap, which is typical for semiconductors.

16.3.1.2.1 Layer-Dependent DOS

As mentioned earlier, the layer thickness strongly influences the DOS. For monolayer TMDs, the DOS gap is prominent, while for few-layer and bulk TMDs, the DOS gap diminishes or shifts.

16.3.1.3 Black Phosphorus

Black phosphorus is another 2D material with a layered structure. It exhibits an anisotropic electronic band structure, with different band gaps along different crystallographic directions. The band gap of black phosphorus depends on the number of layers. As the number of layers increases, the band gap decreases. Black phosphorus has a puckered honeycomb lattice structure, with each layer consisting of two sublattices of phosphorus atoms (A and B). The electronic band structure of black phosphorus is highly anisotropic, meaning that its electronic properties depend on the crystallographic direction [24]. The most notable feature of black phosphorus is its direct band gap in the armchair direction (Γ–X direction of the Brillouin zone). This means that the valence band maximum (VBM) and the conduction band minimum (CBM) occur at the same momentum point in this direction. Monolayer black phosphorus has a relatively wide band gap (typically around 2 eV), which decreases as the number of layers increases. Like the band structure, the DOS of black phosphorus is highly anisotropic. It varies significantly along different crystallographic directions. In the armchair direction, where the direct band gap exists, the DOS near the band edges is relatively high compared to the zigzag direction [25].

242 ■ 2D Metals

16.3.1.4 Silicene and Germanene

Silicene and germanene are 2D analogs of silicon and germanium, respectively. They have a buckled honeycomb structure [25]. Silicene and germanene have been predicted to exhibit the quantum spin Hall effect. Mercury telluride (HgTe) is a prominent 2D material demonstrating the quantum spin Hall effect. Their electronic band structures and density of states (DOS) exhibit unique characteristics due to their different atomic compositions and structural variations.

16.3.1.4.1 Silicene

Silicene is the 2D counterpart of silicon and shares some similarities with graphene. However, its electronic band structure exhibits certain differences: unlike graphene, which has a planar hexagonal lattice, silicene has a buckled structure where silicon atoms in adjacent sublattices are slightly displaced in the vertical direction. This buckling introduces a breaking of symmetry that affects the electronic band structure. Silicene is predicted to exhibit the quantum spin Hall effect, a topological insulator behavior. Its electronic structure has a band gap, and the edge states are topologically protected. It possesses a tunable band gap. Similar to graphene, the DOS of silicene exhibits a vanishing DOS at the Dirac point, resulting from its linear dispersion relationship.

16.3.1.4.2 Germanene

Germanene, like silicene, has a buckled structure due to the vertical displacement of germanium atoms in adjacent sublattices. Germanene has a significant spin–orbit coupling effect, which means the interaction between the electron's spin and its motion is relatively strong. Germanene can exhibit a quantum spin Hall effect similar to silicene, but its topological properties may be more pronounced due to the stronger SOC. The DOS of germanene also exhibits a vanishing DOS at the Dirac point. However, the significant spin–orbit coupling in germanene introduces a spin-splitting effect in the DOS. This means that the DOS for spin-up and spin-down electrons is different, resulting in distinct energy distributions for each spin direction [25].

The 2D material's impressive band gap of 2–3 eV highlights its excellent electronic properties and positions it as a promising candidate for energy storage applications. This advancement underscores its potential as a key player in the quest for more sustainable and powerful energy storage solutions.

16.3.2 Structural Properties

Two-dimensional (2D) metals exhibit various structural properties depending on their specific atomic arrangement and the nature of the metal. These structural properties determine these materials' electronic, mechanical, and thermal behavior. Let's explore some details:

16.3.2.1 Honeycomb Lattice Structure (Graphene)

Graphene is the quintessential 2D metal with a honeycomb lattice structure. The metal atoms (like graphene and carbon atoms) form a hexagonal pattern resembling a honeycomb [26]. Each atom is bonded to three neighboring atoms, creating strong covalent bonds in the plane. This structure contributes to the exceptional mechanical strength and stability of graphene.

16.3.2.2 Buckled Structure (Silicene and Germanene)

Silicene and germanene are two prominent 2D metals with a buckled structure. In a buckled structure, the metal atoms (silicon for silicene and germanium for germanene) arrange themselves in a two-sublattice honeycomb lattice, similar to graphene [25]. However, there is a vertical displacement (buckling) between atoms in different sublattices. This buckling breaks the planar symmetry and can lead to interesting electronic properties, including the quantum spin Hall effect.

16.3.2.3 Layered Structure (Transition Metal Dichalcogenides(TMDs))

TMDs like molybdenum disulfide (MoS_2) and tungsten diselenide (WSe_2) have a layered structure. TMDs consist of layers of metal atoms sandwiched between layers of chalcogen atoms (e.g., sulfur and selenium) [27]. The metal atoms form a trigonal prismatic coordination, and there are van der Waals interactions between adjacent layers.

16.3.2.4 Rhombohedral Lattice Structure (Black Phosphorous)

Black phosphorus is a 2D material with a rhombohedral lattice structure. It consists of layers of phosphorus atoms arranged in a rhombohedral lattice [28]. Each layer is weakly bonded to adjacent layers via van der Waals forces. The anisotropy arises from the difference in bonding within the layers and between the layers.

16.3.3 Thermal Properties

The thermal properties of two-dimensional (2D) metals are of great interest in materials science, as they can significantly differ from their bulk counterparts due to the reduced dimensionality. Let's explore the thermal properties of 2D metals with examples and detailed explanations:

- Thermal conductivity: Thermal conductivity measures a material's ability to conduct heat. In 2D metals, thermal conductivity can be influenced by the reduced dimensionality and lattice structure. Example: graphene.

Graphene, a single layer of carbon atoms in a hexagonal lattice, exhibits exceptionally high thermal conductivity. This is attributed to the strong covalent bonds between carbon atoms and the planar structure. The thermal conductivity of graphene can exceed 3000 W/m·K, which is higher than most bulk materials [29].

- Anisotropy in thermal conductivity: In some 2D metals, the thermal conductivity may be anisotropic, meaning it varies with direction due to the material's lattice structure. Black phosphorus, a layered 2D material, has anisotropic thermal conductivity. The in-plane thermal conductivity is higher (up to ~350 W/m·K) compared to the out-of-plane direction (~20 W/m·K). This anisotropy is due to the weak van der Waals bonds between layers, which restrict heat transfer perpendicular to the layers [30].

- Size-dependent thermal properties: In 2D metals, the size of the material can affect thermal properties due to quantum confinement effects, where heat carriers have quantized energy levels. As 2D metals are scaled down to the nanoscale, their thermal conductivity can change [31].

- Defects and disorder: Defects, impurities, and disorder in 2D metals can significantly impact their thermal properties. Graphene with defects or impurities can exhibit reduced thermal conductivity compared to perfect graphene. Disordered phonon scattering at defect sites can disrupt the efficient phonon heat transfer pathways.

- Electron and phonon coupling: In 2D metals, the interaction between electrons and phonons (lattice vibrations) can influence thermal transport. Example: TMDs. TMDs like molybdenum disulfide (MoS_2) have strong electron-phonon interactions, leading to electron cooling and localized heating effects.

- Quantum size effects: As metal thickness approaches the nanoscale, quantum confinement effects become prominent. In 2D metals, this results in discrete energy levels and a band gap widening. As the number of layers decreases, the quantum confinement effect increases, modifying the electronic structure and band gap, which can be advantageous for energy storage applications. The increased surface-to-volume ratio in nanoscale metals improves charge/discharge kinetics.

16.3.4 Surface Properties

The surface properties of two-dimensional (2D) metals are critical for understanding their behavior and potential applications. The surface properties can differ significantly from their bulk counterparts due to reduced dimensionality. Here, we will provide examples of 2D metals and explain their surface properties in detail.

16.3.4.1 Graphene

Surface reconstruction: Graphene's surface is typically very clean and atomically flat. However, it can undergo surface reconstruction under certain conditions. For example, exposure to hydrogen can lead to the formation of a hydrogenated graphene surface, where hydrogen atoms bond with carbon atoms on the edges, altering its electronic properties [32].

Chemical reactivity: The exposed carbon atoms at the edges of graphene can be highly reactive. Functional groups can be added to these edges to tailor graphene's surface properties.

Adsorption: Graphene's large surface area makes it an excellent platform for adsorption of molecules and atoms. This property is exploited in gas sensors.

16.3.4.2 Platinum Diselenide (PtSe₂)

Surface reconstruction: $PtSe_2$, a 2D transition metal dichalcogenide (TMD), can undergo surface reconstruction, resulting in a periodic arrangement of selenium vacancies [33]. These vacancies can be manipulated to control the material's electronic properties.

Chemical reactivity: The exposed metal (platinum) atoms on the surface of $PtSe_2$ can be catalytically active. This property is useful for applications such as catalysis and energy storage.

16.3.4.3 Bismuthene

Surface states: Bismuthene, a 2D form of bismuth, exhibits surface states that are topologically protected. These surface states host conducting edge states, making bismuthene a potential material for topological insulators and energy storage applications.

Surface reconstruction: Similar to graphene, bismuthene can also undergo surface reconstruction, particularly at its edges [34].

16.3.4.4 Palladium Nanosheets

Surface patterning: Palladium nanosheets, which can be considered 2D due to their ultra-thin nature, can be patterned at the atomic scale. This enables the creation of nanopatterned surfaces with unique properties for applications like plasmonics and energy storage [35].

Catalytic activity: Palladium nanosheets can have highly catalytic surfaces due to their large surface area and the presence of active palladium atoms on the surface.

16.3.4.5 Copper Nanosheets

Surface roughness: Copper nanosheets can exhibit rough surfaces at the atomic scale, which can lead to enhanced catalytic activity. These rough surfaces provide more active sites for reactions, making them valuable for energy storage [36].

Surface oxidation: Copper nanosheets are susceptible to surface oxidation, forming copper oxide layers. These surface oxides can have different electronic properties and reactivity compared to the bulk material, affecting applications such as corrosion resistance.

16.4 APPLICATION OF 2D METALS IN DIFFERENT ENERGY STORAGE SYSTEMS

Two-dimensional (2D) metals have garnered significant attention in various energy storage systems due to their unique properties, such as high electrical conductivity, large surface area, and tunable electronic structures (Table 16.1). Here are some applications of 2D metals in different energy storage systems, along with examples and detailed descriptions (Figure 16.6):

16.4.1 Batteries

16.4.1.1 Li-ion Batteries (LIBs)

2D metals can be used as electrode materials in LIBs to improve energy storage capacity and charge–discharge performance. Molybdenum disulfide (MoS_2) is a 2D material

TABLE 16.1   Applications and Benefits of 2D Metals

Application	Description	Benefits
Anodes	Replace traditional graphite anodes with 2D metal materials (e.g., graphene, MXenes) to enhance lithium storage capacity and charge–discharge rates.	Increased energy density and faster charging.
Catalysts	2D metals like platinum or palladium nanosheets can serve as catalysts in the oxygen reduction reaction (ORR) and oxygen evolution reaction (OER) in lithium-air batteries, improving their overall efficiency.	Higher energy efficiency and longer cycle life.
Current Collectors	Use graphene or other 2D metals as current collectors to reduce weight and volume in LIBs, leading to higher energy density.	Improved energy-to-weight ratio.
Electrodes	Due to their high surface area and excellent electrical conductivity, 2D materials like graphene and MXenes are used as supercapacitor electrodes.	High power density, rapid charge–discharge cycles, and long cycle life.
Pseudocapacitors	2D transition metal dichalcogenides (TMDs) like MoS_2 can be used as pseudocapacitor materials, storing charge through Faradaic reactions.	Increased energy density compared to electrical double-layer capacitors (EDLCs).
Flexible Substrates	2D metals like graphene can serve as flexible substrates for supercapacitors, enabling the development of flexible and wearable energy storage devices.	Enhanced flexibility and integration into wearable electronics.
Hydrogen Adsorption	2D materials like graphene and metal-organic frameworks (MOFs) are used to adsorb and store hydrogen for hydrogen fuel cells and vehicles.	Enhanced hydrogen storage capacity and release kinetics.
Catalysts	2D metals such as palladium and platinum are used as catalysts for hydrogen evolution reactions (HER) and hydrogen oxidation reactions (HOR) in fuel cells.	Improved efficiency and performance of fuel cells.
Anodes	2D metals like MoS_2, phosphorene, and MXenes are explored as anode materials in NIBs and KIBs, providing an alternative to lithium-ion batteries.	Potential for low-cost and abundant resources.
Cathodes	2D materials can also be used as cathodes or cathode additives in NIBs and KIBs to improve their energy storage capabilities.	Enhanced electrochemical performance and energy density.
Photocatalysis	2D metals like titanium dioxide (TiO_2) modified with 2D metal materials are used in photocatalytic systems for solar water splitting and pollutant degradation.	Efficient solar energy conversion and environmental remediation.
Transparent Conductive Electrodes	Graphene and other 2D metals are employed as transparent conductive electrodes in photovoltaic devices (e.g., solar cells) to improve light absorption and charge collection.	Enhanced efficiency and flexibility in solar cell design.

Note: Two-dimensional (2D) metals offer enhanced energy storage capabilities due to their high conductivity, allowing for faster electron mobility and improved performance in energy storage devices.

2D Metals for Energy Storage Applications ■ 247

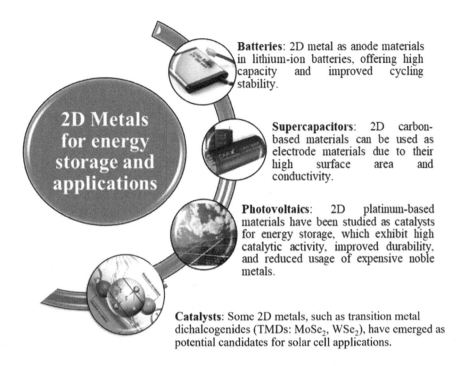

FIGURE 16.6 Applications of 2D metals in energy production (photocatalysts and photovoltaics) and energy-saving devices (batteries and superconductors).

with metallic conductivity that has been explored as an anode material in LIBs. Its layered structure provides ample surface area for lithium-ion adsorption, improving capacity and cycling stability.

16.4.1.2 Na-ion Batteries (SIBs)

2D metals can be explored as electrode materials in SIBs as an alternative to lithium-ion batteries, especially for grid-scale energy storage. Phosphorene, a 2D material composed of phosphorus atoms, has been studied for its potential use in SIBs. It can store sodium ions and exhibit good electrochemical performance.

16.4.2 Supercapacitor

2D metals can serve as electrode materials in supercapacitors to enhance energy and power density. Graphene, a 2D form of carbon, is widely used in supercapacitors. Its high surface area, excellent electrical conductivity, and low resistance at the electrode–electrolyte interface enable rapid charge and discharge, making it ideal for high-performance supercapacitors.

16.4.3 Hydrogen Storage

2D metals can be used for hydrogen storage, crucial for hydrogen fuel cells and the hydrogen economy. Palladium (Pd) nanosheets, when structured into 2D forms, can adsorb a substantial amount of hydrogen due to their high surface area and hydrogen dissociation ability.

16.4.4 *Water Splitting (Electrolysis)*

2D metals can be used as catalysts in water-splitting reactions, particularly in oxygen evolution (OER) and hydrogen evolution reactions (HER). Ruthenium disulfide (RuS_2) is a 2D metal sulfide that has shown excellent catalytic activity for the OER.

16.4.5 Thermoelectric Materials

2D metals can be used as thermoelectric materials for converting waste heat into electricity. Black phosphorus, a 2D material, has been investigated for its thermoelectric properties. It can efficiently convert heat gradients into electrical power and has potential applications in waste heat recovery systems. Low thermal conductivity in 2D materials can be advantageous for thermal management in energy storage devices. These materials exhibit poor heat conduction, so they help minimize heat dissipation during charging and discharging cycles in batteries or capacitors. This property enhances the overall efficiency and safety of energy storage systems. In contrast, high electronic conductivity facilitates the rapid and effective conversion of thermal energy into electrical energy, enhancing the overall performance of the energy storage system.

16.5. FUTURE PERSPECTIVES

Two-dimensional (2D) metals have garnered significant attention in energy storage due to their unique properties and potential to revolutionize various energy storage technologies. Here are some future perspectives on the use of 2D metals in energy storage applications:

16.5.1 High-Performance Electrodes

2D metals like graphene, transition metal dichalcogenides (TMDs), and black phosphorus have shown promise as electrode materials in energy storage devices. Their high electrical conductivity, large surface area, and tunable electronic properties make them suitable for use in batteries and supercapacitors. In the future, we can expect the development of high-performance 2D metal-based electrodes that improve these devices' energy and power density.

16.5.2 Improved Li-ion Batteries

Lithium-ion batteries are the backbone of portable electronics and electric vehicles. 2D metals can enhance the performance of lithium-ion batteries by increasing their capacity, reducing charging times, and extending the cycle life. For example, silicon-based 2D materials can address the issues of silicon anode expansion and contraction, leading to more stable and long-lasting batteries.

16.5.3 Na-ion Batteries

Transition metal dichalcogenides (TMDs), such as MoS_2 and WS_2, represent promising 2D materials for advancing sodium-ion batteries (Na-ion). These materials have a layered structure, allowing facile Na-ion intercalation [37]. The unique electronic properties of TMDs enhance the overall electrochemical performance [37,38] and can serve as efficient catalysts [38]. Incorporating 2D metals in Na-ion batteries holds the potential for addressing the limitations of traditional Li-ion batteries.

16.5.4 Next-Generation Supercapacitors

Supercapacitors are known for their rapid charge/discharge capabilities but have lower energy density compared to batteries. 2D metals, particularly graphene, can be integrated into supercapacitors to improve their energy density while retaining high power density.

16.5.5 Hydrogen Storage

Hydrogen is a clean and efficient energy carrier but requires effective storage solutions. Due to their porous structures and high surface areas, 2D metals such as metal hydrides and metal-organic frameworks (MOFs) can provide high hydrogen storage capacities. Advancements in materials science show promise in enhancing storage capacity and kinetics. Research initiatives, like the European Commission's Hydrogen Valleys and the U.S. Department of Energy's Hydrogen Storage Grand Challenge, highlight the global commitment to overcoming storage barriers. Pioneering technologies like 2D metal-organic frameworks (MOFs) and chemical hydrogen storage are gaining attention.

16.5.6 Flexible and Wearable Energy Storage

The mechanical flexibility of some 2D metals, like graphene, makes them suitable for flexible and wearable energy storage devices. These devices could be integrated into clothing, wearables, and other flexible electronics, offering portable and on-the-go energy solutions.

16.5.7 Sustainable Energy Storage

Sustainability is a key consideration in energy storage. 2D metals can be engineered to use abundant and environmentally friendly materials. For instance, using 2D materials based on earth-abundant elements can reduce the environmental impact of energy storage technologies.

16.5.8 Advanced Energy Conversion Systems

2D metals can also play a role in advanced energy conversion systems, such as fuel cells and thermoelectric generators. Their unique electronic properties and catalytic activity can enhance the efficiency and performance of these systems.

16.5.9 Integration with Renewable Energy Sources

As renewable energy sources like solar and wind become more prevalent, efficient energy storage solutions become increasingly important. 2D metals can help store excess energy generated by renewables, ensuring a stable and reliable energy supply.

16.6 SUMMARY AND CONCLUSIONS

This chapter explored the promising role of two-dimensional (2D) metals in energy storage applications, focusing on their unique properties and potential contributions to advancing energy storage technologies. The chapter covered various 2D metals, including graphene, transition metal dichalcogenides (TMDs), and others, highlighting their electronic, mechanical, and chemical attributes that make them suitable candidates for energy storage systems.

Integrating 2D metals in energy storage applications represents a promising avenue for addressing the growing demand for efficient and sustainable energy storage solutions. Their unique electronic, mechanical, and chemical properties, tunability, durability, and catalytic activity position them as valuable materials for enhancing the performance of batteries, supercapacitors, and other energy storage systems.

As researchers continue to explore the full potential of 2D metals and materials in energy storage, further advancements in energy density, charge/discharge rates, and device lifespan can be expected. However, challenges such as large-scale synthesis, cost-effectiveness, and scalability must also be addressed for practical implementation. Overall, 2D metals hold immense promise in revolutionizing the energy storage landscape, contributing to a sustainable and energy-efficient future.

REFERENCES

1. K.S. Novoselov, A.K. Geim, S.V. Morozov, D. Jiang, Y. Zhang, S.V. Dubonos, I.V. Grigorieva, A.A. Firsov, Electric field in atomically thin carbon films, *Science.* 80 (2004) 666–669.
2. M.C. Argyrou, P. Christodoulides, S.A. Kalogirou, Energy storage for electricity generation and related processes: Technologies appraisal and grid scale applications, *Renew. Sustain. Energy Rev.* 94 (2018) 804–821.
3. H. Chen, T.N. Cong, W. Yang, C. Tan, Y. Li, Y. Ding, Progress in electrical energy storage system: A critical review, *Prog. Nat. Sci.* 19 (2009) 291–312.
4. M. Osada, T. Sasaki, Exfoliated oxide nanosheets: new solution to nanoelectronics, *J. Mater. Chem.* 19 (2009) 2503–2511.
5. J. Nevalaita, P. Koskinen, Stability limits of elemental 2D metals in graphene pores, *Nanoscale.* 11 (2019) 22019–22024.
6. H. Häkkinen, B. Yoon, U. Landman, X. Li, H.-J. Zhai, L.-S. Wang, On the electronic and atomic structures of small AuN- (N = 4–14) clusters: A photoelectron spectroscopy and density-functional study, *J. Phys. Chem. A.* 107 (2003) 6168–6175.
7. P. Koskinen, H. Häkkinen, B. Huber, B. von Issendorff, M. Moseler, Liquid-liquid phase coexistence in gold clusters: 2D or Not 2D?, *Phys. Rev. Lett.* 98 (2007) 15701.
8. T.C.R. Rocha, D. Zanchet, Structural defects and their role in the growth of Ag triangular nanoplates, *J. Phys. Chem. C.* 111 (2007) 6989–6993.
9. Q. Zhang, N. Li, J. Goebl, Z. Lu, Y. Yin, A systematic study of the synthesis of silver nanoplates: Is citrate a "Magic" Reagent? *J. Am. Chem. Soc.* 133 (2011) 18931–18939.
10. J.-W. Yu, X.-Y. Wang, C.-Y. Yuan, W.-Z. Li, Y.-H. Wang, Y.-W. Zhang, Synthesis of ultrathin Ni nanosheets for semihydrogenation of phenylacetylene to styrene under mild conditions, *Nanoscale.* 10 (2018) 6936–6944.
11. H.L. Qin, D. Wang, Z.L. Huang, D.M. Wu, Z.C. Zeng, B. Ren, K. Xu, J. Jin, Thickness-controlled synthesis of ultrathin Au sheets and surface plasmonic property, *J. Am. Chem. Soc.* 135 (2013) 12544–12547.

12. L.-N. Wang, Z.-X. Zhou, X.-N. Li, T.-M. Ma, S.-G. He, Thermal conversion of methane to formaldehyde promoted by gold in $AuNbO_3+$ cluster cations, *Chem. A Eur. J.* 21 (2015) 6957–6961.

13. Y. Kuang, G. Feng, P. Li, Y. Bi, Y. Li, X. Sun, Single-crystalline ultrathin nickel nanosheets array from in situ topotactic reduction for active and stable electrocatalysis, *Angew. Chemie Int. Ed.* 55 (2016) 693–697.

14. Y. Kang, Q. Xue, P. Jin, J. Jiang, J. Zeng, Y. Chen, Rhodium nanosheets–reduced graphene oxide hybrids: A highly active platinum-alternative electrocatalyst for the methanol oxidation reaction in alkaline media, *ACS Sustain. Chem. Eng.* 5 (2017) 10156–10162.

15. M. Faraday, The Bakerian Lecture:Experimental relations of gold (and other metals) to light, *Philos. Trans. R. Soc. London.* 147 (1857) 145–181.

16. X. Lou, H. Pan, S. Zhu, C. Zhu, Y. Liao, Y. Li, D. Zhang, Z. Chen, Synthesis of silver nanoprisms on reduced graphene oxide for high-performance catalyst, *Catal. Commun.* 69 (2015) 43–47.

17. J. Gu, B. Li, Z. Du, C. Zhang, D. Zhang, S. Yang, Multi-atomic layers of metallic aluminum for ultralong life lithium storage with high volumetric capacity, *Adv. Funct. Mater.* 27 (2017) 1700840.

18. T. Wang, Q. He, J. Zhang, Z. Ding, F. Li, Y. Yang, The controlled large-area synthesis of two dimensional metals, *Mater. Today.* 36 (2020) 30–39.

19. T. Wang, M. Park, Q. Yu, J. Zhang, Y. Yang, Stability and synthesis of 2D metals and alloys: A review, *Mater. Today Adv.* 8 (2020) 100092.

20. R. Nandee, M.A. Chowdhury, A. Shahid, N. Hossain, M. Rana, Band gap formation of 2D materialin graphene: Future prospect and challenges, *Results Eng.* 15 (2022) 100474.

21. A. Bostwick, J. McChesney, T. Ohta, E. Rotenberg, T. Seyller, K. Horn, Experimental studies of the electronic structure of graphene, *Prog. Surf. Sci.* 84 (2009) 380–413.

22. R. Ganatra, Q. Zhang, Few-layer MoS_2: A promising layered semiconductor, *ACS Nano.* 8 (2014) 4074–4099.

23. A. Chaves, J.G. Azadani, H. Alsalman, D.R. da Costa, R. Frisenda, A.J. Chaves, S.H. Song, Y.D. Kim, D. He, J. Zhou, A. Castellanos-Gomez, F.M. Peeters, Z. Liu, C.L. Hinkle, S.-H. Oh, P.D. Ye, S.J. Koester, Y.H. Lee, P. Avouris, X. Wang, T. Low, Bandgap engineering of two-dimensional semiconductor materials, *NPJ 2D Mater. Appl.* 4 (2020) 29.

24. Z. Lin, Z. Tian, W. Cen, Q. Zeng, Monolayer black phosphorus: Tunable band gap and optical properties, *Phys. B Condens. Matter.* 657 (2023) 414780.

25. R. Haverkamp, S. Neppl, A. Föhlisch, Near-isotropic local attosecond charge transfer within the anisotropic puckered layers of black phosphorus, *J. Phys. Chem. Lett.* 14 (2023) 8765–8770.

26. T. Arjmand, M.B. Tagani, H.R. Soleimani, Buckling-dependent switching behaviours in shifted bilayer germanene nanoribbons: A computational study, *Superlattices Microstruct.* 113 (2018) 657–666.

27. P. Walimbe, M. Chaudhari, State-of-the-art advancements in studies and applications of graphene: A comprehensive review, *Mater. Today Sustain.* 6 (2019) 100026.

28. R. Gusmão, Z. Sofer, M. Pumera, Black phosphorus rediscovered: From bulk material to monolayers, *Angew. Chemie Int. Ed.* 56 (2017) 8052–8072.

29. M. Potenza, I. Petracci, S. Corasaniti, Transient thermal behaviour of high thermal conductivity graphene based composite materials: Experiments and theoretical models, *Int. J. Therm. Sci.* 188 (2023) 108253.

30. Z. Luo, J. Maassen, Y. Deng, Y. Du, R.P. Garrelts, M.S. Lundstrom, P.D. Ye, X. Xu, Anisotropic in-plane thermal conductivity observed in few-layer black phosphorus, *Nat. Commun.* 6 (2015) 8572.

31. Z. Zhang, Y. Ouyang, Y. Cheng, J. Chen, N. Li, G. Zhang, Size-dependent phononic thermal transport in low-dimensional nanomaterials, *Phys. Rep.* 860 (2020) 1–26.

32. Y. Wang, Z. Wang, Z. Qiu, X. Zhang, J. Chen, J. Li, A. Narita, K. Müllen, C.-A. Palma, Hydrogenation of hexa-peri-hexabenzocoronene: An entry to nanographanes and nanodiamonds, *ACS Nano.* 17 (2023) 18832–18842.

33. J. Chen, J. Zhou, W. Xu, Y. Wen, Y. Liu, J.H. Warner, Atomic-level dynamics of point vacancies and the induced stretched defects in 2D monolayer $PtSe_2$, *Nano Lett.* 22 (2022) 3289–3297.

34. C. Dolle, V. Oestreicher, A.M. Ruiz, M. Kohring, F. Garnes-Portolés, M. Wu, G. Sánchez-Santolino, A. Seijas-Da Silva, M. Alcaraz, Y.M. Eggeler, E. Spiecker, J. Canet-Ferrer, A. Leyva-Pérez, H.B. Weber, M. Varela, J.J. Baldoví, G. Abellán, Hexagonal hybrid bismuthene by molecular interface engineering, *J. Am. Chem. Soc.* 145 (2023) 12487–12498.

35. N. Baig, Two-dimensional nanomaterials: A critical review of recent progress, properties, applications, and future directions, *Compos. Part A Appl. Sci. Manuf.* 165 (2023) 107362.

36. H. Tabassum, X. Yang, R. Zou, G. Wu, Surface engineering of Cu catalysts for electrochemical reduction of CO_2 to value-added multi-carbon products, *Chem Catal.* 2 (2022) 1561–1593.

37. Y. Fang, L. Xiao, Z. Chen, X. Ai, Y. Cao, H. Yang, Recent advances in sodium-ion battery materials, *Electrochem. Energy Rev.* 1 (2018) 294–323.

38. R. Li, L. Zhang, L. Shi, P. Wang, MXene Ti(3)C(2): An effective 2D light-to-heat conversion material., *ACS Nano.* 11 (2017) 3752–3759.

CHAPTER **17**

2D Metals for Energy Conversion Applications

Fatemeh Bahmanzadgan, Fereshteh Pouresmaeil, Shanli Nezami, and Ahad Ghaemi

17.1 INTRODUCTION

2D metals have gained considerable attention in the energy conversion field due to their inherent characteristics such as excellent conductivity, strong thermal stability, and favorable chemical and mechanical resilience. These materials hold a central role in diverse energy conversion technologies, functioning either as catalysts or conductive components within devices designed for harnessing renewable energy sources. This chapter discusses various types of 2D metals, each possessing unique traits that make them suitable for specific applications. The conductivity of 2D metals relies on electron transfers occurring within their structure. Despite their advantages, it's important to recognize that these metals have limitations, compelling researchers to tackle these issues by incorporating effective functional groups into their structures. This chapter provides an extensive examination of the range of 2D metals, emphasizing their suitability and potential in energy conversion.

17.2 TYPES OF 2D METALS

2D noble metal-based catalysts have electrocatalytic properties of controllability by adjusting their size, thickness, composition, and shape. Pd, an often-found face-centered cubic (fcc) noble metal, has succeeded in various catalytic tasks, such as the electrooxidation of organic molecules and oxygen reduction. Meanwhile, Ir and Ru typically set the standard in electrocatalysis for water oxidation [1]. Transition metal dichalcogenides (TMDs) such as molybdenum disulfide (MoS_2) and tungsten diselenide (WSe_2) function as semiconductors and are applied across different energy conversion technologies, like solar cells and electrocatalysis [2]. Layered double hydroxides (LDHs) with their versatile layered

DOI: 10.1201/9781032645001-17

structure are utilized in fuel cells and catalytic processes due to their unique properties [3]. Transition metal oxides (TMOs) like titanium dioxide (TiO_2) and iron oxide (Fe_2O_3) possess varied characteristics suitable for applications in solar cells, batteries, and photocatalysis [4]. MXenes, a group of 2D transition metal carbides and nitrides, show potential in energy storage systems [5]. Metal-organic frameworks (MOFs), renowned for their high porosity, have promising applications in gas storage and separation processes [6]. The graphene-like materials can be categorized into three classes: metal-free analogs, transition metal compounds, and metallic materials [7]. Among these, a newcomer uniquely promising 2D material in the field of energy conversion is the two-dimensional metallic nanomaterials, known as "metallenes," which are defined as metallic nanosheets or nanoplates. These thin-layered materials, known for their unique electronic, mechanical, and thermal properties, have garnered significant attention and development in various applications, particularly energy conversion. The 2D metallenes like bismuthene, antimonene, phosphorene, germanene, etc. provide ultra-fast carrier mobility, outstanding electrical conductivity, super high thermal stability, tunable interlayer band gap, and relatively facile cost-effective preparation methods [8, 9]. These 2D metals and their derivatives demonstrate exceptional qualities essential for diverse energy conversion technologies, spanning from harnessing solar energy to storing and transforming catalytic energy. From the energy storage and conversion point of view, different 2D metals possess characteristics such as carrier mobility and tunable band gap. Metals of group 14 show adjustable band gap and tunable surface electronics by utilizing different approaches and 2D metals of group 15 exhibit high carrier mobility, high theoretical capacity, band gap tunability, and superior stability, making them competitive candidates for energy applications [9]. 2D noble metals, despite their high cost and low availability, demonstrate excellent properties. Their large specific surface area provides abundant active sites on the surface which enhances the atom utilization efficiency and results in suitability for energy conversion applications. Copper, cobalt, nickel, and iron nanosheets are the most studied non-noble 2D metals for this purpose [10]. The physicochemical and electrochemical properties of metals can be optimized by alloying as a consequence of enriching active sites. For Pd-based metals, alloying alters the electronic structure of Pd by downshifting the d-band center [11]. It is possible to optimize the electronic properties of two-dimensional metals by optimizing their morphology [7]. It has been shown by theoretical simulation that group IV plumbenes and metallenes (germanene and stanene) tend to mix SP^2 and SP^3 hybridization, and the electronic properties of 2D metals can be increased by chemical functionalization [12]. Also, two-dimensional materials with high electronic properties can increase light absorption. This characteristic of 2D metals is used in the design of solar cells and optoelectronic and photovoltaic devices. These materials have high stability at high temperatures and can perform well in harsh environmental conditions. 2D metals with high surface area and adjustable electronic properties by changing their morphology, resulting from their unique structural features, can be excellent catalysts for electrochemical reactions in fuel cells [13, 14]. 2D metals demonstrate outstanding electrical conductivity thanks to their electron structure, which allows electrons to move freely [15]. This conductivity is essential for energy-related uses as it facilitates efficient electron transfer, enhancing performance in batteries, supercapacitors, and some electrochemical procedures [16].

In energy conversion and storage devices, 2D metals with a large surface area provide many active sites for reaction and increase the efficiency of these devices. In addition, their unique dimensions show high quantum confinement effects. These materials can be useful in thermoelectric materials by increasing the mobility of energy carriers as a result of their quantum properties. These unique properties of 2D metals make them core materials in energy conversion systems and offer the potential to revolutionize this field by increasing efficiency and performance in various applications.

Some key features of metals bestow upon them the potential to be one of the promising materials in energy conversion processes. For instance, by enhancing the electrochemical reactions, metals can serve as highly efficient catalysts in fuel cells, such as graphene or dichalcogenides of transition metals. Besides, in photovoltaic cells, they improve the charge carrier separation and transport, resulting in efficient conversion of sunlight into electricity. In electrochemical cells, they may act as an electrode for water-splitting reactions to evolve hydrogen and oxygen. In the following sections, the specific applications of 2D metals in electrocatalysis, photovoltaic cells, and thermoelectric energy conversion will be explored in more detail.

17.3 CATALYTIC PROCESSES

There has been an increasing inclination towards the development of sustainable techniques for hydrogen production, with a particular focus on the electrochemical process of water splitting to generate hydrogen and oxygen. The electrochemical cell involves two distinct half-cell reactions, namely the hydrogen evolution reaction (HER) occurring at the cathode and the oxygen evolution reaction (OER) taking place at the anode. This method uses renewable energy to convert water into sustainable hydrogen fuel. Nevertheless, the water-splitting operation necessitates energy and electrocatalysts due to the fact that water electrolysis is not thermodynamically feasible. At present, the electrocatalysts that exhibit the highest efficiency are primarily composed of precious noble metals, specifically platinum-based materials and to a lesser extent, ruthenium-based materials [17].

17.3.1 Oxygen Evolution Reaction (OER)

OER pathways also play a crucial role in catalytic stability. A mechanistic understanding of the reaction pathways can explain how catalysts dissolve during the oxygen evolution [18]. The oxygen evolution mechanism falls into two main categories: the conventional adsorbate evolution mechanism (AEM) and the lattice oxygen oxidation mechanism (LOM) [19]. In AEM, OER's activity strongly correlates with intermediates' adsorption energies. The relationship between these adsorption bonds implies a theoretical limitation on OER activity. Conversely, LOM has garnered significant research interest lately due to its potential to surpass the limitations of AEM. Within the LOM pathway, the oxygen evolution involves the participation of lattice oxygen from catalysts, especially metal oxides, enabling direct O–O coupling. As LOM offers a distinct pathway for O–O coupling, its reaction energy isn't constrained by the adsorption energy scaling seen in the AEM pathway. The reaction pathway for AEM and LOM is indicated in Figure 17.1 based on a synthesis of prior studies to elucidate these pathways [20].

Thus far, the benchmark materials examined as potential NPMCs for the OER are Co- and Ni-based materials (both without support and with support on, e.g., carbon). Among the NPMCs for the OER, Co-based materials are only active in alkaline medium, e.g., CoO_x and CoOOH. The limitation of the traditional OER process based on the four-step mechanism is that it is frequently restricted by its lethargic electron transfer kinetics. The use of noble metal oxides such as IrO_2 and RuO_2 as active catalysts despite their scarcity and high cost is due to their high efficiency and remarkable stability. Therefore, it is very necessary to discover two-dimensional materials to replace Ru and Ir. Catalysts based on graphene without substrate and transition metals have shown many characteristics in addition to high charge transfer. Graphene-like TMDCs are also very active in the field of OER due to their unique structure and electrical properties related to this structure. Studies have shown that the low conductivity of layered TMDCs (WS_2, MoS_2, and $MoSe_2$) is their most important violation. BP is a remarkably functional metal-free layered semiconductor with relatively high transport (≥ 200 cm^2 V^1 s^1) [21].

The particular attention to metal-free electrocatalysts for the OER is because of their high activity, abundance, and inexpensive nature. Single-element doping such as N- and P-doping in graphene overcomes the OER obstacle, yielding η = 10 mA cm^{-2} (0.38 V for N-G241 and 0.33 V for P-G242) compared to that of the IrO_2 benchmark (overpotential: 0.33 V). In the co-doping case, more interesting results and a further improvement in OER activity were achieved. The activity based on electronegativity (EX) and electron affinity (AX) can be represented as ϕ = (EX/EC) × (AX/AC), and Xia et al. showed the binding strength and OER activity of p-block element-doped graphene. Plotting the lower limit of η as a function of ϕ resulted in a volcano shape for mono-elemental doping, which showed that P is the most excellent dopant in the highly active graphene 2D material for the OER. The OER is the reverse reaction of the ORR with 4e transfer. N-Doped graphene, TMs (Ni), metal oxides (Co_3O_4), composites with N-doped carbon nanotubes, and graphitic C have been demonstrated as very active catalysts in the OER process. Qiao et al. reported Ni

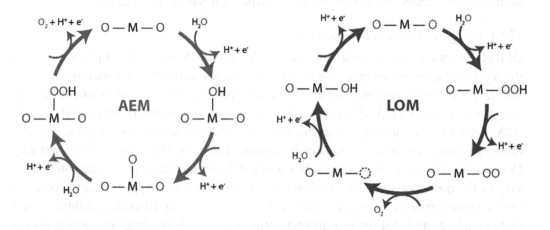

FIGURE 17.1. The reaction pathways of OER [20]. Copyright (2021), Elsevier.

2D Metals for Energy Conversion Applications ■ 257

nanoparticles co-doped in N-doped graphene (Ni/N-G) with OER (η) = 0.32 V, approaching that of the benchmark IrO_2 (0.27 V). These amazingly high activities are credited to the dual active site system.

MXenes are active catalysts with advantages such as high electrical conductivity and a highly hydrophilic surface. The activity of IrO_2 electrocatalysts for OER is very high compared to the activity of metal-free nanoplate electrocatalysts, but nevertheless, the 2D g-N_3C_4 and Ti_3C_2 composite was able to achieve the activity and efficiency corresponding to the activity of carbon-supported IrO_2. The volume of γ-CoOOH 2D plates is almost 20 times higher than that of γ-CoOOH, resulting in twice their activity compared to IrO_2. Therefore, it is said that NNMC nanosheets are more active in OER than IrO_2. Ni-based composites have become superior OER active electrocatalysts with distinct orbital characteristics and 2D planes containing unique 3D electrons. It is also said that these composites are available in very large quantities. From this family, we can refer to Ni-based 2D sheets (Ni–Fe and Ni–Co LDHs) which are very active and can be used in water oxidation, reaching i = 10 mA cm^2 in a small overpotential. Also, compared to IrO_2 and Co–Co LDH, Ni–Fe and Ni–Co LDHs show higher activity. The Ni–V LDH monolayer also showed i = 27 mA cm^2 (57 mA cm^2 after ohmic drop correction) and an overpotential of 350 mV, similar to the most active Ni–Fe LDHs for OER in basic solution. The increased activity of $NiCo_2O_4$, $ZnCo_2O_4$, and Co_3O_4 2D plates is caused by having enclosed O_2 vacancies with low absorption energy for H_2O (smaller TOF slopes, moderate current density, lower overpotential, and higher TOF rotation frequency). 2D porous Co_3O_4 nanosheets showed an electrocatalytic current of up to 342 mA/cm^2 at 1.0 V (vs. Ag/AgCl), about 50 and 30 times higher than bulk Co_3O_4 and the state-of-the-art 20 wt% Pt/C, respectively.

A large amount of unsaturated coordination Co^{3+} ions on the surface provides reactive sites, which speed up the OER process. Similarly, γ-CoOOH sheets with a thickness of 1.4 nm were used as an electrocatalyst for water oxidation, showing high mass activity (66.6 A g^{-1}), which was 20-fold greater than that of bulk γ-CoOOH and 2.4-fold superior to that of the state-of-the-art IrO_2, with the best OER activity among the NNMCs sheets published in the literature. This elevated activity is mainly due to their semi-metallic properties, resulting in improved H_2O electrophilicity and quicker interfacial transfer of electrons between Co ions and adsorbed –OOH intermediate species to make O_2.

The porous composite of g-C_3N_4 and MXene (Ti_3C_2) composed of double plates, in addition to improving the electrochemistry of O_2, provides a higher OER current compared to IrO_2 (20 wt%) and a strong onset potential (1.44 V) in contrast to RHE. Compared to separate Ti_3C_2 and g-C_3N_4 electrodes, this electrode has created a significant synergistic effect. Improving the electrocatalytic performance with BP has attracted the attention of many researchers due to its unique characteristics [22]. Recently, Wang et al. synthesized bulk BP, which had an onset potential of 1.49 V (vs. RHE) and a low Tafel slope of 72.88 mV dec^{-1}, showing comparative electrocatalytic characteristics to the RuO_2 benchmark catalysts for the OER, but its bulk crystal structure still suffers from low active sites. Recently, it was reported that a decrease in the number of layers in 2D materials will result in further exposed electrocatalytic active surface and amplified SSA than that of their bulk counterparts. These favorable results were first observed with the native ultrathin lamellar BP structure, which can open up alternative paths in using 2D layered materials with high electrocatalytic activity. Xiaohui Ren et al. used BP for the OER as a newly rediscovered layered material [23]. Goswami et al. demonstrated that the development of cost-effective, efficient, and durable heterogeneous OER electrocatalysts is possible through the use of non-noble transition metal-containing MOFs with electron-rich bis(5-azabenzimidazole) and organic dicarboxylate bases via hydrothermal synthesis (MOFs 1–4). CV and LSV studies show that these MOFs are very efficient electrocatalysts for the OER process with long-term stability. The overpotential required to produce a current density of 1 mA/cm^2, using Ni (II) containing MOF 2, is the lowest (370 mV), indicating the importance of coordinated water molecules in the 1 coordination sphere around the metal centers and the presence of Ni (II) in the OER process. The highest TOF value of 0.6 s^1 shown by MOF 2 indicates importance [24]. Based on first-principles calculations, Song et al. predicted that the 2D MOF NiIT monolayer could act as a dual-purpose electrocatalyst for efficient overall water splitting. The high electrocatalytic performance of 2D MOFs with overpotentials for V 0.50 (OER) is comparable to common electrocatalysts containing precious metals. The active sites of the NiIT monolayer are on the non-metal atoms instead of the TM atom, i.e., the N atom (for HER) and the C atom that binds to the S atom (for OER). Our simulations highlight a two-dimensional MOF electrocatalyst without precious metals for high-efficiency water splitting as well as a new strategy for catalyst designs [25].

17.3.2. Hydrogen Evolution Reaction (HER)

HER is a two-electron transfer process involving a catalytic intermediate, leading to hydrogen (H) adsorption on a specific site on the electrocatalyst's surface. The energy associated with hydrogen adsorption plays a critical role in determining the overall reaction rate. It is highly desired to have a hydrogen adsorption-free energy value close to zero to achieve effective catalysis for the HER.

Notably, excellent electrocatalytic performances for HER have been achieved using single-atom catalysts such as Pt, Ru, Pd, Co, Mo, Fe, Ni, and W. The unique electronic structure of 2D metals is a crucial factor that enhances the efficiency of the HER [26]. However, these single-metal atom electrocatalysts are susceptible to aggregation due to the high surface energy of individual metal atoms, which can limit their overall stability [17].

The exceptional catalytic performance of electrocatalysts based on TMDs is strongly linked to their layered structure. Consequently, achieving the synthesis of TMDs with controllable layers and large-area uniformity is imperative to unlock their vast range of practical applications. Significant progress has been achieved in producing high-quality monolayer TMD nanosheets (NSs). This section will overview the standard methodologies for synthesizing 2D TMD materials [27]. These methods include top-down approaches, which involve exfoliating bulk TMD counterparts and bottom-up approaches, such as the hydrothermal (solvothermal) method and the vapor-phase deposition route [28, 29].

Their two-dimensional layered structure provides a larger specific surface area, creating numerous active sites for the HER. The in-plane resistivity of 2D TMDs is lower than that of the basal planes, facilitating electron transport to the catalytic edge sites. In recent years, 2D TMD electrocatalysts, including materials such as MoS_2, $MoSe_2$, WS_2, niobium disulfide (NbS_2), and others, have emerged as promising alternatives to address the limitations associated with noble metals and related compounds. These exceptional 2D TMDs possess unique properties that make them highly suitable for various applications, with a particular focus on enhancing their performance in the HER [27]. The distinct d orbitals of various transition metals result in versatile electronic structures in TMDs, leading to diverse catalytic behaviors. TMDs like MoS_2 have remarkably improved catalytic activity over the past decade. Additionally, extensive research has enhanced our understanding of HER catalytic mechanisms in TMDs, offering valuable insights for developing new HER catalysts [27]. Extensive research has been conducted on MoS_2 as an electrocatalyst for the HER. It is the most widely studied TMD for electrochemical hydrogen production. As illustrated earlier, much progress has been made in the synthesis and exfoliation of TMD [30]. The electrocatalytic performance of two-dimensional porous NiCoSe nanosheet arrays in nickel form was investigated for the HER system. It has been shown that the excellent performance with long-term stability is due to the mesoporous structure and the synergistic effect between NiCoSe nanosheet arrays and nickel foam. In addition, to reach a current density of 10 mA cm², it shows the required overpotential of 170 mV for HER [31].

17.3.3 Oxygen Reduction Reaction (ORR)

ORR is a vital electrochemical process that converts oxygen molecules into water. It's a complex, multi-electron reaction involving multiple adsorption and desorption steps. ORR is important in many electrochemical energy conversion and storage technologies. ORR occurs through two pathways in the cathode, depending on the type of acidic or alkaline electrolyte used, O_2 is converted to OH or H_2O. In this way, there is a path of "partial" reduction of 2e, which leads to the formation of absorbed H_2O_2 species, and in another path, complete reduction or "direct" reduction of 4e occurs. Considering that H_2O_2 is more stable than H_2O_2, therefore, the method of reducing 4e is very much considered. In the direct conversion of O_2 to H_2O, by breaking the oxygen–oxygen bond, O_2 absorption occurs on the catalyst surface. Therefore, direct conversion of O_2 to H_2O is a decomposition method. Then, in order to have hydroxyl groups attached to the surface, it is necessary to transfer electrons to the absorbed oxygen atoms. This electron transfer occurs in the form of hydrogen addition. As a result of complete reduction, H_2O is produced while partial reduction of O_2 produces H_2O_2. The ORR (vs. RHE) proceeds via two-step $2e^-$ routes or the direct $4e^-$.

By preventing partial reduction and production of peroxide, maximum energy can be obtained from the reaction. In addition to reducing flow and operational efficiency, partial reduction and production of peroxide leads to the destruction of fuel cell components and unstable performance. 2D metals or two-dimensional metal materials, including substances like graphene, graphene-like materials, and metal-based two-dimensional materials such as MoS_2, have garnered significant attention as potential catalysts. These materials offer substantial surface areas, excellent electrical conductivity, and active sites for oxygen reduction, making them promising candidates for enhancing ORR efficiency [32]. Their adaptable structures and ability to fine-tune their properties make 2D metals valuable for optimizing ORR catalysts [33]. Additionally, research has explored the integration of 2D

MOFs into ORR catalysts, leveraging their stability and electrical conductivity to enhance catalyst performance and durability [34].

The two-dimensional graphene family, due to their distinctive and amazing features such as suitable effective specific surface area, high density and mechanical strength, medium conductivity, and suitable thermal, mechanical, and electrical stability as well as their unique structure, is a suitable alternative to PGM electrocatalyst in ORR to produce. They are sustainable energy. According to the studies that researchers have done in replacing two-dimensional materials for PGM electrocatalysts in ORR, it has been shown that graphene can be an active and low-cost substitute for PGM electrocatalysts. The researchers showed that graphene with B co-doping in N-graphene developed stably and with high N-doping can be of comparable efficiency to benchmark electrocatalysts for ORR. A very high ORR activity of graphene doped with boron or nitrogen has been shown due to its large surface area and high repulsion power in numerous active sites. Graphene doped with boron or nitrogen shows higher ORR activity than platinum-based carbon electrodes in alkaline environments and comparable activity in acidic environments with high stability. The pz atomic orbitals of graphene lead to its chemical stability. These orbitals are hybridized and allow absorption by p–p, p–CH, and p–cation interactions by establishing a stable nonlocal bond in the orbital. By obtaining pentagonal, heptagonal, and octagonal rings, resulting from the removal of heteroatoms from graphene, Yao et al. introduced defects in the structure of graphene, which can strongly affect the ORR properties. In order to eliminate these defects in graphene resulting from the synthesis, the synthesis of non-doped and edge-doped graphene by Ar plasma treatment is considered. MXenes are a new family of two-dimensional materials that are promising candidates in the field of electrocatalysis. $Ti_3C_2X_2$, one of the most widely used MXenes in the field of electrocatalytic application, has two very important advantages. First of all, this material has high resistance in acidic and oxidative environments and mechanical stress. Second, it effectively prevents corrosion, while it can be an effective support for carbon-based electron transfer channels. For example, Ag shows relatively high ORR performance and strength in an alkaline medium, but its reported electrocatalytic behavior is not satisfactory. Another brilliant program to further enhance the catalytic activity is to prepare more conductive supports along with some effective bimetallic catalysts. Therefore, Peng et al. synthesized MXene–Ag composites by directly mixing $AgNO_3$ and MXene alkaline solution (alk–MXene, $Ti_3C_2(OH/ONa)_2$) containing polyvinylpyrrolidone (PVP) at room temperature. This unique alk–MXene–$Ag_{0.9}Ti_{0.1}$ composite nanowire with a width of 42 ± 5 nm showed increased conductivity and synergistic effect of high ORR activity by increasing the active sites and showed the best ORR activity with $E_{1/2}$ = 0.782 V vs. RHE at 1600 rpm and η = 0.921 V vs. RHE, the best structure stability and reversibility after 1000 cycles compared to the commercial Ag/C catalyst. Deng et al. investigated the catalytic potential of single-atom catalysts (SACs) within 2D transition metal-based tetracyanoquinodimethane (TM-TCNQ) monolayers. These monolayers, hosting a dense and periodic distribution of single transition metal atoms, proved remarkably efficient in ORR and OER. Notably, Fe-TCNQ and Ni-TCNQ emerged as highly promising catalysts, with Fe-TCNQ demonstrating anticipated activity surpassing that of Pt in ORR, indicating its potential as an exceptionally active catalyst in this domain [35]. Liu et al. developed a novel set of versatile electrocatalysts composed mainly of metallic nickel or cobalt along with a minor proportion of their oxides attached to nitrogen-doped reduced graphene oxide (rGO). This involved creating Co-CoO/N-rGO and Ni-NiO/Ni-N-RGO through a pyrolysis process involving graphene oxide and either cobalt or nickel salts. The Ni-NiO/N-rGO catalyst demonstrated superior electrocatalytic performance for OER when compared to commercial IrO_2. On the other hand, Co-CoO/N-rGO exhibited heightened activity for ORR comparable to Pt/C, particularly in the context of zinc–air batteries [36]. Wei et al. examined a series of 2D MOFs with TMN_2O_2 structures (TM = Cr, Mn, Fe, Co, Ni) to gauge their capability in catalyzing ORR. Among them, the CoN_2O_2 configuration emerged as a standout, demonstrating superior electrocatalytic performance for both ORRs [37].

17.4 PHOTOVOLTAICS

Two-dimensional metal nanomaterials have the potential to play a significant role in increasing the efficiency of solar cells and photovoltaic energy conversion processes. Their compelling characteristics render a great potential to improve the performance of photovoltaic devices. Despite the great importance and the considerable impact that 2D metals can have on increasing the efficiency of solar cells, there is still relatively limited research in this field, mainly due to the challenges in their fabrication and utilization in photovoltaic devices, which is as a result of the high energy barrier to overcome and the fact that the metal monolayers are of low thermodynamic stability [7]. However, the promising results from the existing research are a confirmation of their effectiveness in improving photovoltaic energy conversion technologies.

Among different experimentally and theoretically investigated metallenes, antimonene stands out as a highly stable band gap semiconductor, and antimonene-based 2D materials are intriguing for researchers due to their superior physicochemical properties and promising applications in optoelectronics devices. A high-quality antimonene oxide-based low dimensional materials were prepared using a liquid exfoliation method in butanol and were applied for perovskite solar cells based on copper-doped nickel oxide hole transfer layer, by improving power conversion efficiency from 15.7% to near 18% [38]. The application of antimonene, as a 2D metal, leads to an effective charge extraction and a decreased recombination loss, which helps to achieve improvement in photovoltaic performance. In order to achieve high-performance nonfullerene organic solar cells (NF-OSCs), a bilayer hole extraction layer (HEL) with solution-processed molybdenum trioxide (MoO_3) and two-dimensional (2D) material of antimonene was developed. By inserting the antimonene layer, the power conversion efficiency (PCE) of devices with MoO_3 HEL was increased from 8.92% to 11.30% in OSCs with nonfullerene systems of PBDB-T-2F:IT-4F, which was much higher than that of the devices with PEDOT:PSS HEL (10.59%) [39].

One of the challenges that limits the practical application of quantum-dot-sensitized solar cells (QDSSCs) is their high recombination effect and low stability, which lowers their efficiency. Antimonene quantum dots (AMQDs) were applied as an effective photoactive materials in QDSSCs and their advantages, in association with a strong light-matter interaction, moderate energy band gap in the visible range, and antioxidation properties, resulted in a promoted photoelectric conversion efficiency (PCE) of 3.07%. The as-fabricated solar cell demonstrated long-term stability, where the PCE value maintains to more than 90% of the initial value after 1000 h [40].

The short layer distance and high binding energy of antimonene, as a monolayer 2D material, makes its preparation very challenging, while its band gap tunability and semiconductive properties should be confirmed.

In one of the studies for verification of semiconductor properties of monolayer antimonene nanosheets, the obtained semiconductive antimonene nanosheets (SANs) exhibited photoluminescence (PL) band gap at about 2.33 eV and PL lifetime of 4.3 ns, which shows that SANs were ideal for the hole extraction layer in planar inverted perovskite solar cells (PVSCs). Owing to their fast hole extraction and efficient hole transfer at the perovskite/hole transport layer interface, they could significantly enhance the device performance.

In another approach, uniform layers of antimonene were prepared, using a pregrinding and subsequent sonication-assisted liquid-phase exfoliation process. Pregrinding makes a shear force along the layer surfaces, forming large thin antimony plates. These plates then easily are exfoliated into smooth, large antimonene, avoiding long sonication times and antimonene destruction. The resulting antimonene verified the tunable band gap from 0.8 eV to 1.44 eV. Subsequently, when the antimonene was used as a hole transport layer in perovskite solar cells, 30% hole extraction and current enhancement occurred [41].

17.5 THERMOELECTRIC CONVERSION

Thermoelectric energy conversion is a process that converts heat energy into electrical energy (and vice versa) using materials with specific thermoelectric properties [42]. It represents a groundbreaking technology that exploits a fundamental physical phenomenon called the Seebeck effect [43]. This phenomenon facilitates the immediate conversion of temperature variances into electrical energy. Fundamentally, a temperature difference between dissimilar materials, often semiconductors, causes electron motion and generates an electric current. This exceptional characteristic sets thermoelectric materials apart from conventional approaches to power generation, which depend on mechanical components or chemical processes. These materials possess unique properties, such as high electrical conductivity and customizable thermal properties. These properties make them promising candidates for enhancing thermoelectric energy conversion. 2D metals' high mechanical strength and stability at high temperatures make them suitable for use in harsh environmental conditions, further enhancing their utility in energy conversion systems. 2D transition metal dichalcogenides (TMDCs), exemplified by MX_2 (with M = Mo, W, Ti and X = S, Se, Te), have gained significant attention recently due to their layered structure and outstanding characteristics. Their semiconductor nature, strong chemical stability, and exceptional mechanical and physical attributes have driven extensive exploration. These materials have undergone thorough investigation in various domains, including optoelectronics, energy conversion, and applications in fields like cancer therapy. TMDCs have notably attracted considerable interest in photodetectors, thermoelectric uses, and gas-sensing technologies. Their noteworthy traits involve relatively high electrical conductivity combined with lower thermal conductivity, positioning TMDCs as promising options for high-performance thermoelectric devices. Their potential extends significantly to creating wearable heating/cooling tools and efficient power generators. Patel et al. utilized monolayers WSSe and WSTe. Both materials showcased direct band gaps, remarkable Seebeck coefficient, and electrical conductivity, resulting in a notably high-power factor. Additionally, these Janus monolayers exhibited significantly lower lattice thermal conductivity in comparison to WS_2. As a result, WSTe is a promising thermoelectric material due to its exceptional performance, positioning the Janus monolayer as an advantageous choice for effective thermoelectric conversion [44]. However, some metallenes, like aluminene, are just theoretically predicted to have high thermal stability [45]. In some cases, the thickness of metallenes affects their thermal stability [46].

17.6 CONCLUSIONS

2D metals have exhibited unique properties in the field of energy conversion namely exceptional electrical conductivity, expansive specific surface area, thermal stability, remarkable flexibility, and the ability to fine-tune their electronic structure. In this chapter we explored some features that are related to energy conversion applications and provided a foundation for realizing how 2D metal can significantly enhance efficiency in energy conversion systems. Initially, the main catalytic processes in electrochemical energy conversion such as OER, HER, and ORR were investigated by highlighting the effect of 2D metals on their efficiency. The potential of 2D metal to catalyze these reactions can revolutionize sustainable energy production. After all, the light-absorbing capabilities, appropriate band gap, and electrical conductivity of 2D metal which leads to more efficient solar energy conversion were examined. Despite the intriguing characteristics of 2D metals, their practical applications, especially the energy-related ones, are seriously limited. These limitations particularly are due to the challenges in synthesis methods. The exfoliation method, which is of more interest to researchers in this field, is not only challenging for achieving one or a few layers of metal nanosheets but also may be very expensive. Moreover, the amount of work on the evaluation of 2D metals for energy storage and conversion is extremely limited owing to the synthesis routes, and it necessitates more attention in academic environments. In conclusion, this chapter provided an overview of the potential that this new category of two-dimensional materials has in energy conversion, specifically by demonstrating their effect on various energy conversion systems. The study of 2D metal presents a promising path towards more efficient and environmentally friendly energy conversion technologies.

REFERENCES

1. Y. Li, S. Guo, Noble metal-based 1D and 2D electrocatalytic nanomaterials: Recent progress, challenges and perspectives, *Nano Today* 28 (2019) 100774.
2. X. Wu, H. Zhang, J. Zhang, X.W. Lou, Recent advances on transition metal dichalcogenides for electrochemical energy conversion, *Advanced Materials* 33 (2021) 2008376.
3. K. Fan, P. Xu, Z. Li, M. Shao, X. Duan, Layered double hydroxides: Next promising materials for energy storage and conversion, *Next Materials* 1 (2023) 100040.
4. L.G. Gerling, S. Mahato, C. Voz, R. Alcubilla, J. Puigdollers, Characterization of transition metal oxide/silicon heterojunctions for solar cell applications, *Applied Sciences* 5 (2015) 695–705.

5. P. Lokhande, A. Pakdel, H. Pathan, D. Kumar, D.-V.N. Vo, A. Al-Gheethi, A. Sharma, S. Goel, P.P. Singh, B.-K. Lee, Prospects of MXenes in energy storage applications, *Chemosphere* 297 (2022) 134225.
6. R. Nivetha, S. Sharma, J. Jana, J.S. Chung, W.M. Choi, S.H. Hur, Recent advances and new challenges: Two-dimensional metal–organic framework and their composites/derivatives for electrochemical energy conversion and storage, *International Journal of Energy Research* 2023 (2023) 711034, 47 pages.
7. M. Xie, S. Tang, B. Zhang, G. Yu, Metallene-related materials for electrocatalysis and energy conversion, *Materials Horizons* 10 (2022) 407–431.
8. Y. Wang, K. Chen, H. Hao, G. Yu, B. Zeng, H. Wang, F. Zhang, L. Wu, J. Li, S. Xiao, J. He, Y. Zhang, H. Zhang, Engineering ultrafast charge transfer in a bismuthene/perovskite nanohybrid, *Nanoscale* 11 (2019) 2637–2643.
9. Z. Wu, J. Qi, W. Wang, Z. Zeng, Q. He, Emerging elemental two-dimensional materials for energy applications, *Journal of Materials Chemistry A* 9 (2021) 18793–18817.
10. B. Jiang, Y. Guo, F. Sun, S. Wang, Y. Kang, X. Xu, J. Zhao, J. You, M. Eguchi, Y. Yamauchi, H. Li, Nanoarchitectonics of metallene materials for electrocatalysis, *ACS Nano* 17 (2023) 13017–13043.
11. X. Chu, J. Li, W. Qian, H. Xu, Pd-based metallenes for fuel cell reactions, *Chemical Record* 23 (2023).
12. Y. Liu, K.N. Dinh, Z. Dai, Q. Yan, Metallenes: Recent advances and opportunities in energy storage and conversion applications, *ACS Materials Letters* 2 (2020) 1148–1172.
13. A. Hayat, M. Sohail, A. El Jery, K.M. Al-Zaydi, S. Raza, H. Ali, Z. Ajmal, A. Zada, T. Taha, I.U. Din, Recent advances, properties, fabrication and opportunities in two-dimensional materials for their potential sustainable applications, *Energy Storage Materials* 59 (2023) 102780.
14. V. Kochat, A. Samanta, Y. Zhang, S. Bhowmick, P. Manimunda, S.A.S. Asif, A.S. Stender, R. Vajtai, A.K. Singh, C.S. Tiwary, P.M. Ajayan, Atomically thin gallium layers from solid-melt exfoliation, *Science Advances* 4 (2018) e1701373.
15. T. Dutta, N. Yadav, Y. Wu, G.J. Cheng, X. Liang, S. Ramakrishna, A. Sbai, R. Gupta, A. Mondal, Z. Hongyu, Electronic properties of 2D materials and their junctions, *Nano Materials Science* 6 (2024) 1–23.
16. J. Pang, R.G. Mendes, A. Bachmatiuk, L. Zhao, H.Q. Ta, T. Gemming, H. Liu, Z. Liu, M.H. Rummeli, Applications of 2D MXenes in energy conversion and storage systems, *Chemical Society Reviews* 48 (2019) 72–133.
17. R. Sukanya, D.C. da Silva Alves, C.B. Breslin, Recent developments in the applications of 2D transition metal dichalcogenides as electrocatalysts in the generation of hydrogen for renewable energy conversion, *Journal of The Electrochemical Society* 169 (2022) 064504.
18. J. Suntivich, K.J. May, H.A. Gasteiger, J.B. Goodenough, Y. Shao-Horn, A perovskite oxide optimized for oxygen evolution catalysis from molecular orbital principles, *Science* 334 (2011) 1383–1385.
19. K. Khan, A.K. Tareen, M. Aslam, Y. Zhang, R. Wang, Z. Ouyang, Z. Gou, H. Zhang, Recent advances in two-dimensional materials and their nanocomposites in sustainable energy conversion applications, *Nanoscale* 11 (2019) 21622–21678.
20. F.-Y. Chen, Z.-Y. Wu, Z. Adler, H. Wang, Stability challenges of electrocatalytic oxygen evolution reaction: From mechanistic understanding to reactor design, *Joule* 5 (2021) 1704–1731.
21. H. Zhong, C.A. Campos-Roldán, Y. Zhao, S. Zhang, Y. Feng, N. Alonso-Vante, Recent advances of cobalt-based electrocatalysts for oxygen electrode reactions and hydrogen evolution reaction, *Catalysts* 8 (2018) 559.
22. F. Song, X. Hu, Exfoliation of layered double hydroxides for enhanced oxygen evolution catalysis, *Nature Communications* 5 (2014) 4477.

23. H. Liang, F. Meng, M. Cabán-Acevedo, L. Li, A. Forticaux, L. Xiu, Z. Wang, S. Jin, Hydrothermal continuous flow synthesis and exfoliation of NiCo layered double hydroxide nanosheets for enhanced oxygen evolution catalysis, *Nano Letters* 15 (2015) 1421–1427.

24. A. Goswami, D. Ghosh, V.V. Chernyshev, A. Dey, D. Pradhan, K. Biradha, 2D MOFs with Ni (II), Cu (II), and Co (II) as efficient oxygen evolution electrocatalysts: Rationalization of catalytic performance vs structure of the MOFs and potential of the redox couples, *ACS Applied Materials & Interfaces* 12 (2020) 33679–33689.

25. X. Song, J. Wang, S. Qi, Y. Fan, W. Li, M. Zhao, Bifunctional electrocatalytic activity of bis (iminothiolato) nickel monolayer for overall water splitting, *The Journal of Physical Chemistry C* 123 (2019) 25651–25656.

26. M. Chhowalla, H.S. Shin, G. Eda, L.-J. Li, K.P. Loh, H. Zhang, The chemistry of two-dimensional layered transition metal dichalcogenide nanosheets, *Nature Chemistry* 5 (2013) 263–275.

27. Q. Fu, J. Han, X. Wang, P. Xu, T. Yao, J. Zhong, W. Zhong, S. Liu, T. Gao, Z. Zhang, 2D transition metal dichalcogenides: Design, modulation, and challenges in electrocatalysis, *Advanced Materials* 33 (2021) 1907818.

28. J. You, M.D. Hossain, Z. Luo, Synthesis of 2D transition metal dichalcogenides by chemical vapor deposition with controlled layer number and morphology, *Nano Convergence* 5 (2018) 1–13.

29. S. Alam, M.A. Chowdhury, A. Shahid, R. Alam, A. Rahim, Synthesis of emerging two-dimensional (2D) materials–Advances, challenges and prospects, *FlatChem* 30 (2021) 100305.

30. T.F. Jaramillo, K.P. Jørgensen, J. Bonde, J.H. Nielsen, S. Horch, I. Chorkendorff, Identification of active edge sites for electrochemical H_2 evolution from MoS_2 nanocatalysts, *Science* 317 (2007) 100–102.

31. Y. Zhou, Y. Chen, M. Wei, H. Fan, X. Liu, Q. Liu, Y. Liu, J. Cao, L. Yang, 2D MOF-derived porous NiCoSe nanosheet arrays on Ni foam for overall water splitting, *CrystEngComm* 23 (2021) 69–81.

32. H. Xu, J. Zhu, Q. Ma, J. Ma, H. Bai, L. Chen, S. Mu, Two-dimensional MoS_2: Structural properties, synthesis methods, and regulation strategies toward oxygen reduction, *Micromachines* 12 (2021) 240.

33. R. Iqbal, S. Ali, G. Yasin, S. Ibraheem, M. Tabish, M. Hamza, H. Chen, H. Xu, J. Zeng, W. Zhao, A novel 2D Co3 (HADQ) 2 metal-organic framework as a highly active and stable electrocatalyst for acidic oxygen reduction, *Chemical Engineering Journal* 430 (2022) 132642.

34. U. Khan, A. Nairan, J. Gao, Q. Zhang, Current progress in 2D metal–organic frameworks for electrocatalysis, *Small Structures* 4 (2023) 2200109.

35. Q. Deng, J. Zhao, T. Wu, G. Chen, H.A. Hansen, T. Vegge, 2D transition metal–TCNQ sheets as bifunctional single-atom catalysts for oxygen reduction and evolution reaction (ORR/OER), *Journal of Catalysis* 370 (2019) 378–384.

36. X. Liu, W. Liu, M. Ko, M. Park, M.G. Kim, P. Oh, S. Chae, S. Park, A. Casimir, G. Wu, Metal (Ni, Co)-metal oxides/graphene nanocomposites as multifunctional electrocatalysts, *Advanced Functional Materials* 25 (2015) 5799–5808.

37. X. Wei, S. Cao, H. Xu, C. Jiang, Z. Wang, Y. Ouyang, X. Lu, F. Dai, D. Sun, Novel two-dimensional metal organic frameworks: High-performance bifunctional electrocatalysts for OER/ORR, *ACS Materials Letters* 4 (2022) 1991–1998.

38. J. He, F. Zhang, Y. Xiang, J. Lian, X. Wang, Y. Zhang, X. Peng, P. Zeng, J. Qu, J. Song, Preparation of low dimensional antimonene oxides and their application in Cu:NiOx based planar p-i-n perovskite solar cells, *Journal of Power Sources* 435 (2019) 226819.

39. W. Lan, X. Gao, Y. Liu, J. Gu, M. Zhao, B. Wei, F. Zhu, An antimonene modified hole extraction layer for high efficiency PEDOT:PSS-free nonfullerene organic solar cells, *Organic Electronics* 93 (2021).

40. C. Zhang, Y. Li, P. Zhang, M. Qiu, X. Jiang, H. Zhang, Antimonene quantum dot-based solid-state solar cells with enhanced performance and high stability, *Solar Energy Materials and Solar Cells* 189 (2019) 11–20.

41. X. Wang, J. He, B. Zhou, Y. Zhang, J. Wu, R. Hu, L. Liu, J. Song, J. Qu, Bandgap-tunable preparation of smooth and large two-dimensional antimonene, *Angewandte Chemie International Edition* 57 (2018) 8668–8673.

42. D. Enescu, Thermoelectric energy harvesting: Basic principles and applications, *Green Energy Advances* 1 (2019) 38.

43. H. Jouhara, A. Żabnieńska-Góra, N. Khordehgah, Q. Doraghi, L. Ahmad, L. Norman, B. Axcell, L. Wrobel, S. Dai, Thermoelectric generator (TEG) technologies and applications, *International Journal of Thermofluids* 9 (2021) 100063.

44. A. Patel, D. Singh, Y. Sonvane, P. Thakor, R. Ahuja, High thermoelectric performance in two-dimensional Janus monolayer material WS-X (X= Se and Te), *ACS Applied Materials & Interfaces* 12 (2020) 46212–46219.

45. C. Kamal, A. Chakrabarti, M. Ezawa, Aluminene as highly hole-doped graphene, *New Journal of Physics* 17 (2015) 083014.

46. K.G. Steenbergen, N. Gaston, Thickness dependent thermal stability of 2D gallenene, *Chemical Communications* 55 (2019) 8872–8875.

CHAPTER **18**

2D Metals for Sensors and Actuators

Hadiseh Masoumi, Azam Aslani,
Ahad Ghaemi, and Ram K. Gupta

18.1 INTRODUCTION

In this current era, 2D metals have garnered significant attention because of their unique intrinsic properties such as high conductivity, good thermal stability, and suitable chemical and mechanical stability. Each of these 2D metals has unique characteristics, which makes them feasible in some cases. The conductivity of 2D metals is attributed to electron transfers in their structure. It is essential to mention that each of these 2D metals has also several shortcomings that force scholars to solve these challenges by incorporating efficient functional groups into their skeleton. This chapter attempts to explain the different kinds of 2D metals in detail.

18.2 TYPES OF 2D METALS

Recognition of the kinds of 2D metals can help us to know the characteristics and properties for analyzing the behavior of these materials about sensing performance. Besides, learning this case can guide us to decide which ones are suitable for gas, heavy metal, and biomolecules.

18.2.1 Transition Metal Dichalcogenides (TMDs)

TMDs can be described in the general form of MX2 where M belongs to transition metals like Ti, Zr, Hf, V, Nb, Ta, Mo, W, Tc, and Re, and X is often S, Se, and Te. The electrical properties of TMDs are stronger than those of other 2D materials such as graphene. The thickness of films in 2D TMDs is in the range of 6–7 Å. The electrical properties of these components depend on the coordination number of M and the electrons in the orbital of

DOI: 10.1201/9781032645001-18

267

d. The electrical nature of TMDs can be tuned from semiconductors (MoS$_2$, MoSe$_2$, WS$_2$, and WSe$_2$) to metallic (NbTe$_2$, TaTe$_2$) to superconductors (NbS$_2$, NbSe$_2$). 2D TMDs are flexible in terms of chemical architecture and have adjustable band gaps, which makes them desirable for fabricating the sensing devices. The researchers are still involved with multiple challenges in this case including selectivity, response, response duration, recovery period, and resistance. Among these factors, response duration is the most crucial issue. Moreover, as the plane of TMDs is exposed to the atmospheric oxygen or water for a long duration, their skeleton gets damaged, ultimately mitigating the resistance. Thus, it is advised to incorporate metal oxides into the structure of TMDs in order to stimulate the diffusion of target moieties onto the surface of TMDs. Indeed, metal oxides can improve the sensing function by manipulating the geometrical properties of these materials. The resistance ability is defined as the tendency of charges of the target molecules (gas, heavy metal, bioparticles, etc.) to withdraw the electrons of the sensing components [1].

18.2.2 2D MOF Nanosheets

2D MOF nanosheets are commonly utilized for the detection of water pollutants such as heavy metal ions, antibiotics, organic particles, and bacteria because of their high porosity and large surface area. The most conventional sensors in which these materials act as sensing agents are chemical ones, including electrochemical, colorimetric, and luminescent sensors. As an example in the application of 2D MOF in recognizing the metallic ions, it can be assigned to the use of Ru-MOF for determining the mercury ions. With respect to Figure. 18.1, the Ru-MOF precipitates in water as a yellow powder in the absence of mercury ions, and once the mercury ions with the concentration range of 25 pM to 50 Nm are injected into the solution, they interact with Ru-MOF immediately in such a way

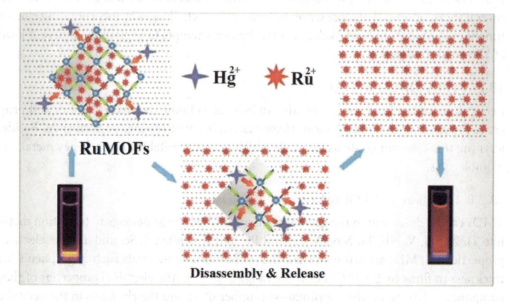

FIGURE 18.1 The interaction of Ru-MOF with mercury ions. Adapted with permission from [2]. Copyright (2018) Springer.

that Ru-MOF decomposes and a considerable amount of luminescent guest moieties like $Ru(bpy)_3^{2+}$ (bpy: 2,2-bipyridine-5,5-dicarboxylic acid) are distributed toward the solution which alters the appearance of solution to yellow again and converts to red color under the exposure of UV radiation. Indeed, the type of metal in the core of 2D MOF causes the adherence of organic particles onto their surface, which assists the sensing performance [2].

18.3 GAS SENSING

18.3.1 NO_x and NH_3

Li et al. [3] were the first researchers who detected NO gas using MoS_2 as the sensor at the normal temperature. They concluded that the two layers of MoS_2 can detect NO gas with a response level at 0.8 ppm. It was observed that using the single layer of 2D metals is not stable. Late et al. [4] worked on this topic and explained the reason for instability, which was owing to the low density of carriers. It can be derived that the response of the 2D metals relies on charge transfer. In fact, using more layers of 2D metals causes more charge transfer.

The position of 2D metals can affect the gas response. For example, vertically aligned MoS_2 reveals a higher response to NO_x gas than the horizontally aligned MoS_2, which is due to the higher activity of edge sites of the vertically 2D metals compared to the terrace sites of the horizontally 2D metals [5].

Doping has also a considerable influence on the performance of 2D metals for the detection of NO_2 gas, which is ascribed to the significant change in band alignment and carrier transport behavior. For example, Choi et al. [6] doped a low concentration of Nb atoms in the $MoSe_2$ network, and it was observed that the low amount of Nb atoms can enhance the sensing performance dramatically.

A number of the 2D metals have a higher ability for sensing the NH_3 gas. Sakhuja et al. [7] produced WS_2 nanosheets for the detection of NH_3 gas. WS_2 revealed a high response of 1805% in 3 ppm of NH_3 gas at a short time of response and recovery, which were calculated at 147 s and 14 s, respectively. Ko et al. [8] added Ag nanowires into the structure of WS_2, and it was found that the sensitivity was enhanced by 12 fold relative to the pristine WS_2. This discrepancy can be ascribed to the synergetic interaction of catalytic and n-type doping effects.

18.3.2 H_2

H_2 is known as a poisonous and combustible gas. Therefore, the detection of this gas is crucial. Many 2D metals and composites of 2D metals were synthesized for this aim. It has been elucidated that the composites of 2D metals show a higher response and sensitivity to H_2 gas. In this part, some works in this field are explained. In this regard, MoS_2 exhibited a high response at a short time of 143 s in 25 °C. The reason was an exposure of the H_2 atmosphere to the edge places of MoS_2. In order to improve the ability of 2D metals for H_2 detection, some components were grafted to them such as Pd, Si, and GaN. It is important to note that some materials reduce the performance of 2D metals such as ZnO. ZnO covers the edge sites of 2D metals which prevents the exposure of H_2 gas to the edge sites of 2D metals, and it cannot sense this gas. Agrawal et al. [9] grafted Pd into the network of the MoS_2 for detecting H_2 gas.

18.3.3 H_2S, CO, and CO_2

Several 2D metals were selected as potential sensors for detecting H_2S, CO, and CO_2, like $MoSe_2$, Ag-$MoSe_2$-rGO (rGO: reduced graphene oxide), WS_2, 1T-TiS_2, WS_2-Au, and MoS_2-Ni. Jha et al. [10] used $MoSe_2$ for detecting H_2S gas at 200°C. The response was obtained at 15.87–53.04% at a concentration range of 50 ppb to 5.45 ppm with a response time and recovery time of 15 and 43 s, respectively. The sensing mechanism for the interaction between $MoSe_2$ and H_2S is charge transfer. They synthesized $MoSe_2$ via the exfoliation method. Luo et al. [11] could sense H_2S gas at 25°C using Ag-$MoSe_2$-rGO via the hydrothermal synthesis method. The response was attained at 40.60% in 10 ppm of H_2S. Sakhuja et al. [12] fabricated a 1T-TiS_2 nanosheet for sensing 4 ppm of H_2S gas with a high response of 395% at a response time of 19.7 s and a recovery time of 48 s. They found that the interaction between the sensor and H_2S is physical, which returns to the van der Waals forces. Kim et al. [13] added Au particles into the structure of WS_2 for sensing 0.2 ppm of CO gas at 25°C at a lesser recovery and high response period. The energy use was very low in the sensing of this gas (28.6 μW).

18.3.4 Gas Sensing Mechanism

One of the main reasons for the good gas sensing capabilities of 2D metals is their adjustable band gap. The band gaps of MoS_2, WS_2, and SnS_2 are in the range of 1–2 eV because of their low carrier concentration. In fact, some properties such as layer-dependent band gap, high surface-to-volume ratio, and an excellent sorption capacity distinguish 2D metals from other components in gas sensing. In gas sensing, WS_2 has attracted more attention than MoS_2 because WS_2 has ambipolar field-modulation properties. Among the 2D metals, $MoTe_2$ exhibits higher sensitivity toward the gases due to its greater bond length and lower binding energy. In addition, SnS_2 has a larger electronegativity than other 2D metals; thus, it has a higher adsorption potential and correspondingly more gas response than other 2D metals.

Charge transfer is another parameter that impacts the gas sensing mechanism, which depends on the surface skeleton and electrical features. The contact of gas molecules with the 2D metals can be implied by the adsorption energy and the value of charge transfer. In this regard, molecular dynamic simulations can give us useful results about the adsorption position and orientation of gas particles. The essential energy for adsorbing of gas particles onto the surface of the 2D metals can be calculated by Eq. (18.1):

$$E_{ad} = E_{(2D+gas)} - \left\lfloor E_{2D} + E_{gas} \right\rfloor \tag{18.1}$$

In Eq. (18.1), $E_{(2D+gas)}$ refers to the sum energy of a layer consisting of a 2D metal and gas particle. The higher adsorption energy of a certain gas molecule relative to the others can imply the higher selectivity of that gas than others.

The number of 2D layers in a sensor has a significant effect on the sensing performance. Late et al. [4] studied this case. They examined two and five layers of MoS_2 for sensing the NO_2 and NH_3. It was observed that five layers of MoS_2 have a higher response for sensing both gases because more gas particles can move toward the inner layers of 2D metals.

The other parameter that can influence the gas response is the presence of oxygen in the ambient. According to the literature, the presence of oxygen promotes the response. The reactions of NH_3 with sulfur vacancies in the absence of oxygen atoms are shown below:

$$SnS_2 \rightarrow SnS(s) + \frac{1}{x} S_x(g) + V_S^x \tag{18.2}$$

$$2NH_3(ads) + V_S^X \leftrightarrow N_2(g) + 6H^+ + V_S^{++} + 8e^- \rightarrow (3) \tag{18.3}$$

$$X = 2\text{--}8.$$

Regarding Eq. (18.2), SnS_2 is degraded to sulfur vacancies in the absence of an oxygen atom. Then, the gas molecule like NH_3 reacts with the sulfur vacancies and produces electrons (Eq. (18.3)), which reduces the resistance.

18.4 HEAVY METAL SENSING

Heavy metal ions are highly toxic carcinogens and cause severe health problems in humans. Recently, several human health-related threats have been observed worldwide because of the environmental contamination caused by heavy metals.

18.4.1 Arsenic Ion Detection

Arsenic ions, especially pentavalent and trivalent arsenite, can remain in the water sources for a long time and lead to inevitable diseases such as skin disease and kidney, lung, and bladder failures. Thus, it is urgent to apply sensors with the base of 2D metals for detecting these perilous ions. There have been relatively few studies conducted on detecting arsenic ions using 2D metals. Some reports are indicated in this section. Rajkumar et al. detected arsenic ions using a gold layer as the sensor and indium tin oxide as the electrodes, with the lower detection value of 0.2 µM. Babar et al. used Au/GNE for detecting trace amounts of arsenic ions at a concentration of 0.08 ppb. Gao et al. reported the detection of trivalent arsenite using iron oxide as sensors at a normal temperature with a low detection limit of 0.0008 ppb [14].

18.4.2 Chromium Ion Detection

Chromium ions are abundant heavy metal ions that cause severe damage to humans and the ecosystem, prompting governments to mitigate these ions from water sources. Some works in diminishing this ion are explained here. Wang et al. fabricated electrodes using carbon nanotubes (CNTs) for detecting trace amounts of Cr^{6+} in contaminated water, which reveals greater sensitivity in comparison to carbon electrodes, with a detection limit of 5 ppb. Rosolina et al. altered the carbonaceous electrode with the help of single-walled carbon nanotubes in the vicinity of a thin layer of pyridine for the elimination of Cr^{6+} ions under square wave voltammetry with the detection limit of 0.8 µg l[µl]. Liu et al. proved that the modification of electrodes with graphene can be suitable for detecting Cr^{6+} at a detection limit of 1.5×10^{-7} mol L[µl] using pulse cathodic stripping voltammetry [14].

18.4.3 Cadmium Ion Detection

Cadmium ions are the most toxic metal ions that should be detected to prevent incurable diseases. There isn't a large amount of work in this case. Graphene, SnO_2, AuNP, and Bi/GaN were largely utilized in the detection of cadmium ions. For instance, Zhang et al. utilized AuNP as sensors with functionalized electrodes for the detection of cadmium ions. In the case of graphene, the limit detection and sensitivity have been determined at 0.0054 µM and 124.03 µA/µm, respectively. Li et al. synthesized Nafion/graphene composites for detecting low concentrations of cadmium with promising reusability. Si et al. constructed reduced graphene with coats of gold particles, which exhibits high selectivity in the range of 50 nM to 300 µM [14].

18.4.4 Copper Ion Detection

To avoid the entry of Cu^{2+} into the environment, some detection procedures were utilized by different research teams. Pesavento et al. employed a highly sensitive surface plasmon resonance fiber optic sensor with the covering of a gold layer for the fast and economic detection of copper ions in drinking water. Cui et al. proposed an electrochemical method by using copper-catalyzed oxidation of cysteamine for detecting copper ions in the range of 1–1000 nM with a detection limit of up to 0.48 nM. Zhu et al. took advantage of the strong affinity between histidine and Cu^{2+} ions and the highly conductive behavior of multi-walled carbon nanotubes to fabricate an electrochemical sensor for detecting Cu^{2+} ions in water. Tang et al. constructed CdTe nanowires using a one-step view for the detection of the Cu^{2+} ions selectively at 0.078 µm [14].

18.4.5 Mercury Ion Detection

Mercury ion is recognized as a carcinogenic particle even at a low concentration (1 ppb); thus, detection of this ion is essential. Graphene and MoS_2 were the most common 2D metals in detecting mercury ions. Graphene oxide has been employed as the field-effect transistor (FET) for detecting the mercury ions at a low detection limit of 2.5×10^{-8} M. It exhibited a rapid response time of 10 s. For attaining such lower detection limits and higher selectivity, functionalization of the graphene oxide has been accomplished with thioglycolic acid and AuNPs. Additionally, Sett and Bhattacharya modified rGO layers with AuNP with the detection limit and sensitivity at 1 ppb and 0.3 µA/ppb, respectively. It was found that the semiconducting properties of MoS_2 has an important efficiency in the detection limit. The high binding dependency between sulfur and mercury, ultra-high surface area to volume ratio, and the electronic features of the MoS_2 can be significantly influenced by their environment. Zhou et al. proposed the construction of FET-based sensors with MoS_2 coated with DNA/AuNPs for detecting mercury ions.

18.4.6 Sensing Mechanism

The mechanism of sensors for detecting heavy metal ions widely relies on adsorption/desorption, redox reactions, chelate formation, and charge transport. The interaction of sensing devices and metallic ions alters the potential of the gate, which affects conductivity. Also, the alteration of conductivity is observed in the current according to Eq. (18.4).

$$I_{DS} = \frac{\varepsilon_n \mu W}{2dL} \left(2\left((V_G - V_T) V_{DS} - V_{DS}^2 \right) \right) \tag{18.4}$$

In Eq. (18.4), $\mathscr{E}_n$, V_G, W, L, μ, V_T, and d refer to the total permittivity of oxide or insulating layers, gate potential, width of the gate, length of the gate, carrier mobility in the channel, threshold voltage of the device, and spacing between the channel and the surface, respectively.

Zhou et al. used a sensor with the base of MoS_2 and DNA for detecting heavy metal ions. In their work, the metallic ions considered adhere to the DNA (Figure 18.2a). As envisaged, the conductivity and finally the current between the source and drain have changed. Similarly, the cadmium ions bind to glutathione on the surface of the AlGaN/GaN forming the complex Cd-GSH and finally altering the current between the source and drain. Figure. 18.2b displays the impact of UV radiation on the sensing behavior of the MoS_2/AlGaN composite for the detection of mercury ions. Regarding this figure, electron–hole pairs of n-type MoS_2 were excited due to the UV exposure since the absorption energy of UV light exceeds the optical bandgap of MoS_2 of 1.2 eV. Therefore, electrons were transferred from the valence to the conduction region, which creates more binding sites for mercury ions.

18.5 SYNTHESIS PRINCIPLES

Selecting the types of synthesis for producing sensors with the base of 2D metals is important because the synthesis method directly relates to the cost and production quality. Therefore, it is essential to survey the entire synthesis methods for finding the practical way [15, 16]. A large number of 2D metals are constructed by the hydrothermal and solvothermal synthesis methods [17].

18.5.1 Ligand-Assisted Growth

In a wet reaction environment, capping ligands are mostly utilized in the formation of 2D metals. The structure of a ligand has two parts. The first part is called the head, which consists of an active functional moiety, and the second part is an inactive molecule that plays the role of a stabilizer. The active part can create a link with the atoms of metal. Molecules with a larger chain are highly recommended for selecting the second part of a ligand. The 2D metals with the nanosize are generated along this synthesis procedure. In this case, Xu

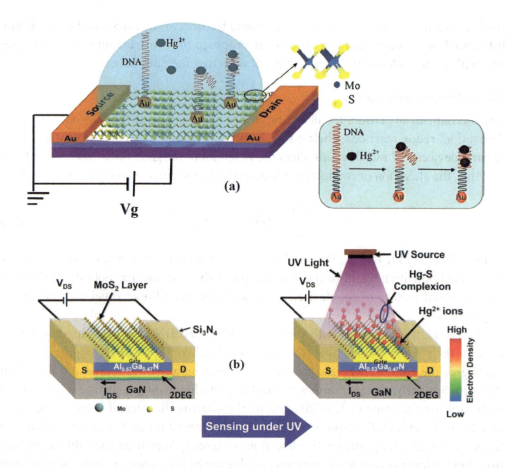

FIGURE 18.2 (a) Interaction of mercury ions and MoS₂/DNA composite. (b) Effect of UV radiation on the electron transport of MoS₂/AlGaN composite. Adapted with permission from [14]. Copyright (2021) Elsevier.

et al. [18] suggested a surfactant with a base of pyridinium as a capping agent for building Pd nanolayers at a mild temperature (35°C). H₂PdCl₄ was employed as an intermediate in their work. As shown in Figure 18.3, the morphology of the produced 2D metals is envisaged to be hexagonal and triangular. Besides pyridinium, poly(vinylpyrrolidone) (PVP), cetyltrimethylammonium bromide (CTAB), and oleylamine are the remaining suitable surfactant candidates.

18.5.2 Small-Molecule-Mediated Growth

The molecules or ions with small sizes are often exploited as additives for making the shapes of crystalline to the built network. Small molecules can enhance the adsorption capacity more effectively than larger ones. CO and halide ions are the most commonly used small particles in synthesizing 2D metals. For instance, the CO molecule is used in constructing the nanosheets with the base of Pd. In the research of Zheng et al., the nanosheets with the base of Pd were produced with a thickness of 1.8 nm using CO at a pressure of 1

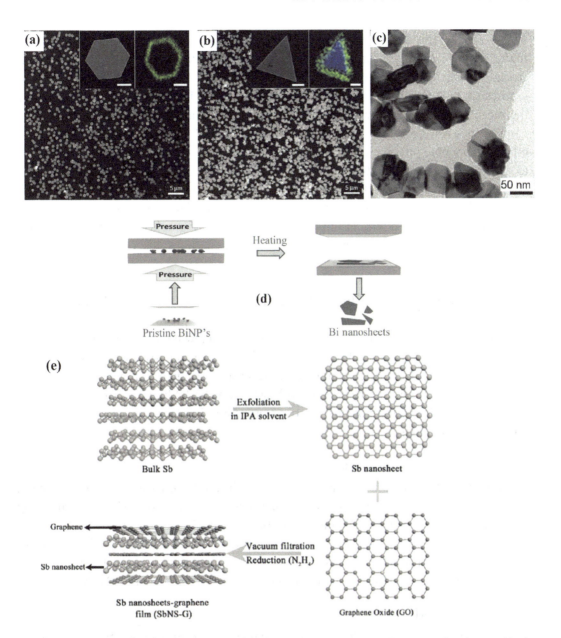

FIGURE 18.3 The morphology of 2D metals with (a) hexagonal and (b) triangular shapes. (c) The morphology of Pt nanosheets without using the surfactant. (d) The formation of Bi nanosheets using mechanical compression. (e) Formation of Sb nanosheet. Adapted with permission from [19]. Copyright (2020) Elsevier.

bar. Moreover, palladium acetylacetonate, PVP, and a halide salt have been mixed with an appropriate solvent like benzyl alcohol. It should be noted that each 2D metal can accept a certain halide, and a certain halide cannot always act as an additive for the entire 2D metal. In other words, some halides may have a poor interaction with certain metals such as chloride ions with gold. In the case of Au, bromide and iodide ions act selectively in the adsorption of this metal.

18.5.3 2D Template-Confined Growth

In the current era, scholars utilize 2D materials as a template for producing 2D metals. It is crucial to explain that it isn't a novel idea because this approach was examined before for synthesizing 2D metals with the base of Pt and Pd in the interlayers of graphite. Among the recently published manuscripts, graphene was used as a template for constructing nanosheets with the base of gold in the presence of PVP and CTAB. Besides graphene, other components can be utilized as a template like indium oxide and MoS_2. It is possible to synthesize 2D metals in the absence of surfactant, as demonstrated in the work of Yang et al. [20]. With respect to their investigation, $PtCl^{2\mu}$ adsorbed on the plane of graphene oxide by the oxygen atoms of the functional group in the first stage. In the latter stage, Pt nanoparticles interacted with H_2 molecules on the plane of graphene oxide, which generates Pt nanosheets with certain shapes (Figure 18.3c).

18.5.4 Mechanical Compression

The mechanical compression for synthesizing 2D metals is folding and rolling. Figure 18.3d displays the formation of Bi nanosheets achieved by pressing raw Bi nanoparticles with applied heat. This method is appropriate only for a few metals including Ag, Al, Au, Ni, Pt, In, and Bi. In this case, Wu et al. [21] constructed a composite of Ag–Al by rolling and folding Ag–Al layers. The obtained thickness was lower than 10 nm. In the same way, the Al–graphene composite has been synthesized via the folding and rolling of Al and graphene layers. The failure of this synthesis procedure is a lack of ability to control the geometry of 2D metals.

18.5.5 Solution-Based Exfoliation

In this method, the solid phase of metal is dissolved in a suitable solvent as the liquid phase. For instance, Yang et al. introduced the powder of Sb in propyl alcohol (PA) for generating nanosheets of Sb (Figure 18.3e). The size of the obtained Sb nanosheet was 400 nm with a thickness of 4 nm [22]. Matsumoto et al. [23] prepared nanosheets with the base of Pt electrochemically through exfoliating the layers of Pt. Similarly, nanosheets with the base of Ru have also been fabricated during the reduction of the exfoliated RuO_2 layers.

18.6 MODIFICATION OF SURFACES

Modification of the surface of the 2D metals can increase the number of active sites, thereby enhancing particle adsorption and improving sensing performance. In this section, it was attempted to indicate the concepts for the modification procedures of 2D metals. These conventional techniques have been disclosed below.

18.6.1 Oxide Layer

The interface of 2D metals is changed through mechanical deformation (Figure 18.4a) of the oxide (scraping, mixing, vibrating, and swelling), chemical dissolution of the oxide (Figure. 18.4b), and electrochemical method (Figure 18.4c) till the additives can join to them [24]. In the chemical procedure, the oxide is dissolved in acidic or basic environments.

2D Metals for Sensors and Actuators ■ 277

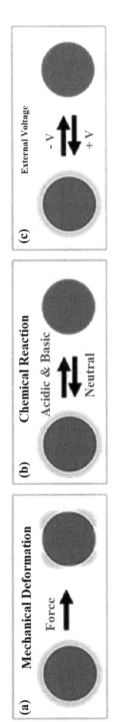
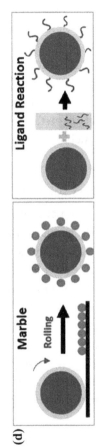
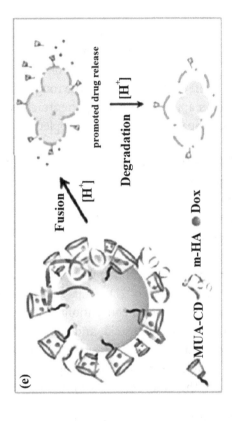

FIGURE 18.4 (a) Mechanical, (b) chemical, and (c) electrochemical techniques. Schematic of (d) marble and ligand reaction. (e) Application of ligand modification of 2D metals in drug delivery. Adapted with permission from [25]. Copyright (2021) Wiley.

In the physical procedure, the oxide films are broken by exerting the external force. For instance, shearing liquids into complex particles (SLICE) is a strategy that divides the metals into particles for blending with the other components especially functional groups such as thiol. In the chemical procedure, the degradation of oxide films occurred at a higher rate in the basic media rather than in the acidic media. However, the oxide films don't have adequate resistance at a low (<3) or high pH (>3). In the electrochemical procedure, the interface mechanism of metals is varied by injecting the voltage.

18.6.2 Marbles

The "marble" refers to a procedure that homogeneously covers the surface of a 2D metal by rolling it on the powdery particles (Figure 18.4d). These powdery moieties are mostly micro- or nanoparticles like polytetrafluoroethylene (PTFE) with a size of 35 μm, ferronickel, and polyethylene with micro-size particles. The addition of some powdery particles like a blend of ferronickel and polyethylene induces the magnetic and waterproofing features to the 2D metals. The covers of the powdery moieties inhibit the 2D metals from attaching to the surfaces owing to the roughness of the covering, which is dedicated by the Cassie–Baxter approach. Avoiding the attachment of 2D metals to the surfaces is desired in sensors and apparatuses that need droplet fluidity, such as lab-on-a-chip devices.

18.6.3 Ligand Binding

The interaction of ligands with the surface of the 2D metals is a procedure for modifying them. In this method, the polymers or small components are chemically or physically bound to 2D metals. In the case of physical binding, mechanical agitation or sonication is applied. For instance, the stability of 2D metals is promoted in water and biological buffer via the incorporation of poly(1-octadecene-alt-maleic anhydride) (POMA) in water. Furthermore, modifying the surface of 2D metals with covalent or hydrogen bonds enhances the resistance of 2D metals such as adherence of thiol groups of 1-dodecanethiol to the 2D metal. The most conventional chemical ligands include dodecylphosphonic acid, 11-phosphonoundecanoic acid, and silanes. The introduction of ligands enables narrowing the size distribution of particles in 2D metals (Figure 18.4d). This technique is widely applied in drug delivery in which drugs adhere to the surface of 2D metals and are transferred toward the cancer cells because of variations in the pH (Figure 18.4e) [26].

18.7 BIOMOLECULES SENSING

18.7.1 Glucose and Nucleic Acid

Glucose enzymes such as glucose oxidase (GOD) have been utilized for making sensitive sensors in glucose recognition. Pumera's team discovered that the heterogeneous electron transfer (HET) of WSe_2 is stronger than WS_2, MoS_2, and $MoSe_2$ [27]. This phenomenon is attributed to the metallic 1t phase, which can augment the electrocatalytic potential in sensors for detecting glucose molecules. One of the main issues in the sensors of glucose is a weak electron transfer between the electrode and GOD which decreases the sensitivity. Several scientists attempted to solve this challenge by incorporating some metals in the matrix of 2D metals. For example, Su et al. coated MoS$_2$ with the layers of AuNPs to immobilize GOD, which promoted electron transfer activity. In addition, they could detect glucose molecules in the range of 10–300 μM [28]. Wu et al. suggested an approach to boosting the conductivity and electron transfer of 2D metals via electrochemical reduction for glucose detection [29]. Another method for improving the sensitivity and stability of 2D metals is adding peroxide mimics into the sensors of glucose recognition. For instance, Cai et al. coated MoS$_2$ with Pt$_{74}$Ag$_{26}$ as the peroxide mimics during an easy synthesis route [30]. It was observed that MoS$_2$–Pt$_{74}$Ag$_{26}$ exhibits excellent colorimetric detection of fructose, lactose, and mannitol in the range of 1–10 μM. One of the main problems in the sensors of glucose is the susceptibility of enzymes to temperature, pH, and humidity [31]. In order to boost the stability of biosensors in the various ambient conditions, CuS with diverse structures was used.

The mechanism of the interaction is the van der Walls bond between DNA and the basal plane of MoS$_2$ at a DNA concentration of 10^{-10}–10^{-4} μM. Loo et al. suggested a functionalized disposable electrical printed (DEP) device for sensitive detection of DNA. In fact, the incorporation of MoS$_2$ sheets on modified DEP could alter the voltammetric signal with various intensities, which relies on the amount of DNA. It was reported that elevating the surface area of 2D metals boosted the sensitivity of biosensors. Figure 18.5 displays the production steps of microRNA sensing performance. The structure of MoS$_2$ was modified with AuNPs via electrodeposition for stable binding of the SH-RNA probe that can hybridize microRNA-155 through hydrogen bonding. Besides, toluidine blue was loaded onto the MoS$_2$–AuNPs network for the selective hybridization of microRNA-155, which can determine microRNA-155 at a low concentration of 0.32 fM.

18.7.2 Dopamine, Ascorbic Acid, and Uric Acid

Dopamine, ascorbic acid, and uric acid are small molecules that play an important role in the nervous system. The lack of dopamine in the brain can generate neurological illnesses like Parkinson's and schizophrenia [33]. Because of the lower concentration of dopamine (0.01–1 μM) compared to ascorbic acid and uric acid and the proximity of their oxidation values, the sensitive and specific recognition of dopamine among various components is still an issue. For rectifying the selectivity of dopamine, Wu et al. reduced MoS$_2$ (rMoS$_2$) in the [Fe(CN)$_6$]$^{3-/4-}$ and [Ru(NH$_3$)$_6$]$^{2+/3+}$ redox couples by finding a sharp peak-to-peak separation potential (ΔE_p) of ~85 mV and ~88 mV, respectively [29]. This sensor could detect selectively dopamine in the combination of uric acid and ascorbic acid because of

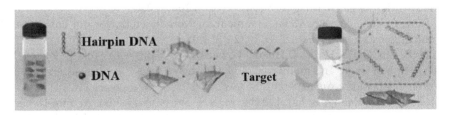

FIGURE 18.5 Hybridization chain reaction. Adapted with permission from [32]. Copyright (2019) Elsevier.

the stronger signal of dopamine relative to other acids. In other words, the attraction force generates between the negative charge of rMoS$_2$ and the positive charge of dopamine, while the repulsion force generates between the negative charge of rMoS$_2$ and the negative charge of uric and ascorbic acids. An important issue in determining dopamine in a basic environment is the polymerization of dopamine, which decreases sensitivity and resistance. Xia et al. grafted glutathione to prevent the polymerization of dopamine at pH 8.5 [34].

18.8 CONCLUSIONS

The current chapter was presented to explain the application of 2D metals in sensing toxic particles. Generally speaking, TMDs have a superior potential in terms of sensitivity, response time, and recovery time relative to other 2D metals. However, some issues exist that limit the utilization of TMDs on a larger scale such as consuming the expensive pristine materials, difficult synthesis routes, and releasing the perilous vapors during their synthesis.

According to the conducted probes, the following suggestions can be considered for augmenting the performance of 2D metals in the future:

- Simulation of the sensing performance of 2D metals with some software like molecular dynamics (MD), artificial neural network (ANN), and Design-Expert (DOE) can assist researchers in estimating some possible challenges prior to examination at the laboratory.

- Grafting novel components to the structure of 2D metals for providing their composites.

- Introducing photo-reactive components into the network of 2D metals.

- Evaluating other species of 2D metals that haven't been studied till now.

- Studying the surface chemistry of TMDs and the electron transfer mechanism during material–analyte interactions.

- Designing a simple synthetic route with a high yield for constructing 2D metals using low-cost intermediates for producing them on a commercial scale.

REFERENCES

1. E. Lee, Y.S. Yoon, D.J. Kim, Two-dimensional transition metal dichalcogenides and metal oxide hybrids for gas sensing, *ACS Sens.* 3 (2018) 2045–2060.
2. X. Fang, B. Zong, S. Mao, Metal–organic framework-based sensors for environmental contaminant sensing, *Nanomicro. Lett.* 10 (2018) 1–19.
3. H. Li, Z. Yin, Q. He, H. Li, X. Huang, G. Lu, D.W.H. Fam, A.I.Y. Tok, Q. Zhang, H. Zhang, Fabrication of single-and multilayer MoS$_2$ film-based field-effect transistors for sensing NO at room temperature, *Small.* 8 (2012) 63–67.
4. D.J. Late, Y.-K. Huang, B. Liu, J. Acharya, S.N. Shirodkar, J. Luo, A. Yan, D. Charles, U.V. Waghmare, V.P. Dravid, Sensing behavior of atomically thin-layered MoS$_2$ transistors, *ACS Nano.* 7 (2013) 4879–4891.

5. S.Y. Cho, S.J. Kim, Y. Lee, J.S. Kim, W.B. Jung, H.-W. Yoo, J. Kim, H.-T. Jung, Highly enhanced gas adsorption properties in vertically aligned MoS_2 layers, *ACS Nano.* 9 (2015) 9314–9321.
6. S.Y. Choi, Y. Kim, H.S. Chung, A.R. Kim, J.D. Kwon, J. Park, Y.L. Kim, S.H. Kwon, M.G. Hahm, B. Cho, Effect of Nb doping on chemical sensing performance of two-dimensional layered $MoSe_2$, *ACS Appl. Mater. Interfaces.* 9 (2017) 3817–3823.
7. N. Sakhuja, R.K. Jha, N. Bhat, Tungsten disulphide nanosheets for high-performance chemiresistive ammonia gas sensor, *IEEE Sens. J.* 19 (2019) 11767–11774.
8. K.Y. Ko, J.G. Song, Y. Kim, T. Choi, S. Shin, C.W. Lee, K. Lee, J. Koo, H. Lee, J. Kim, Improvement of gas-sensing performance of large-area tungsten disulfide nanosheets by surface functionalization, *ACS Nano.* 10 (2016) 9287–9296.
9. A. Agrawal, R. Kumar, S. Venkatesan, A. Zakhidov, Z. Zhu, J. Bao, M. Kumar, M. Kumar, Fast detection and low power hydrogen sensor using edge-oriented vertically aligned 3-D network of MoS_2 flakes at room temperature, *Appl. Phys. Lett.* 111 (2017) 1–6.
10. R.K. Jha, J.V. Costa, N. Sakhuja, N. Bhat, $MoSe_2$ nanoflakes based chemiresistive sensors for ppb-level hydrogen sulfide gas detection, *Sens. Actuators B Chem.* 297 (2019) 1–34.
11. Y. Luo, D. Zhang, X. Fan, Hydrothermal fabrication of Ag-decorated $MoSe_2$/reduced graphene oxide ternary hybrid for H_2S gas sensing, *IEEE Sens. J.* 20 (2020) 13262–13268.
12. N. Sakhuja, R.K. Jha, R. Chaurasiya, A. Dixit, N. Bhat, 1T-phase titanium disulfide nanosheets for sensing H_2S and O_2, *ACS Appl. Nano Mater.* 3 (2020) 3382–3394.
13. J.-H. Kim, A. Mirzaei, H.W. Kim, S.S. Kim, Realization of Au-decorated WS_2 nanosheets as low power-consumption and selective gas sensors, *Sens. Actuat. B Chem.* 296 (2019) 1–40.
14. A. Nigam, N. Sharma, S. Tripathy, M. Kumar, Development of semiconductor based heavy metal ion sensors for water analysis: A review, *Sens. Actuator A Phys.* 330 (2021) 1–24.
15. H. Masoumi, A. Aslani, A. Ghaemi, H. Farokhzad, Engineering and chemistry aspects of the well-known conductive polymers as sensors: Characterization, mechanism, synthesis, scale-up: A review, *Sens. Int.* 4 (2023) 1–24.
16. H. Masoumi, A. Ghaemi, H.G. Gilani, Synthesis of polystyrene-based hyper-cross-linked polymers for Cd (II) ions removal from aqueous solutions: Experimental and RSM modeling, *J. Hazard. Mater.* 416 (2021) 1–19.
17. A.P. Dral, J.E. ten Elshof, 2D metal oxide nanoflakes for sensing applications: Review and perspective, *Sens. Actuat. B Chem.* 272 (2018) 369–392.
18. D. Xu, Y. Liu, S. Zhao, Y. Lu, M. Han, J. Bao, Novel surfactant-directed synthesis of ultrathin palladium nanosheets as efficient electrocatalysts for glycerol oxidation, *Chem Comm.* 53 (2017) 1642–1645.
19. T. Wang, M. Park, Q. Yu, J. Zhang, Y. Yang, Stability and synthesis of 2D metals and alloys: A review, *Mater. Today Adv.* 8 (2020) 1–18.
20. S. Yang, P. Qiu, G. Yang, Graphene induced formation of single crystal Pt nanosheets through 2-dimensional aggregation and sintering of nanoparticles in molten salt medium, *Carbon.* 77 (2014) 1123–1131.
21. H. Liu, H. Tang, M. Fang, W. Si, Q. Zhang, Z. Huang, L. Gu, W. Pan, J. Yao, C. Nan, 2D metals by repeated size reduction, *Adv. Mater.* 28 (2016) 8170–8176.
22. J. Gu, Z. Du, C. Zhang, J. Ma, B. Li, S. Yang, Liquid-phase exfoliated metallic antimony nanosheets toward high volumetric sodium storage, *Adv. Energy Mater.* 7 (2017) 1–8.
23. A. Funatsu, H. Tateishi, K. Hatakeyama, Y. Fukunaga, T. Taniguchi, M. Koinuma, H. Matsuura, Y. Matsumoto, Synthesis of monolayer platinum nanosheets, *Chem. Comm.* 50 (2014) 8503–8506.
24. N. Syed, A. Zavabeti, J.Z. Ou, M. Mohiuddin, N. Pillai, B.J. Carey, B.Y. Zhang, R.S. Datta, A. Jannat, F. Haque, Printing two-dimensional gallium phosphate out of liquid metal, *Nat. Commun.* 9 (2018) 1–10.

25. K.Y. Kwon, V.K. Truong, F. Krisnadi, S. Im, J. Ma, N. Mehrabian, T.-i. Kim, M.D. Dickey, Surface modification of gallium-based liquid metals: Mechanisms and applications in biomedical sensors and soft actuators, *Adv. Intell. Syst.* 3 (2021) 1–11.

26. Y. Lu, Q. Hu, Y. Lin, D.B. Pacardo, C. Wang, W. Sun, F.S. Ligler, M.D. Dickey, Z. Gu, Transformable liquid-metal nanomedicine, *Nat. Commun.* 6 (2015) 1–10.

27. N. Rohaizad, C.C. Mayorga-Martinez, Z.k. Sofer, M. Pumera, 1T-phase transition metal dichalcogenides (MoS_2, $MoSe_2$, WS_2, and WSe_2) with fast heterogeneous electron transfer: application on second-generation enzyme-based biosensor, *ACS Appl. Mater. Interfaces.* 9 (2017) 40697–40706.

28. S. Su, H. Sun, F. Xu, L. Yuwen, C. Fan, L. Wang, Direct electrochemistry of glucose oxidase and a biosensor for glucose based on a glass carbon electrode modified with MoS_2 nanosheets decorated with gold nanoparticles, *Mikrochim. Acta.* 181 (2014) 1497–1503.

29. S. Wu, Z. Zeng, Q. He, Z. Wang, S.J. Wang, Y. Du, Z. Yin, X. Sun, W. Chen, H. Zhang, Electrochemically reduced single-layer MoS_2 nanosheets: Characterization, properties, and sensing applications, *Small.* 8 (2012) 2264–2270.

30. S. Cai, Q. Han, C. Qi, Z. Lian, X. Jia, R. Yang, C. Wang, $Pt_{74}Ag_{26}$ nanoparticle-decorated ultrathin MoS_2 nanosheets as novel peroxidase mimics for highly selective colorimetric detection of H_2O_2 and glucose, *Nanoscale.* 8 (2016) 3685–3693.

31. X. Lin, Y. Ni, S. Kokot, Electrochemical and bio-sensing platform based on a novel 3D Cu nano-flowers/layered MoS_2 composite, *Biosens. Bioelectron.* 79 (2016) 685–692.

32. L. Lan, Y. Yao, J. Ping, Y. Ying, Ultrathin transition-metal dichalcogenide nanosheet-based colorimetric sensor for sensitive and label-free detection of DNA, *Sens. Actuat. B Chem.* 290 (2019) 565–572.

33. C. Xue, Q. Han, Y. Wang, J. Wu, T. Wen, R. Wang, J. Hong, X. Zhou, H. Jiang, Amperometric detection of dopamine in human serum by electrochemical sensor based on gold nanoparticles doped molecularly imprinted polymers, *Biosens. Bioelectron.* 49 (2013) 199–203.

34. X. Xia, X. Shen, Y. Du, W. Ye, C. Wang, Study on glutathione's inhibition to dopamine polymerization and its application in dopamine determination in alkaline environment based on silver selenide/molybdenum selenide/glassy carbon electrode, *Sens. Actuat. B Chem.* 237 (2016) 685–692.

CHAPTER 19

2D Metals for Biomedical Applications

Elnaz Fekri and Mir Saeed Seyed Dorraji

19.1 INTRODUCTION

Metals are known as one of the most useful groups of materials. These materials are used in various fields due to their unique structural, electrical, thermal, and chemical properties. Metallenes, a class of emerging materials composed of pure metal atoms with a relatively simple structure, have recently attracted much attention. The concept of metallenes has been proposed in recent years, but there is still no clear definition for it. However, they are generally defined as two-dimensional nanomaterials which consist of pure metal atoms. According to the properties of metals, metallenes can be divided into two main categories: layered and non-layered metals. Some of them are layered metallic materials, such as antimonene and bismuthene, having a six-membered rhombic atomic structure [1]. Some other metals such as magnesium are non-layered materials.

19.2 PROPERTIES AND CLASSIFICATION

19.2.1 Properties

When the thickness of nanosheets is very small in the dimensions of one atom or a few atoms, unique properties appear. For example, metallenes have a high density of metal atoms that can act as catalytically active sites. The wide and unique physicochemical properties of metallenes, including catalytic, plasmonic, photoluminescent, and magnetic properties, have made them one of the most promising groups of materials for various applications.

DOI: 10.1201/9781032645001-19

283

19.2.1.1 Catalytic Properties

The high density of metal atoms in metallenes leads to an increase in the efficiency of the optimal use of metal atoms. Studies show that metallenes have high activity, selectivity, and stability in heterogeneous catalytic processes.

In addition, metallenes have attracted much attention in electrocatalytic reactions such as hydrogen evolution reaction (HER), oxidation–reduction reaction (ORR), and oxygen evolution reaction (OER).

In photocatalytic processes, metallenes can effectively act as co-catalysts or heterojunctions based on two-dimensional metal nanoplates and two-dimensional semiconductor nanoplates [2]. Metallenes are also useful in biomedicine due to their attractive catalytic properties.

19.2.1.2 Plasmonic Properties

Localized surface plasmon resonance (LSPR) is one of the attractive features of metal nanoparticles and has been the focus of researcher's attention in many applications such as optical sensing of biomolecules, photothermal treatment, spectral signal enhancement, and optical wave enhancement [3]. LSPR is a collective oscillation of conduction band electrons that is excited by the electromagnetic field.

LSPR characteristics of metal nanoparticles are very sensitive to factors such as composition, size, and morphology.

The near-infrared (NIR-LSPR) properties of metal nanostructures such as gold, copper, and silver make them useful agents for the photothermal treatment of cancer using NIR lasers. Huang et al. reported the synthesis of free hexagonal palladium nanosheets with a thickness of less than ten atomic layers. The synthesized blue nanosheets showed a distinct and tunable surface plasmon resonance peak in the near-infrared region [3].

In addition, metallenes are often of interest in disease research due to their amazing optical and photothermal conversion properties.

In terms of biological applications, the photothermal performance of multi-element metals is better than that of mono-metals [4].

19.2.1.3 Photoluminescence Properties

Photoluminescence (PL) is a common physical phenomenon of two-dimensional semiconductor materials. Hussain et al. successfully synthesized high crystallinity Bi nanosheets (BiNSs) with a thickness of about 2 nm. These nanosheets showed an obvious Pl effect, along with wide light emission in the visible region This remarkable PL response of ultrathin BiNSs is attributed to the effects of carrier confinement and crystal defects/dislocations associated with the trapping process of electrons and excitons in ultrathin BiNSs [5].

According to Zhang et al., it was confirmed that the obtained semiconducting antimonene nanosheets exhibit indirect band gap characteristics, with a (PL) band gap of about 2.33 eV and a PL lifetime of 4.3 ns and emits green fluorescence under excitation at 470 nm [6].

19.2.1.4 Magnetic Properties

The magnetic properties of metals, including anisotropic magnetism and isotropic super-paramagnetism, are largely determined by their size, structure, and surface electrons. Lang et al. synthesized a series of nickel nanosheets with size-dependent superparamagnetism. Small nickel nanoplates have weaker magnetic properties than nanoplates with large edge lengths. However, its coercive force increases with decreasing size, which increases its anisotropic magnetization [7].

In addition, Zhao et al. used electron beam irradiation to spread an atomic layer of iron in the holes of porous graphene and found that the two-dimensional iron nanosheets exhibited large vertical magnetic anisotropy [8].

19.2.2 Classification

In general, the metallenes that have been reported to date can be mainly classified into three groups: (1) main group metallenes such as galena (Ga-ene) and germanene (Ge-ene); (2) transition metal-based metallenes, including noble metallenes (e.g., Au, Pd. Etc.) and non-noble metallenes (e.g., Fe, Co, Ni, etc.); and (3) alloy-based metals [9].

According to the number of elements, metallenes can also be classified as monometallene, bimetallene, and trimetallene. After 2015, with the discovery of element-based main group metallens, research on metallens rapidly emerged. In addition, polymetallic alloy nanosheets have also received much attention in recent years.

19.2.2.1 Monometallenes

Monometallens are two-dimensional nanosheets containing one type of metallic element.

Monometallens have high strength due to the singularity and symmetry of their crystal structures and can be classified into metallenes based on the main element (Sb, Bi, etc.) and metallenes based on transition elements (Pd, Co, etc.). The experimental observations of Huang et al. confirmed ultrathin gold nanosheets exhibiting a pure hcp phase with a thickness of about 2.4 nm. From these observations, it can be concluded that in very small dimensions (for example, thickness less than 6 nm for AuSS), gold structures are stabilized in the hcp phase [10].

By promoting their anisotropic growth, Peng et al. prepared single-crystal hexagonal antimony sheets with large lateral dimensions (0.5–1.5 mm) and very thin thickness (5–30 nm) [11].

19.2.2.2 Bimetallenes

Bimetallenes are metal nanosheets that are composed of two types of metal elements. Suprathin PdMo nanosheets were synthesized using a one-pot wet chemical method by Liu et al. Bimetallic PdMo alloy in the form of a highly curved metal nanosheet with sub-nm thickness was used as an efficient electrocatalyst for non-alkaline ORR and OER electrolytes [12].

286 ■ 2D Metals

19.2.2.3 Trimetallenes

Recently, trimetallenes consisting of three metallic elements have also been reported. Qin et al. used a mature dimetallene synthesis method to obtain new trimetals called porous PdWM (M = Nb, Mo, and Ta). The porous and wrinkled structure can accelerate electron transfer and mass transfer [13].

19.3 SYNTHESIS STRATEGIES

The preparation of metallenes with the desired thickness, compositions, and crystalline phases has important effects on their physicochemical properties and catalytic performance. Strategies for the synthesis of metallenes can be divided into two categories: (1) top-down approaches include mechanical cleavage, ultrasonic exfoliation, electrochemical exfoliation, and plasma-assisted processes, which use mechanical forces to disrupt the bonds between layers [14]. (2) Bottom-up approaches, known by various wet chemical methods, include epitaxial molecular beam and chemical vapor deposition (CVD), which are chemically facilitated by in-plane addition or self-assembly of atoms or molecules [15].

Many metallenes have a strong tendency to bind non-oriented metals and form good-quality 3D crystals.

Structural symmetry creates strong crystal structures, which result in nanostructures with a high aspect ratio. Therefore, it is difficult to separate bulk intermetallic layers.

However, the addition of surfactants and ligands to facilitate the reduction of surface binding energy is one of the methods that can be used to prepare atomically thin metals.

19.3.1 Top-down Approaches

19.3.1.1 Mechanical Cleavage

During mechanical cleavage (MC), no chemical reaction occurs and is known as the most common method for making two-dimensional nanomaterials with good crystallinity and high purity. However, the small production scale and relatively low performance are among the disadvantages of using this method [16].

Ares et al. used the MC method for the preparation of antimonene for the first time. They prepared antimonene-layered nanosheets (AM NSs) and were transferred to the Si/SiO$_2$ substrate through adhesive tape layering. After washing with acetone and methanol, all the samples were baked at 180°C to remove the remaining solvents. (AM NSs) had side dimensions less than 240 nm and thickness less than 8 nm [17].

19.3.1.2 Ultrasonic Exfoliation

Recently, the use of ultrasonic waves in material synthesis, especially for the fabrication or modification of various nanomaterials, has attracted the attention of researchers. Simplicity, high efficiency, and short reaction time are the reasons for the popularity of sonochemistry. The essential physical phenomenon for the sonochemical process, mainly due to acoustic cavitation, involves the formation, growth, and explosive collapse of microbubbles in the liquid. The resulting hot spots/microjets generate extremely high temperatures of ~5000 K and pressures of ~150 MPa in one nanosecond. The bubbles collapse on the sheet surface due to rapid pressure changes, crushing the sheets dispersed in the solvent. These extreme reaction conditions are usually not possible through conventional synthesis techniques.

Parameters such as ultrasound frequency, input power, type of ultrasonic probe, and the choice of suitable solvent have a great impact on the quality and amount of ultrasonic exfoliation.

Gibaja et al. showed that a stable suspension of micron-sized antimonene can be obtained under sonication in a 4:1 isopropanol/water mixture without adding any surfactant [18].

19.3.1.3 Electrochemical Exfoliation

Electrochemical exfoliation is a widely used method for the large-scale synthesis of monolayer or multilayer metallenes. In this method, the electrolyte ions enter the two-dimensional structure under an applied voltage, which leads to the separation of layered parts limited by van der Waals interactions. The applied voltage and the types and concentration of the electrolyte are parameters whose control affects the quality of exfoliated products [19].

Electrochemical exfoliation of bismuth nanosheets was performed by Wu et al. in a two-electrode system. In this system, small bismuth ingots were used as raw materials, graphite rods as anodes, copper strips as cathodes, and tetrapropylammonium bromide solution in acetonitrile as electrolyte [20].

19.3.1.4 Plasma-Assisted Processes

Plasma-assisted processes are used to synthesize high-quality and ultrathin metal layers by plasma immersion in a metal substrate.

The basic steps of this method are divided into three steps: substrate preparation, plasma immersion, and annealing.

Tsai et al. successfully synthesized antimony layers. The InSb (001) substrates were first immersed in N_2 plasma generated using a radio frequency system (13.56 MHz) with a power of 50–200 W for 30–60 min, and then for 30–60 min in the machine was cooked. N_2/H_2 environment (1/10, v/v) at 450°C [21].

19.3.1.5 Other Top-down Methods

Solid-melt exfoliation is one of the emerging methods. Kuchat et al. developed a facile exfoliation synthesis of 2D Ga with an atomically thin layer structure from the molten phase [22].

The principle is to separate the interfacial solid layer from the solid-melt interface of the low-melting metal material into a thin layer of metal atoms.

In addition to exfoliation by solid melting, dealloying strategies have also been considered.

Wang et al. used an in situ dealloying method under transmission microscopy to fabricate single-atom-thick gold layers (0.6 nm) and free-standing and stable nanoribbons from gold–silver alloys [23].

288 ■ 2D Metals

19.3.2 Bottom-up Approaches

19.3.2.1 Epitaxial Molecular Beam

Molecular beam epitaxy is now widely used for the synthesis of nanosheets. Thickness, crystal orientation, and doping level of metals are controllable parameters in this technique. Niu et al. used the molecular beam epitaxy technique to directly synthesize high-quality 2D antimonene on a dielectric copper oxide substrate formed from copper (111) [24].

19.3.2.2 Vapor Deposition Technology

Vapor deposition technology is a well-known technology, in which a metal or composite is deposited on a surface by physical and chemical processes that occur in the gas phase. Vapor deposition techniques are classified into two categories: chemical vapor deposition (CVD) and physical vapor deposition (PVD) [25].

In a typical process, a preselected substrate is placed in a furnace chamber where one or more gas/vapor precursors are circulated, and the precursors react/decompose on the surface of the substrate.

By using the ultra-high-vacuum chemical vapor deposition (UHV-CVD) system, Tsai et al. synthesized germanium nanosheets with high crystallinity on Si (100) wafer substrates by an annealing process [26].

PVD is the process of ion deposition on a collective surface using physical methods under vacuum conditions. This method has many advantages over other methods for preparing two-dimensional materials. For example, the deposition rate can be controlled by adjusting parameters such as gas pressure, gas flow rate, and heating rate.

19.3.2.3 Wet Chemical Methods

Wet chemical synthesis methods, such as solvothermal synthesis, have been widely used to prepare ultrathin two-dimensional nanomaterials with high yields.

These methods enable the large-scale production of two-dimensional metals with controlled structures and abundant surface functional groups. The four common methods of wet chemical synthesis of metallenes are:

(1) **Ligand-confined growth:**
Ligand-limited growth is a process in which ligands with metal ions/atoms are used to control crystal growth. Choosing the right ligands is an important factor that affects the final products. Ligands are generally divided into two categories: small molecule ligands and organic ligands [27].

(2) **Two-dimensional templated synthesis:**
In the two-dimensional template method, two-dimensional nanomaterials are mainly used as templates to prevent the growth of two-dimensional nanosheets along the three-dimensional direction.

(3) **Solvothermal synthesis:**
Solvothermal solution synthesis is based on the classic hydrothermal route at a relatively high temperature and pressure, which improves the reactivity and solubility of the reactants.

(4) *Soft colloidal-templated synthesis*:

The synthesis of soft colloidal templates of metallenes is inspired by composite nanosheets of other transition metals.

Zhang et al. reported a soft template method for the synthesis of antimonene and bismuthene nanosheets in colloidal solutions.

The dissolution of $SbCl_3$ in alkylphosphonic acids can form a layered structure that allows the preparation of antimonene nanosheets with a rhombic crystal structure [28].

19.4 BIOLOGICAL EFFECTS OF METALLENES

Despite the biomedical potential of metallenes, the use of two-dimensional metals in biological processes and their clinical translation still faces challenges due to potential physiological toxicity and other risk factors. Therefore, there is a need to evaluate the potential risks of metallenes at the cellular and animal levels.

The biological effects of metallenes such as toxicity, cellular uptake, pharmacokinetics, nanobiological interactions, and interactions with specific cellular pathways are related to their composition, structure, and physicochemical properties such as dispersion and stability.

19.4.1 Toxicity

The development of metallenes for biomedical applications is at an early stage.

However, researchers believe that understanding the toxicity of metals is very important for biological and biomedical applications.

Reactive nitrogen species (RNS) and reactive oxygen species (ROS) such as hydroxyl radicals, superoxide anion radicals, hydrogen peroxide, nitric oxide, and peroxynitrite can be considered as reaction products catalyzed by metallenes. Although ROS and RNS can be useful in biological systems, sometimes they can be harmful to living systems. Among the beneficial effects of ROS at low concentrations, the induction of cellular physiological responses or defense against infectious agents can be mentioned.

On the contrary, at high concentrations, ROS causes damage to cellular structures including lipids, membranes, and proteins. Oxidative damage accumulates during the cell life cycle and damage to the DNA and RNA structure and oxidation of unsaturated fatty acids in biological membranes leads to diseases such as cancer, arteriosclerosis, arthritis, and neurological disorders [29].

Alteration of pH and cell permeability and interference with biomembrane integrity by methallenes is a common toxic mechanism.

Perforation of the cytoplasmic membrane and release of cell contents, enzymes, proteins, and nucleic acids lead to direct cell death, while specific damage to the mitochondrial membrane ultimately leads to mitochondrial apoptosis [30].

Furthermore, after intravenous injection, metallens may affect various types of blood cells (red blood cells, leukocytes, and platelets) and proteins (hemoglobin and plasma proteins) in the blood and cause their aggregation or change their ability to change shape [31].

Rupture of red blood cells (hemolysis) can lead to anemia with severe complications. Hemolytic activity is directly related to the size and surface properties of metals.

290 ■ 2D Metals

Non-hemolytic interactions with blood cells and protein components may result in pro-coagulant activity with possible risk of thrombotic and embolic complications.

Therefore, it is worth noting that while metals exhibit various physiotherapeutic activities in vivo, it is important to ensure that they do not affect the normal activity of living organisms.

19.4.2 Cellular Uptake

Due to their small size, nanoparticles (NPs) can easily enter cells and also translocate in cells, tissues, and organs. However, cellular uptake and targeting of nanoparticles can be controlled by adjusting the physicochemical properties of nanoparticles such as size, shape, and surface properties such as surface hydrophobicity/hydrophilicity. Hence, knowledge of the basic mechanisms involved in cellular uptake is very important to assess and track the fate of NPs [32].

Cellular uptake of metallenes is one of the significant issues in their biological activity. Cellular uptake is determined by the interaction between the molecular membrane and the plasma membrane.

Macropinocytosis is a form of endocytosis that is carried out by the plasma membrane, trapping medium droplets in micron-sized vesicles.

Macropinocytosis involves the non-selective uptake of macromolecules with a diameter greater than 0.2 μm. Nutrients, solutes, and antigens are factors influencing the absorption of metals. And it works independently of specific binding between ligands and receptors [33].

NPs in the size range of 250 nm to 3 μm have been shown to have optimal phagocytosis in vitro, while nanoparticles with a size range of 120–150 nm are internalized via clathrin- or caveolin-mediated endocytosis.

Metal NP shape also plays a pivotal role in the uptake and trafficking pathways of NPs.

Chithrani et al. studied the effect of the shape of colloidal nanoparticles on the uptake of HeLa cells. The result showed that spherical AuNPs absorb much more than rod AuNPs [34].

Intracellular transport of metallenes is an intermediate step in cellular uptake. After NPs are internalized by cells, they first encounter membrane-bound intracellular vesicles called early endosomes (EE) [27]. The early endosome transports the cargo to the desired cellular destination. Early endosomes become late endosomes (LE) through the process of maturation and differentiation. LE then fuse with lysosomes to form endolysosomal vesicles, and hydrolytic enzymes present in these vesicles degrade the trapped nanoparticles.

The function of LE is to transport metallenes to the lysosome for degradation.

19.4.3 Pharmacokinetics

Pharmacokinetic and biodistribution studies of metallenes introduced into the body are necessary to help evaluate safety and efficacy as well as estimate clinical doses and species differences of such products. There is a close relationship between the biodistribution profile and physical and chemical properties of nanoparticles. Therefore, identifying whether nanoparticles can accumulate in target and non-target tissues is critical for designing adequate safety studies. In vivo pharmacokinetics of nanomaterials provide important insights into toxicity. Before applying metallenes in vivo, the ADME process (absorption, distribution, metabolism, excretion) must be evaluated. The tests to evaluate the characteristics of nanoparticles need to investigate the parameters that affect the biodistribution of drugs containing nanoparticles and possibly their safety. One of the most important and documented parameters is the measurement of particle size, size distribution, shape, and state of aggregation [35].

19.4.3.1 Biomacromolecule Interaction

The distribution of metals in living organisms depends on the permeability of the membranes of organs, tissues, and the circulatory system. When nanomaterials enter the body through various ways such as inhalation, oral consumption, contact with the skin, wounds, and injection, the nanomaterials including metals flow throughout the circulatory system and body fluids and are internalized in cellular compartments and enter the cell. They reach different organs.

Some metals, such as Pd, are very stable and tend to accumulate in physiological environments and do not dissipate rapidly.

They may undergo a wide variety of reactions in vivo, such as dissolution, aggregation, absorption of proteins, and changes in osmotic pressure [36].

This may cause toxicity. Metallenes may interfere with cellular metabolic processes by disrupting the integrity of the cell or cell membrane. Also, by producing harmful ROS, they can disrupt the function of tissues and organs.

However, the formation of metallene aggregates can significantly affect their biodistribution in the body. They tend to accumulate nonspecifically in organs of the reticuloendothelial system (RES) (such as the liver and spleen) over a long period of time and are difficult to break down or remove from the body [37].

The passive accumulation of nanoparticles at tumor sites is attributed to the enhanced permeability and retention effect (EPR) that results from the rich vasculature and incomplete lymphatic system in the tumor region. If they are not small enough to be excreted by the kidney or large enough to be quickly recognized and trapped by the (RES), they tend to remain in circulation for a long time [38].

19.4.3.2 Biodegradation

Biodegradation refers to the ability of living organisms to break down drugs or substances into smaller components that can be absorbed by various organs or transported through metabolic organs such as the liver and kidney.

The biodegradability of metals is an important indicator of their biological safety. Metallenes have good crystallinity with some defects and are difficult to decompose in physiological environments. For in vivo biomedical applications, developing methods to ensure effective degradation of 2D nanomaterials and thus reducing toxicity is an urgent issue.

Biodegradation is a two-step process: degradation and purification. The clearance pathway of biodegradable nanoparticles mainly involves hepatic clearance, which is primarily

292 ■ 2D Metals

determined by the hydrodynamic size (HD) of the particles. Nanomaterials with a (HD) less than 20 nm can be rapidly absorbed by the liver and kidneys and metabolized from the blood, while nanomaterials with a large size (HD > 420 nm) can be filtered in the liver sinuses. Other physicochemical characteristics of nanomaterials such as morphology, surface functional groups, and surface charge also have a significant effect on their cleaning rate. For example, Chen et al. reported that small glutathione-modified Pd nanosheets (NSs) were efficiently eliminated through renal excretion, and indeed this modification effectively improved the clearance rate of Pd NSs [39].

19.5 BIOMEDICAL APPLICATIONS

Due to their unique magnetic, optical, thermal, catalytic, and electrical properties, metallenes have recently attracted attention in important biomedical applications. Metallenes (such as stannane, bismuthene, and PdMo) find their wide applications in biosensors, bioimaging, antibacterial, anti-inflammatory, and antitumor.

19.5.1 Antitumor Applications

Traditional methods of tumor treatment such as radiotherapy, chemotherapy, and surgery have disadvantages such as easy recurrence, high side effects, and low survival rate. In addition, traditional methods often face challenges due to the lack of specificity and drug resistance.

Metal nanoparticles may have the potential to overcome problems associated with conventional chemotherapy. According to reports, metal nanoparticles play a useful and powerful role in cancer therapy and antitumor applications and provide advantages such as better targeting, gene silencing, photothermal therapy, dynamic therapy, and drug delivery. Metal nanoparticles are also used as a diagnostic tool for imaging cancer cells. Metal NP-based therapeutic systems not only provide simultaneous diagnosis and treatment but also enable controlled and targeted drug release, which has revolutionized cancer treatment and management [40].

The selection and specificity of tumor targeting is a very important aspect.

Metal nanoparticles target tumor cells through two strategies: passive targeting and active targeting. The EPR effect is a passive phenomenon in which nanoparticles of certain sizes have a greater tendency to accumulate in tumor tissue than in normal tissues [41].

With unique features such as a large specific surface area and high loading capacity of the two-dimensional structure, metallenes are a promising drug delivery system (DDS) that can be an effective alternative to conventional chemotherapy [42].

For instance, polydopamine/polyethylene glycol (Ti@PDA-PEG)-based multifunctional drug delivery system based on Ti NSs showed good performance in chemotherapy [43].

19.5.2 Biosensors

With properties such as high surface-to-volume or mass ratio and atomic thinness, metallenes seem interesting for use in biosensing applications and have a strong response to surface adsorption events.

These properties and features such as optical phenomena, tunable band structure, and favorable electronic properties make them one of the suitable candidates for biosensor applications. Metallenes have also been considered in the field of biosensors due to their fluorescence-quenching properties.

Due to their special affinity for nucleic acid materials such as DNA and RNA, a series of metallene-based nucleic acid biosensors have been developed. García-Mendiola et al. functionalized antimonene with an oligonucleotide as a DNA sensor. The functionalization process resulted in DNA-modified multilayer antimonene, which, after deposition of gold plate-printed electrodes, created a simple and efficient DNA electrochemical sensing platform. The selectivity of this device not only allows the identification of a specific DNA sequence but also directly detects the mutation of this gene associated with breast cancer in clinical samples [44].

Due to its high carrier mobility and high stability at room temperature, bismuthene is promising in the fields of biomedicine and biosensors. Zhou et al. developed a bismuthene-based ultrathin detection platform based on the specific detection of RNA (miRNA), the detection of which can reach 60 pM. Excitingly, this platform provides sensitive detection of RNA molecules for early stages of cancer [45].

In addition to the specific detection of nucleic acid materials, metals are also very selective and sensitive to some vitamins, and small organic molecules of living organisms.

19.5.3 Bioimaging

One of the methods used in the diagnosis of diseases is medical imaging. This method uses various techniques and processes to visually examine the function of body organs and tissues (physiology) for clinical purposes and is also used to diagnose and treat diseases.

Recently, metallenes have been investigated for fluorescence, photoacoustic imaging (PAI), and computed tomography (CT) imaging. Li et al. synthesized poly(vinylpyrrolidone)-protected bismuth nanodots (PVP-Bi nanodots) through a very facile strategy both in the laboratory and on a large scale. Due to the high X-ray attenuation ability of Bi element, PVP-Bi nanodots showed excellent performance in (X-ray CT imaging) [46].

Luke et al. developed a novel stable contrast agent for PAI-guided SLN biopsy: silica-coated gold nanoparticles (Si-AuNPs). Si-AuNP showed high thermal stability when exposed to pulsed and continuous wave laser irradiation. This makes them suitable for in vivo PAI [47].

Metallenes are known as potential contrast agents for in vitro and in vivo imaging, especially CT and PAI imaging.

However, the advantages of metallens over clinical contrast agents should be demonstrated in further studies.

19.5.4 Anti-inflammation

Due to the sonodynamic and photodynamic properties of some metallenes, they are used to induce pro-inflammatory effects for tumor treatment.

Also, metallenes can be used as a drug delivery platform to promote the release of anti-inflammatory drugs and increase the therapeutic effect of anti-inflammatory drugs. For example, rheumatoid arthritis (RA) is a drug-resistant disease caused by an overactive immune system.

Methotrexate (MTX) is a drug used for the clinical treatment of RA. However, toxicity due to poor selectivity to inflammatory cells has presented challenges to the use of MTX. To reduce the side effects of MTX and improve its therapeutic effect, Chen et al. used a new therapeutic strategy by synthesizing a novel nanotherapy, which can target inflammatory cells and control the release of MTX. Novel hexagonal palladium nanosheets were used as a near-infrared (NIR) photothermal agent with arginine glycine aspartic acid (RGD) peptides on the surface to enhance the targeting ability of the nanosheets to inflammatory cells [26].

Metallenes can be used as anti-inflammatory agents to treat inflammation as well. Inflammatory bowel disease (IBD) is caused by the excessive production of ROS and affects normal intestinal function. Zhang et al. used a type of orally zero-valent molybdenum nanodots (ZVMNs) to treat IBD by scavenging ROS. In addition, RNA sequencing revealed that ZVMNs can protect colon tissues against oxidative stress by inhibiting the nuclear factor kB signaling pathway and reducing the production of excessive pro-inflammatory factors [48].

19.5.5 Antibacterial

An overgrowth of harmful bacteria causes bacterial infection. Different types of bacteria can cause different symptoms. Two-dimensional metallenes with different properties have potential antibacterial effects.

Two-dimensional silver nanosheets show more antibacterial ability in many cases. Li et al. used silver nanosheets with a thickness of 1.5 nm to make antibacterial membranes. Ag nanosheet@TiO_2 was introduced into the titania cell and the prepared membrane had a molecular weight cutoff of 22,163 Da and a high permeability of 236 L/(m^2h bar) [25]. Using a one-step strategy, He et al. developed two-dimensional AuPd alloy sheets as imaging-guided photonic nanoantibiotics. AuPd 2D alloy nanosheets with very small thickness (~1.5nm) showed outstanding thermal effects and excellent ROS production as well as favorable biocompatibility. AuPd nanosheets can kill 100% of Gram-positive (*Staphylococcus aureus*) and Gram-negative (*Escherichia coli*) bacteria when irradiated with 808 nm laser at 1 W/cm^2 for 5 min. It also showed a good therapeutic effect in the wounded mouse model. This work expands the biomedical applications of two-dimensional metallenes in the field of photonic nanoantibiotics [49].

19.6 CONCLUSION AND PERSPECTIVE

In this chapter, two-dimensional metals ranging from transition group to main group metals were summarized in terms of their structure, properties, and types, and examples of each case were mentioned. Synthesis strategies, biological effects, and biomedical applications were also investigated. Research on 2D metals has clearly made great progress in recent years. However, there are still challenges from the viewpoints of synthesis strategies, biological effects, biomedical applications, and clinical translation, which require further investigation.

19.6.1 Synthesis Strategies

Considering the inherent isotropic properties of metals and the high surface energy of two-dimensional morphologies, the synthesis of metallenes still faces challenges. New types of metallenes, such as high-entropy metallenes, are likely to be suitable for biomedical applications. High-entropy alloys consisting of five or more metallic elements have been gradually reported. Due to the interaction between multiple metallic elements, high-entropy alloys are likely to have interesting properties for various applications including biomedical applications. However, the high-entropy two-dimensional alloy has not been investigated to date. More extensive research is needed.

In addition, due to the difference in the properties of different metals, there is no standard method for preparing high-quality 2D metals, and their large-scale preparation remains challenging.

19.6.2 Biological Effects

One of the main obstacles limiting the use of 2D metals in medicine is their potential toxicity. Current research shows that the distribution of metals in living organisms has toxic effects on major organs and blood.

However, the precise effects of metallenes on the nervous and immune systems, which play a vital role in the effectiveness of treatment, require extensive systematic research. In addition, during pharmacokinetic research, great importance should be given to the accumulation of methallenes in the diseased area.

Currently, the accumulation of metals in tumor regions is mainly dependent on the EPR effect.

However, the EPR effect may not be effective enough in the clinic. To increase the efficiency of metal accumulation, the effectiveness can be increased by modifying factors such as size distribution, colloidal dispersion, and surface modifications such as target molecules (such as antibodies).

In addition, in nanobiological interactions, when metals are exposed to the physiological environment, during processes such as oxidation, accumulation, instability, and uncontrolled degradation, it can directly lead to interactions between metals and biological interfaces such as proteins, DNA, cell membrane, and cytoskeleton. These underlying mechanisms of nanobiological interactions need further clarification.

19.6.3 Biomedical Applications

Metallenes have wide applications in biomedicine. Despite effective research on therapeutic effects, the analysis of the underlying therapeutic mechanisms is still unclear.

It is also necessary to develop new metal-based diagnostic and therapeutic methods, such as their use as probes for positron emission tomography (PET) or PET/MRI, to help guide therapy.

In addition, researchers should develop more types of animal models with similar characteristics to humans with homologous genes. Furthermore, disease models should be developed for metallen-based diagnosis and therapy, including disease diagnosis, neuro-degenerative diseases, and regenerative medicine (e.g., stem cells).

19.6.4 Clinical Translation

Various types of metals have been used in medical and diagnostic treatments, including arthritis (gold), bipolar disorder (lithium), anemia and low blood pressure (iron), and broad-spectrum antibiotics (bismuth), antibacterial (silver), and antitumor (platinum). However, 2D materials, including metallenes, have rarely entered clinical trials. Therefore, the barriers and potential for clinical translation of metallenes require further research.

REFERENCES

1. W. Tao, N. Kong, X. Ji, Y. Zhang, A. Sharma, J. Ouyang, B. Qi, J. Wang, N. Xie, C. Kang, Emerging two-dimensional monoelemental materials (Xenes) for biomedical applications, *Chemical Society Reviews*, 48 (2019) 2891–2912.
2. S. He, J. Chai, S. Lu, X. Mu, R. Liu, Q. Wang, F. Chen, Y. Li, J. Wang, B. Wang, Solution-phase vertical growth of aligned $NiCo_2O_4$ nanosheet arrays on Au nanosheets with weakened oxygen–hydrogen bonds for photocatalytic oxygen evolution, *Nanoscale*, 12 (2020) 6195–6203.
3. X. Huang, S. Tang, X. Mu, Y. Dai, G. Chen, Z. Zhou, F. Ruan, Z. Yang, N. Zheng, Freestanding palladium nanosheets with plasmonic and catalytic properties, *Nature Nanotechnology*, 6 (2011) 28–32.
4. Y. Chen, Z. Fan, Z. Zhang, W. Niu, C. Li, N. Yang, B. Chen, H. Zhang, Two-dimensional metal nanomaterials: Synthesis, properties, and applications, *Chemical Reviews*, 118 (2018) 6409–6455.
5. N. Hussain, T. Liang, Q. Zhang, T. Anwar, Y. Huang, J. Lang, K. Huang, H. Wu, Ultrathin Bi nanosheets with superior photoluminescence, *Small*, 13 (2017) 1701349.
6. F. Zhang, J. He, Y. Xiang, K. Zheng, B. Xue, S. Ye, X. Peng, Y. Hao, J. Lian, P. Zeng, Semimetal–semiconductor transitions for monolayer antimonene nanosheets and their application in perovskite solar cells, *Advanced Materials*, 30 (2018) 1803244.
7. Y. Leng, Y. Wang, X. Li, T. Liu, S. Takahashhi, Controlled synthesis of triangular and hexagonal Ni nanosheets and their size-dependent properties, *Nanotechnology*, 17 (2006) 4834.
8. J. Zhao, Q. Deng, A. Bachmatiuk, G. Sandeep, A. Popov, J. Eckert, M.H. Rümmeli, Freestanding single-atom-thick iron membranes suspended in graphene pores, *Science*, 343 (2014) 1228–1232.
9. C. Cao, Q. Xu, Q.-L. Zhu, Ultrathin two-dimensional metallenes for heterogeneous catalysis, *Chemical Catalysis*, 2 (2022) 693–723.
10. X. Huang, S. Li, Y. Huang, S. Wu, X. Zhou, S. Li, C.L. Gan, F. Boey, C.A. Mirkin, H. Zhang, Synthesis of hexagonal close-packed gold nanostructures, *Nature Communications*, 2 (2011) 292.
11. L. Peng, S. Ye, J. Song, J. Qu, Solution-phase synthesis of few-layer hexagonal antimonene nanosheets via anisotropic growth, *Angewandte Chemie International Edition*, 58 (2019) 9891–9896.
12. M. Luo, Z. Zhao, Y. Zhang, Y. Sun, Y. Xing, F. Lv, Y. Yang, X. Zhang, S. Hwang, Y. Qin, PdMo bimetallene for oxygen reduction catalysis, *Nature*, 574 (2019) 81–85.
13. Y. Qin, H. Huang, W. Yu, H. Zhang, Z. Li, Z. Wang, J. Lai, L. Wang, S. Feng, Porous PdWM (M = Nb, Mo and Ta) trimetallene for high C1 selectivity in alkaline ethanol oxidation reaction, *Advanced Science*, 9 (2022) 2103722.
14. S. Zhou, X. Liu, Y. Lin, D. Wang, Spontaneous growth of highly conductive two-dimensional single-crystalline $TiSi_2$ nanonets, *Angewandte Chemie*, 120 (2008) 7795–7798.

15. N. Abid, A.M. Khan, S. Shujait, K. Chaudhary, M. Ikram, M. Imran, J. Haider, M. Khan, Q. Khan, M. Maqbool, Synthesis of nanomaterials using various top-down and bottom-up approaches, influencing factors, advantages, and disadvantages: A review, *Advances in Colloid and Interface Science*, 300 (2022) 102597.

16. H. Liu, A.T. Neal, Z. Zhu, Z. Luo, X. Xu, D. Tománek, P.D. Ye, Phosphorene: An unexplored 2D semiconductor with a high hole mobility, *ACS Nano*, 8 (2014) 4033–4041.

17. P. Ares, F. Aguilar-Galindo, D. Rodríguez-San-Miguel, D.A. Aldave, S. Díaz-Tendero, M. Alcamí, F. Martín, J. Gómez-Herrero, F. Zamora, Mechanical isolation of highly stable antimonene under ambient conditions, *Advanced Materials*, 28 (2016) 6332–6336.

18. C. Gibaja, D. Rodriguez-San-Miguel, P. Ares, J. Gómez-Herrero, M. Varela, R. Gillen, J. Maultzsch, F. Hauke, A. Hirsch, G. Abellán, Few-layer antimonene by liquid-phase exfoliation, *Angewandte Chemie*, 128 (2016) 14557–14561.

19. Z. Huang, H. Hou, Y. Zhang, C. Wang, X. Qiu, X. Ji, Layer-tunable phosphorene modulated by the cation insertion rate as a sodium-storage anode, *Advanced Materials*, 29 (2017) 1702372.

20. D. Wu, X. Shen, J. Liu, C. Wang, Y. Liang, X.-Z. Fu, J.-L. Luo, Electrochemical exfoliation from an industrial ingot: Ultrathin metallic bismuth nanosheets for excellent CO_2 capture and electrocatalytic conversion, *Nanoscale*, 11 (2019) 22125–22133.

21. H.-S. Tsai, C.-W. Chen, C.-H. Hsiao, H. Ouyang, J.-H. Liang, The advent of multilayer antimonene nanoribbons with room temperature orange light emission, *Chemical Communications*, 52 (2016) 8409–8412.

22. V. Kochat, A. Samanta, Y. Zhang, S. Bhowmick, P. Manimunda, S.A.S. Asif, A.S. Stender, R. Vajtai, A.K. Singh, C.S. Tiwary, Atomically thin gallium layers from solid-melt exfoliation, *Science Advances*, 4 (2018) e1701373.

23. X. Wang, C. Wang, C. Chen, H. Duan, K. Du, Free-standing monatomic thick two-dimensional gold, *Nano Letters*, 19 (2019) 4560–4566.

24. T. Niu, Q. Meng, D. Zhou, N. Si, S. Zhai, X. Hao, M. Zhou, H. Fuchs, Large-scale synthesis of strain-tunable semiconducting antimonene on copper oxide, *Advanced Materials*, 32 (2020) 1906873.

25. R. Platz, S. Wagner, Intrinsic microcrystalline silicon by plasma-enhanced chemical vapor deposition from dichlorosilane, *Applied Physics Letters*, 73 (1998) 1236–1238.

26. H.-S. Tsai, Y.-Z. Chen, H. Medina, T.-Y. Su, T.-S. Chou, Y.-H. Chen, Y.-L. Chueh, J.-H. Liang, Direct formation of large-scale multi-layered germanene on Si substrate, *Physical Chemistry Chemical Physics*, 17 (2015) 21389–21393.

27. C. Lu, R. Li, Z. Miao, F. Wang, Z. Zha, Emerging metallenes: Synthesis strategies, biological effects and biomedical applications, *Chemical Society Reviews*, 52 (2023) 2833–2865.

28. J. Zhang, S. Ye, Y. Sun, F. Zhou, J. Song, J. Qu, Soft-template assisted synthesis of hexagonal antimonene and bismuthene in colloidal solutions, *Nanoscale*, 12 (2020) 20945–20951.

29. M. Valko, C. Rhodes, J. Moncol, M. Izakovic, M. Mazur, Free radicals, metals and antioxidants in oxidative stress-induced cancer, *Chemico–Biological Interactions*, 160 (2006) 1–0.

30. H.L. Karlsson, P. Cronholm, Y. Hedberg, M. Tornberg, L. De Battice, S. Svedhem, I.O. Wallinder, Cell membrane damage and protein interaction induced by copper containing nanoparticles—Importance of the metal release process, *Toxicology*, 313 (2013) 59–69.

31. K.M. de la Harpe, P.P. Kondiah, Y.E. Choonara, T. Marimuthu, L.C. du Toit, V. Pillay, The hemocompatibility of nanoparticles: A review of cell–nanoparticle interactions and hemostasis, *Cells*, 8 (2019) 1209.

32. P. Foroozandeh, A.A. Aziz, Insight into cellular uptake and intracellular trafficking of nanoparticles, *Nanoscale Research Letters*, 13 (2018) 1–12.

33. X. Xie, J. Liao, X. Shao, Q. Li, Y. Lin, The effect of shape on cellular uptake of gold nanoparticles in the forms of stars, rods, and triangles, *Scientific Reports*, 7 (2017) 3827.

34. K.A. Willets, R.P. Van Duyne, Localized surface plasmon resonance spectroscopy and sensing, *Annual Review of Physical Chemistry*, 58 (2007) 267–297.

35. B.S. Zolnik, N. Sadrieh, Regulatory perspective on the importance of ADME assessment of nanoscale material containing drugs, *Advanced Drug Delivery Reviews*, 61 (2009) 422–427.
36. Y. Huang, X. Chen, S. Shi, M. Chen, S. Tang, S. Mo, J. Wei, N. Zheng, Effect of glutathione on in vivo biodistribution and clearance of surface-modified small Pd nanosheets, *Science China Chemistry*, 58 (2015) 1753–1758.
37. B. Wang, X. He, Z. Zhang, Y. Zhao, W. Feng, Metabolism of nanomaterials in vivo: Blood circulation and organ clearance, *Accounts of Chemical Research*, 46 (2013) 761–769.
38. Y. Nakamura, A. Mochida, P.L. Choyke, H. Kobayashi, Nanodrug delivery: Is the enhanced permeability and retention effect sufficient for curing cancer? *Bioconjugate Chemistry*, 27 (2016) 2225–2238.
39. X. Chen, S. Shi, J. Wei, M. Chen, N. Zheng, Two-dimensional Pd-based nanomaterials for bioapplications, *Science Bulletin*, 62 (2017) 579–588.
40. Y.-W. Jiang, G. Gao, H.-R. Jia, X. Zhang, X. Cheng, H.-Y. Wang, P. Liu, F.-G. Wu, Palladium nanosheets as safe radiosensitizers for radiotherapy, *Langmuir*, 36 (2020) 11637–11644.
41. A. Sharma, A.K. Goyal, G. Rath, Recent advances in metal nanoparticles in cancer therapy, *Journal of Drug Targeting*, 26 (2018) 617–632.
42. X.-J. Liang, C. Chen, Y. Zhao, P.C. Wang, Circumventing tumor resistance to chemotherapy by nanotechnology, *Multi-Drug Resistance in Cancer*, 596 (2010) 467–488.
43. X. Yuan, Y. Zhu, S. Li, Y. Wu, Z. Wang, R. Gao, S. Luo, J. Shen, J. Wu, L. Ge, Titanium nanosheet as robust and biosafe drug carrier for combined photochemo cancer therapy, *Journal of Nanobiotechnology*, 20 (2022) 154.
44. T. Garcia-Mendiola, C. Gutierrez-Sanchez, C. Gibaja, I. Torres, C. Buso-Rogero, F. Pariente, J. Solera, Z. Razavifar, J.J. Palacios, F. Zamora, Functionalization of a few-layer antimonene with oligonucleotides for DNA sensing, *ACS Applied Nano Materials*, 3 (2020) 3625–3633.
45. S.R. Rani, I. Kainthla, S. Dongre, L. D'Souza, R.G. Balakrishna, Recent advances of eco-friendly 2D monoelemental bismuthene as an emerging material for energy, catalysis and biomedical applications, *Journal of Materials Chemistry C*, 11 (2023) 6777–6799.
46. P. Lei, R. An, P. Zhang, S. Yao, S. Song, L. Dong, X. Xu, K. Du, J. Feng, H. Zhang, Ultrafast synthesis of ultrasmall poly (vinylpyrrolidone)-protected bismuth nanodots as a multifunctional theranostic agent for in vivo dual-modal CT/photothermal-imaging-guided photothermal therapy, *Advanced Functional Materials*, 27 (2017) 1702018.
47. G.P. Luke, A. Bashyam, K.A. Homan, S. Makhija, Y.-S. Chen, S.Y. Emelianov, Silica-coated gold nanoplates as stable photoacoustic contrast agents for sentinel lymph node imaging, *Nanotechnology*, 24 (2013) 455101.
48. G.S. Firestein, Evolving concepts of rheumatoid arthritis, *Nature*, 423 (2003) 356–361.
49. S. He, G. Zhu, Z. Sun, J. Wang, P. Hui, P. Zhao, W. Chen, X. Jiang, 2D AuPd alloy nanosheets: One-step synthesis as imaging-guided photonic nano-antibiotics, *Nanoscale Advances*, 2 (2020) 3550–3560.

Index

Actuators, 26
Alkaline fuel cell (AFC), 204, 205, 219
Atomic force microscopy (AFM), 25, 26, 37, 38, 82, 132, 162, 165, 192
Atomic layer deposition (ALD), 78, 110, 213, 220

Batteries, 18, 27, 29, 208, 211, 234, 236, 245, 247–250, 254, 255, 261
Battery, 234–236
Biomedical, 1, 118, 289, 291, 292, 294, 295
Black phosphorus (BP), 2, 20, 76, 78, 241, 243, 244, 248, 256–258
Bottom-up, 3, 4, 12, 114, 127, 134, 162, 178, 179, 190, 238, 258, 286

Carbon dioxide reduction reaction (CO_2RR), 7, 9, 10, 139, 140, 151
Chemical vapour deposition (CVD), 19, 24–26, 35, 70, 72, 79, 80, 84, 107, 110, 111, 179, 286, 288

Defect, 1, 2, 10, 19, 22, 25, 26, 69, 92, 93, 97, 100, 112, 125, 129, 131, 145, 148, 151, 158, 159, 161, 166, 190, 197, 198, 200, 201, 229, 238, 244, 260, 284, 291
Degradation, 20, 127, 140, 156, 162, 165–168, 182–184, 190, 230, 278, 290, 291, 295
Density functional theory (DFT), 51, 52, 54, 55, 59, 99, 100, 127, 131, 132, 151, 183, 184, 186, 198, 199, 228
Direct ethanol fuel cell (DEFC), 219
Direct methanol fuel cell (DMFC), 204, 219

Electrocatalysis, 2, 7, 10, 11, 27, 34, 124, 125, 127, 138, 146, 148, 211, 221, 222, 253, 255, 260
Electrocatalytic, 5, 8, 10, 13, 126, 139, 140, 145, 148, 151, 152, 211, 212, 214, 223, 224, 227, 253, 257–261, 278, 284

Electronics, 1, 12, 16–18, 25–29, 66, 76–78, 81, 92, 96, 97, 107, 108, 113, 118, 157, 184, 248, 249, 254, 261
Energy-dispersive X-ray spectroscopy (EDS), 12, 38, 39, 194
Energy Storage, 9, 27–29, 63, 76, 77, 96, 151, 157, 230, 234–237, 242, 244, 245, 247–250, 254, 263
Exfoliation, 4, 5, 21, 22, 32, 36, 43, 48, 84, 87, 108, 128, 164, 178, 179, 181, 190, 191, 238, 252, 259, 261, 263, 270, 276, 286, 287

Field effect transistors (FET), 17, 77, 81–83, 91, 103, 272, 273
Fourier transform infrared spectroscopy (FTIR), 26, 42, 132, 164, 225
Fuel cell, 8, 27, 77, 139, 148, 203–212, 214, 219, 221, 223–226, 230, 231, 236, 247, 249, 254, 255, 259

Graphene, 1, 2, 4, 16, 18–21, 24, 25, 27, 34, 35, 47–49, 51, 52, 54–58, 60, 63, 65, 67, 76–78, 80, 85, 86, 93, 96, 97, 99–102, 107, 109, 110, 112, 113, 123, 132, 167, 168, 179, 189, 192, 203, 207–210, 212, 229, 234, 238–245, 247–249, 254–257, 259–261, 267, 270, 272, 276, 285
Graphene oxide (GO), 19, 35, 67, 76, 113, 167, 168, 191, 192, 207, 209, 210, 238, 261, 270, 272, 276
Graphitic carbon nitride (g-C3N4), 8, 19, 21, 47, 63, 78, 122, 123, 125, 128, 130–132, 197, 198, 204, 234, 257

Hall effect, 26, 96, 242, 243
Heavy metal, 28, 166, 182, 184, 186, 267, 268, 271, 273
Heterostructure, 17, 77, 84, 96, 97, 100, 103, 104, 118, 125, 130, 131, 167, 196, 199–201, 229
Hexagonal boron nitride (h-BN), 1, 16, 17, 19, 21, 78, 85, 99, 103, 204, 238

300 ■ Index

Hybridization, 92, 161, 162, 164, 174, 175, 182, 194, 254, 279
Hydrogen evolution process (HER), 8, 138, 140–145, 211–214, 258, 259

Inductively coupled plasma atomic emission spectroscopy (ICP-AES), 39, 40
Inductively coupled plasma mass spectrometry (ICP-MS), 39, 40

Layered double hydroxide (LDH), 1, 2, 4, 19, 20, 78, 142, 145, 174, 175, 182, 185, 191, 234, 253, 257

Magnetic random-access memory (MRAM), 92, 97
Magnetic tunnel junction (MTJ), 92, 102, 103
Metallenes, 2–5, 8, 37, 39, 76–78, 96, 97, 100, 101, 123–128, 130, 134, 189–192, 194–201, 254, 261, 263, 283–296
Metallization, 3, 6, 32, 35, 36, 40, 190, 192
Metal-organic framework (MOF), 1, 2, 108, 114, 118, 174, 176, 177, 183, 234, 249, 254, 258, 260, 261, 268, 269
Metal oxide-semiconductor field-effect transistor (MOSFET), 91
Molecular beam epitaxy (MBE), 36, 37, 288
Molten carbonate fuel cell (MCFC), 204, 206
MXenes, 78, 107, 108, 113, 118, 174, 175, 182, 204, 207, 210, 211, 214, 234, 254, 257, 260

Nanomaterials (NMs), 1, 19, 63, 65, 68, 69, 78, 138, 166, 178, 179, 182, 189, 190, 207, 224, 225, 234, 254, 261, 283, 286, 288, 291, 292
Nanoparticles(NPs), 1, 78, 107, 125, 148, 162, 165–167, 177, 189, 208, 209, 211, 213, 214, 220–222, 229, 238, 257, 276, 278, 284, 290–293
Nanosheets (NSs), 1, 5, 8, 33, 36, 64, 69, 71, 78–82, 84–86, 107, 111, 114, 123, 125, 128, 129, 131, 146, 148, 150, 151, 157, 159, 161, 162, 165, 174, 175, 177–179, 181–185, 189–192, 194, 197–199, 203, 208, 212–214, 221–225, 229, 234, 245, 247, 254, 257, 258, 262, 263, 268, 269, 274, 276, 283–289, 292, 294
Nanowires (NWs), 1, 35, 64, 177, 212, 213, 269, 272

Optoelectronics, 26, 28, 66, 107, 108, 113, 118, 157, 261
Organic solar cell (OSC), 115, 261

Oxygen evolution reaction (OER), 7, 8, 142, 145, 146, 211, 214, 248, 255–258, 260, 261, 263, 284, 285
Oxygen reduction reaction (ORR), 8, 59, 77, 124, 127, 146, 209, 211, 212, 256, 259–261, 263, 285

Perovskite, 19, 78, 107, 108, 112, 115, 116, 118, 175, 261, 262
Perovskites solar cell (PeSC), 115
Phosphoric acid fuel cell (PAFC), 204, 205
Photodetector, 28, 65, 77, 108, 110–112, 116–118, 262
Photoluminescence spectroscopy (PL), 26, 42, 66, 67, 195, 262, 284
Physical vapour deposition (PVD), 23, 24, 288
Proton exchange membrane fuel cell (PEMFC), 204, 205, 208, 209
Pulsed laser deposition (PLD), 4, 24, 36, 37, 110

Quantum dot (QD), 10, 128, 161, 162, 261

Raman, 11, 19, 25, 26, 40, 41, 43, 225
Reduced graphene oxide (rGO), 19, 113, 167, 209, 261, 270, 272

Scanning electron microscopy (SEM), 23, 25, 37–39, 82, 131, 132
Sensors, 18, 20, 26–29, 63, 64, 69, 76, 77, 114, 115, 244, 268, 270–273, 278, 279, 292, 293
Sheet resistance measurement (SRM), 26
Solid oxide fuel cell (SOFC), 204–207
Spin field-effect transistor, 103
Spintronics, 65, 66, 68, 79, 92, 96, 97, 100, 103, 104
Sputtering, 23, 24, 213
Supercapacitor, 29, 79–81, 115, 231, 237, 247–250, 255

Template, 4, 32, 34, 35, 49, 52, 57, 59, 78, 159, 180, 190–192, 212, 238, 276, 288, 289
Thermoelectric, 248, 249, 255, 262, 263
Top-down, 3–5, 108, 113, 127, 178, 179, 190, 238, 258, 286, 287
Transition metal carbide (TMC), 175, 210
Transition metal dichalcogenide (TMDC), 4, 16, 18, 19, 29, 47, 48, 51, 55, 63, 66, 67, 78, 96, 107, 109, 157, 176, 204, 234, 238, 241, 243, 245, 248, 250, 253, 256, 262, 263, 267, 280
Transition metal oxide (TMO), 19, 107, 134, 174, 175, 179, 234, 238, 254

Transition metal sulfides (TMD), 1, 2, 4, 16, 18–20, 24, 25, 29, 47, 48, 51, 55, 63, 66, 67, 69, 78, 84, 85, 96, 107–110, 112, 113, 118, 157, 174, 176, 179, 204, 234, 241, 243–245, 248–250, 253, 256, 258, 259, 262, 266, 267, 280

Transmission electron microscope (TEM), 5, 12, 26, 35–41, 43, 131, 132, 164, 165, 192

Tunneling magnetoresistance (TMR), 92, 103

van der Waals (vdW), 4, 17, 18, 20, 36, 47, 48, 64, 65, 72, 77–79, 100, 103, 107, 110, 125, 128–130, 164, 176, 178, 190, 199, 238, 243, 244, 270, 287

Water splitting, 8, 27, 122, 125, 133, 138, 141, 156, 168, 169, 182, 185, 190, 194, 196, 199, 211, 214, 248, 255, 258

X-ray absorption spectroscopy (XAS), 11, 39

X-ray diffraction (XRD), 11, 25, 26, 40, 43, 162, 163, 165, 225

X-ray photoelectron spectroscopic (XPS), 25, 26, 39, 42, 132, 165